Demonstrational Optics

Demonstrational Optics

Part 2: Coherent and Statistical Optics

Oleg Marchenko

St. Petersburg State University
St. Petersburg, Russia

Sergei Kazantsev

Paris Observatory
Paris, France

and

Laurentius Windholz

Technical University of Graz
Graz, Austria

 Springer

ISBN-13: 978-1-4419-4081-0   e-ISBN-13: 978-0-387-68327-0

Printed on acid-free paper.

9 8 7 6 5 4 3 2 1

springer.com

Contents

Preface

This book is the second part of a two-volume textbook which emphasizes the experimental demonstration of optical effects and properties. The physical background of the phenomena within is thoroughly discussed in order to ensure a didactic approach to the field of optics. Besides experiments, computer simulations of optical phenomena based on the statistical properties of light are performed. Thus, this second volume can be seen also as a book which gives an introduction to statistical optics.

In the first volume (Wave and Geometrical Optics) wave properties of light, polarization phenomena, light waves in media, optical anisotropy and geometrical optics are discussed. The second volume (Coherent and Statistical Optics) discusses interference and diffraction phenomena, Fourier optics, the history of light quanta, detection of photons, properties of white light, and the correlation of light amplitudes and light intensities.

First, interference and diffraction phenomena are treated in a classical manner, involving a large number of demonstrational experiments, followed by an introduction to Fourier optics. The development of the radiation laws and the idea of light quanta is treated thoroughly. A large part of this book is devoted to thermal radiation. All phenomena under consideration are in some way connected with the fact that thermal radiation has properties of partial coherence. All interference and diffraction phenomena show an inherent dependency of the fringe contrast on the dimensions and the spectral composition of the light source used. Further, coherence phenomena lead to measuring procedures for the intensity of light by means of photodetectors. To understand the detection of weak intensities it is necessary to be familiar with statistical properties of the photoeffect. Thus computer simulations are included in order to illustrate statistical properties of photoeffect and radiation. Then interference as the case of amplitude correlation is treated in terms of correlation functions. Finally, the correlation of intensities is treated. Again, all phenomena are illustrated by experiments and computer simulations.

The authors would like to thank the authorities of the Technical University of Graz for support.

Introduction

The first two chapters are devoted to traditional subjects such as the interference and diffraction of light waves. Classical interference schemes with wave-front splitting such as the double-slit YOUNG interferometer, LLOYD's mirror, and FRESNEL's bi-prism and bi-mirrors are discussed in the beginning of Chapter 1. Next, interferometers with amplitude splitting of the MICHELSON-type and multiple-beam interferometers are discussed. Along with the description of all these schemes, the role of all experimental parameters is emphasized, and interference patterns corresponding to the experiments are presented.

In Chapter 2 diffraction phenomena are discussed mainly using the idea of FRESNEL zones in order to explain all observations.

Chapter 3 treats diffraction phenomena in terms of two-dimensional FOURIER transforms. As an example of such a mathematical approach, the operational principle of a thin positive lens is discussed in detail. Questions of spatial filtration are illustrated by examples, and ABBE's theory of the resolving power of a microscope is explained. Also, the phase-contrast ZERNIKE microscope is considered and illustrated by demonstrational experiments.

Chapter 4 reviews the basic laws of thermal equilibrium radiation in the context of a black body. Here, PLANCK's distribution is deduced in a way similar to that applied by PLANCK in his works on problems of the equilibrium radiation, which was based on the statistical definition of entropy. Emphasizing EINSTEIN's hypothesis of light, including the photoeffect, spontaneous and induced radiation and the population of atomic states is considered.

Chapter 5 contains a statistical treatment of basic principles of photodetection, explained with the help of monochromatic waves. The introduction of a shortest interval needed for creation a photoelectron, together with the probability describing elementary photoemission, allows the use of the POISSON stochastic process as a simple model of initiation of a photoelectron. Several examples using computer models illustrate the properties of shot noise. Under monochromatic radiation, requirements for the propagation of photons are treated in terms of POISSON statistics. The important property of light quanta, that they form interference patterns even under the propagation of single photons, is illustrated by experiments.

A description of quasi-monochromatic light in terms of random fields is given in Chapter 6. In contrast to monochromatic waves, quasi-monochromatic light fields undergo intensity fluctuations. A useful representation of quasi-monochromatic waves by quadrature components is introduced. Using this representation, fundamental laws of GAUSSian light are established with respect to absolutely incoherent light radiation, so-called white light. Computer models for the simulation of a GAUSSian distribution of the quadrature components and for the exponential distribution of instantaneous intensities are discussed.

Chapter 7 is devoted to the correlation of electromagnetic fields responsible for interference and diffraction phenomena. The spatial and temporal correlation of optical fields is thoroughly investigated, and relationships between the field correlation functions and the visibility of interference fringes are established. These properties are illustrated with the help of interference schemes of YOUNG and MICHELSON type. The development of the correlation properties of quasi-monochromatic waves is discussed in the context of a spatial mode. Emphasis is given here to the physical sense of the degree of degeneracy, which allows the estimation of the total number of light quanta within the spatial mode. The problem of the interference of single photons in classical interference experiments is solved by further analysis of the light energy radiated by thermal optical sources. As in the previous chapters, computers models of spatially and temporally correlated GAUSSian light are discussed.

The final chapter, Chapter 8, is devoted to the correlation of light intensities detected at different places within one coherence volume. The detection of quasi-monochromatic waves is discussed in connection with the properties of correlation functions of intensity and of correlation functions of intensity fluctuations. The statistics of the observed photocurrent leads to computer models of inertialess and inertial photodetectors. To verify predictions of the MANDEL equation a computer model of the BOSE-EINSTEIN statistics, dealing with the statistics of photons within one spatial mode, is considered. Pioneering experimental works concerning the correlation of intensities, like the observation of optical beats and the stellar interferometer of intensities, are discussed here in conjunction with computing models simulating the observed results.

Chapter 1

INTERFERENCE OF LIGHT WAVES

Optical interference is a direct proof of the wave nature of light. The interference principle for light waves promoted by T.YOUNG (1773-1829) was corroborated excellently by his famous double-slit experiment. During further development of wave optics, a large number of new experimental methods to investigate the interference of light beams were proposed. Most of these classical experiments carry the names of their creators: LLOYD mirror, bi-prism and bi-mirrors of FRESNEL, NEWTON rings, MICHELSON-, LUMMER–GEHRCKE-, and FABRY–PEROT interferometers and many others. It is justified to relate the features of these classical optical experiments to the methods of classical natural science. In fact, through the creation of each particular interference instrument, a certain concrete physical problem was solved. For example, in his first double slit experiment YOUNG was able to estimate the wavelength of a light wave. Further, as optical experimental technique was developed to more sophisticated instruments, the precise measurement of physical quantities became a leading motivation for the construction of interferometers. For this purpose, different types of MICHELSON interferometers were used to establish a more accurate length standard. To solve the basic problem of light propagation through vacuum, such interferometers were used to try to detect the "aether wind" (this experiment later played a crucial role in confirming the laws of relativity theory). On the basis of MICHELSON's interference scheme, the stellar interferometer was constructed to measure the angular diameters of stars for the first time. A large number of concrete spectrometric measurements have been carried out with the help of FABRY–PEROT- and LUMMER–GEHRCKE interferometers. Summarizing, interference of light is a part of optics which greatly contributed to the investigations of the nature of light. Additionally, the methods to obtain interference patterns and the resulting instrumental techniques were applied to measure precisely different physical quantities.

1.   Classical interference schemes

Let us consider the distribution of intensity resulting from the superposition of two beams which originate from a monochromatic point source S (Fig.1.1). Practically, this experiment can be realized by splitting a primary beam of light into two beams, passing through two pinholes S_1 and S_2 in an opaque screen. According to HUYGENS's *principle* (CH.HUYGENS, 1629-1695), the pinholes can be regarded as secondary monochromatic point sources, each emitting a spherical wave. If both sources are illuminated by one monochromatic incident wave, they are always emitting waves with a fixed phase difference. Such secondary point sources are called *mutually coherent*. The beams from the secondary sources are superimposed within the region behind the screen with the pinholes, and an interference pattern is formed in this region. Let the (x, y)-plane be the plane of observation, parallel to the screen with the pinholes (x–axis parallel to S_1S_2 direction), and let R_0 be the distance between this screen and the (x, y)-plane as shown in Fig.1.1.

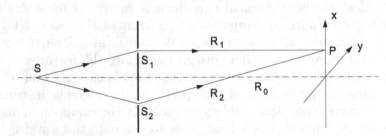

Figure 1.1.   Interference of two secondary waves from mutually coherent sources S_1, S_2. The sources are illuminated by a monochromatic point source S.

Let a monochromatic spherical wave from the primary point source S create the secondary spherical waves at S_1 and S_2 at the moment t_0, and let the amplitude of these waves at S_1 and S_2 be E_0. The secondary spherical waves at point P of the (x, y)-plane can be represented by the complex functions E_1 and E_2:

$$E_1 \sim \frac{E_0}{R_1} \exp\{-i2\pi\nu_0(t_0 + R_1/c)\} \quad ,$$

$$E_2 \sim \frac{E_0}{R_2} \exp\{-i2\pi\nu_0(t_0 + R_2/c)\} \quad ,$$

where E_0 is the field strength at unit distance from S_1 or S_2, and $R_1 = \overline{S_1P}$, $R_2 = \overline{S_2P}$ are the distances between point P and the pinholes. The total field at point P is given by the function

$$E_P = E_1 + E_2 \sim \frac{E_0}{R_1} \exp\{-i2\pi\nu_0(t_0 + R_1/c)\}+$$

$$+ \frac{E_0}{R_2} \exp\{-i2\pi\nu_0(t_0 + R_2/c)\} \quad .$$

Let x be the x-coordinate of point P. We assume both R_1 and R_2 to be large compared to x, so that the amplitude factors E_0/R_1 and E_0/R_2 can be substituted by E_0/R_0. Hence, the total field is given by

$$E_P \approx \frac{E_0}{R_0} \exp\{-i2\pi\nu_0t_0\} \left[\exp\{-ikR_1\} + \exp\{-ikR_2\}\right] \quad ,$$

where $k = 2\pi\nu_0/c = 2\pi/\lambda_0$. The intensity I_P of light at point P is calculated to be proportional to $E_P E_p^*$:

$$I_P \sim E_P E_P^* \approx \frac{E_0^2}{R_0^2} \left[\exp\{-ikR_1\} + \exp\{-ikR_2\}\right] \cdot$$

$$\cdot \left[\exp\{-ikR_1\} + \exp\{-ikR_2\}\right] = 2\frac{E_0^2}{R_0^2}[1 + \cos(k\Delta R)] \quad , \qquad (1.1)$$

where $\Delta R = R_2 - R_1$ is the difference of the geometrical path lengths from S_2 and S_1 to point P. The intensities of the interfering waves at P are both proportional to E_0^2/R_0^2. Hence, it follows (1.1) that the total intensity can be written as

$$I_P = 2I_0 + 2I_0 \cos k\Delta R = 2I_0(1 + \cos k\Delta R) \quad , \qquad (1.2)$$

where $2I_0 \cos k\Delta R$ is called the *interference term*. This expression represents the intensity dependency on the path difference, or the dependency on the phase difference $\delta = k\Delta R$. Evidently, a periodically varying intensity pattern will be obtained with maxima of intensity

$$I_{max} = 4I_0 \qquad \text{when} \quad \delta = 0, \pm 2\pi, \pm 4\pi \dots \quad , \qquad (1.3)$$

as well as with minima of intensity

$$I_{min} = 0 \qquad \text{when} \quad \delta = \pm\pi, \pm 3\pi \dots \quad . \qquad (1.4)$$

The total intensity within the superposition region of two coherent beams (each having intensity I_0) varies periodically with the phase difference and changes between maxima equal to $4I_0$ and minima equal to zero.

Let us now assume that a light wave from a monochromatic point source is split by a beam splitter into two waves with amplitudes E_1 and E_2, respectively. Further, let these waves interfere at a point of observation P. Assuming that the phase difference between the interfering waves is $\Delta\varphi$, let us derive a formula for the intensity at this point. The complex amplitude E at point P may be represented as follows:

$$E_P = E_1 \exp(i\varphi) + E_2 \exp(i\varphi) \exp(i\Delta\varphi) ,$$

where φ is an initial phase of the incident waves. φ must be the same for both interfering waves because both waves initiate from one monochromatic point source. The intensity calculated by means of $E_P E_P^*$ can be written as

$$I_P \sim E_P E_P^* = E_1 E_1^* + E_2 E_2^* + E_1 E_2^* \exp(i\Delta\varphi) + E_2 E_1^* \exp(i\Delta\varphi) .$$

We introduce the intensities of the interfering waves, $I_1 \sim E_1 E_1^*$ and $I_2 \sim E_2 E_2^*$, respectively. Since the initial phase is assumed to be the same for both waves, the magnitudes $E_1 E_2^*$ and $E_2 E_1^*$ are real magnitudes and can be represented in terms of the intensities as follows:

$$E_1 E_2^* = E_2 E_1^* \sim \sqrt{I_1 I_2} .$$

Using the intensities I_1 and I_2 we can write

$$I_P = I_1 + I_2 + 2\sqrt{I_1 I_2} \cos(\Delta\varphi) . \qquad (1.5)$$

If we assume $I_1 = I_2 = I_0$, we get the same equation as before, (1.2), and the intensity changes with varying $\Delta\varphi$ between $I_{max} = 4I_0$ and $I_{min} = 0$.

Let us define the *contrast* or the *visibility* V of the interference pattern in the form

$$V = \frac{I_{\max} - I_{\min}}{I_{\max} + I_{\min}} , \qquad (1.6)$$

where we take the intensities at two neighbouring extreme values of the function I: one value is associated with a maximum of I, $I_{\max}$, and the other with a minimum $I_{\min}$. For the case $I_1 = I_2 = I_0$ we obtain the highest possible contrast, $V = 1$.

When the intensities are unequal, the contrast of the interference pattern is lower. According to (1.5) $I_{\max} = I_1 + I_2 + 2\sqrt{I_1 I_2}$, and $I_{\min} = I_1 + I_2 - 2\sqrt{I_1 I_2}$. When substituting $I_{\max}$ and $I_{\min}$ by these expressions we find V as:

$$V = \frac{2\sqrt{I_1 I_2}}{I_1 + I_2} . \qquad (1.7)$$

Sometimes it is more convenient to introduce the ratio of the intensities, $x = I_1/I_2$. Then we get

$$V = \frac{2\sqrt{I_1/I_2}}{I_1/I_2 + 1} = \frac{2\sqrt{x}}{x + 1} \quad .$$

The right-hand side of the last expression has a maximal value of 1 for $x = 1$ ($I_1 = I_2$). If $I_1 \neq I_2$, the contrast V is always smaller than 1.

We note that the interference law represented in terms of the superposition of two monochromatic waves in formula (1.1) holds, if the assumption of monochromatic point sources S_1 and S_2 is true. In turn, this implies that both superimposed waves must have the same frequency ν_0. Thus the secondary point sources have to be initiated from a monochromatic point source S. In this case the phase difference for a certain point P is constant. Conversely, this condition of a constant phase difference allows one to say that two sources S_1 and S_2 are mutually coherent. Interference of two mutually coherent waves represented by the expressions (1.1) and (1.2) is additionally based on the assumption that the waves have the same polarization direction. The reason is that these waves are caused by radiation of a monochromatic wave of a certain polarization at S. We have seen in Part I (Chapter 3) that a monochromatic wave with elliptical polarization may be represented by a superposition of two waves which are linearly polarized in two mutually orthogonal directions. In the general case of an elliptical polarization of the original monochromatic wave, the interference can therefore be well described by expressions analogous to (1.1) and (1.2).

1.1 Young's double slit experiment

The well known interference experiment of YOUNG, schematically considered above, can be realized under the following experimental conditions (Fig.1.2). A beam from a mercury-helium lamp is focused by a lens onto a narrow vertical slit. This slit is placed on the first focal plane of the objective O_1 (focal length 10 cm). A double slit is placed after this objective. The two single slits forming the double slit are separated by 0.8 mm. A second objective O_2 (focal length 60 cm) is mounted close to the double slit.

In our arrangement, the expanding beam from the first slit is transformed to a parallel beam by the first objective; this parallel beam illuminates the double slit. Two light beams expanding from the double slit pass through objective O_2 and form an interference pattern in its focal plane. Since the interference pattern is located on its second focal plane, it is caused by parallel rays, emitted by the double slit throughout

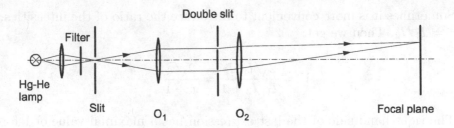

Figure 1.2. YOUNG's double slit interference experiment. The width of the primary slit is 0.03 mm. The double slit consists of two identical slits (width 0.02 mm), separated by 0.8 mm. The primary slit is placed in the focal plane of objective O_1 (focal length 10 cm). The focal length of the second objective O_2 is 60 cm.

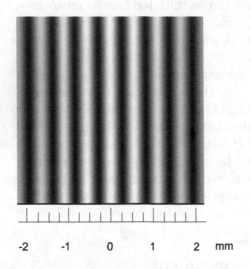

Figure 1.3 Interference fringes observed with YOUNG's interference scheme for the yellow lines of a Hg-He lamp ($\overline{\lambda} = 580$ nm).

the angular space (this arrangement corresponds to the observation of so-called FRAUNHOFER diffraction, see Chapter 2).

The interference pattern consists of bright and dark bands called *interference fringes* (Fig.1.3). We observe an equidistant system of linear fringes oriented parallel to the double-slit. The separation between two neighbouring bright fringes is the *fringe spacing.*

Let two parallel rays from the centers of slits S_1 and S_2 fall on the objective under such a (small) angle θ that these rays are focused at a point P corresponding to the first minimum of the intensity (Fig.1.4). If $h/2$ is the position of P then the fringe spacing is equal to h. It is evident that $\theta \approx h/(2f)$ and $\Delta R = b\theta$, where b is the separation between the centers of the slits and f is the focus length of the objective. One gets $\Delta R = bh/(2f)$. Since the coordinate $h/2$ corresponds to the path difference $\Delta R = \lambda/2$, the fringe spacing can be found as

$$h = f\frac{\lambda}{b} = \frac{\lambda}{\alpha} \quad , \tag{1.8}$$

with $\alpha = b/f$. One can treat the relationship (1.8) in terms of the angular separation α between the two sources S_1 and S_2 (note that the angle b/f has to be small to use the approximation $\sin\alpha \approx \alpha$). Concurrently, the angle α is the angle between the interfering rays. The smaller the separation b is (and the smaller the angle between the interfering rays), the larger is the fringe spacing.

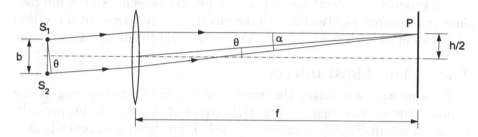

Figure 1.4. Relationships between the fringe spacing and the angle between two interfering rays.

It is useful to introduce the angular spacing of the fringes in addition to its linear dimension. In the case under consideration, the angular spacing is the angle between lines joining two adjacent minima of the intensity distribution with the central point of the objective. We can therefore write the angular spacing of the fringes in a form:

$$\beta = \frac{h}{f} = \frac{\lambda}{b} \quad . \tag{1.9}$$

Let us estimate the fringe spacing. It is found from Fig.1.3 that six bright fringes occupy 2.75 mm, hence the fringe spacing is approximately 0.45 mm. One can assume that two points sources located at the centers of the slits produce interference in just the same way as any other pair of points located in the slits; this is true for very narrow slits. If the separation of these points is 0.8 mm and the focal length of the objective O_2 is 60 cm, then the angle between the interfering rays α in (1.8) is equal to $b/f \approx 1.3 \cdot 10^{-3}$ rad. For the bright yellow lines in the spectrum of a mercury–helium lamp we have a mean wavelength $\bar{\lambda} = 580$ nm. From (1.8), we find for the fringe spacing $h \approx 0.435$ mm, in good agreement with the observed data. For the given fringe spacing h and the focal length 60 cm of the objective we find for the angular spacing of the fringes from (1.9):

$$\beta = \frac{h}{f} \approx 7 \cdot 10^{-4} \text{ rad} \quad .$$

Twelve bright and dark fringes compose the principal part of the interference pattern shown in Fig.1.3 which are concentrated within an angle of only $8 \cdot 10^{-3}$ rad.

The idea of a point monochromatic source used in the treatment of the interference from two mutually coherent waves can also be applied in the case of the real experiment. Each small element along the vertical direction of the primary vertical slit can be regarded as one point source which produces its interference pattern over the screen. All the interference patterns are identical due to the identity of the geometrical position of any element with respect to the double slit and the observation screen.

1.2   The Lloyd mirror

Another way of splitting the wave front of a light wave coming from a primary source was suggested by H.LLOYD (1800-1881). In his arrangement, shown in Fig.1.5, a mirror is used. Light from a mercury-helium lamp focused by a lens on a horizontal narrow slit falls on a horizontal plane surface mirror placed after the slit. The slit is parallel to the plane of the mirror and close to its surface plane which caused the light to be reflected at nearly grazing incidence. As before, an interference pattern is formed by an objective within its focal plane. For any point in the focal plane, the interference results from two rays: one is reflected by the mirror, and the other passes directly from the slit to the point. Interference fringes obtained by means of the LLOYD *mirror* are shown

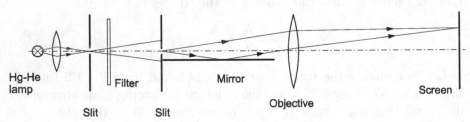

Figure 1.5. Experimental arrangement using LLOYD's mirror. The width of the primary slit is 0.06 mm, the width of the second slit 0.025 mm, the distance between the slits 40 cm, the focal length of the objective 60 cm, the length of the mirror 7 cm.

in Fig.1.6. With increasing distance from the mirror surface (its position corresponds to 0 mm in Fig.1.6) the contrast and the brightness of the fringes decreases. This decrease is caused by diffraction of light due to the finite width of the slit (see Chapter 2). This phenomenon takes place

in all experiments considered here. We note that the ray reflected from
the mirror changes its phase by π due to reflection from an optically
thick medium. Thus the dark fringes occur when the path difference
ΔR between two interfering rays is equal to λ, unlike formula (1.3). The
first fringe is dark ($\Delta R = 0$), the second one, corresponding to the path
difference $\lambda/2$, is bright, and so on. The fringes in Fig.1.6 are obtained
under the conditions: primary slit width 0.06 mm, secondary slit width
0.025 mm, distance between the slits 40 cm, focal length of the objective
60 cm, and length of the mirror 7 cm. 16 bright fringes occupy 7 mm,
which gives a distance between neighbouring maxima of approximately
0.5 mm; the angular fringe spacing is therefore about $8 \cdot 10^{-4}$ rad.

Fig.1.7 illustrates the geometrical considerations for creating two mutually coherent sources in the interference scheme with the LLOYD mirror. The interfering beams come from two point sources S_1 and S_2: one
is the slit, the other is its virtual image formed by the mirror.

Two parallel rays from S_1 and S_2 propagate at a small angle θ with
respect to the mirror plane. These rays are then focused by the objective
at point P, having the coordinate $f\theta$ in the direction normal to the slit.
The distance b between S and S_1 increases with increasing distance $b/2$
of the slit from the mirror surface.

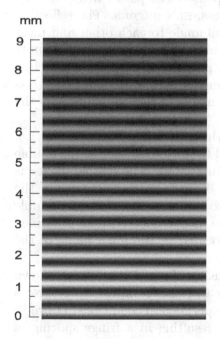

Figure   1.6 Interference
pattern   using   LLOYD's
mirror for the yellow lines
of a Hg-He lamp ($\bar{\lambda} = 580$
nm). The brightness of
the fringes decreases with
increasing distance from
the mirror plane.

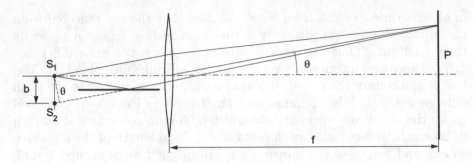

Figure 1.7. Geometrical consideration for creating interference fringes with the LLOYD mirror. The slit S_1 of width a and its virtual image S_2 separated by b act as two mutually coherent sources.

1.3   Fresnel's mirrors

In a famous interference scheme suggested by A.FRESNEL (1788-1827), two mutually coherent sources are created by means of two mirrors. In the experiment described below, two plane glass plates fitted side by side (distance ca. 0.1 mm) compose FRESNEL's *mirrors*. The reflecting surfaces of the plates are at a very small angle to each other and must touch each other (distance less than 0.1 mm). This arrangement permits the creation of two virtual images of a primary source which are close to each other. To avoid reflections from the back surfaces of the glass plates practically opaque black glass is used. Another possibility is the use of two rectangular surface mirrors.

Such a double mirror is the basis of the experiment shown in Fig.1.8. A narrow vertical slit positioned parallel to the joining side of the double mirror is illuminated by a mercury-helium lamp. The double mirror is placed 25 cm from the slit. The light beam from the slit falls on the double mirrors at the angle of 15°. Two beams reflected by the double mirror pass through an objective. The interference pattern is formed by nearly parallel rays and therefore located in the focal plane of the objective.

The system of interference fringes consists of equidistant bands shown in Fig.1.10,b. These fringes are similar to those obtained by YOUNG's double slit interference scheme. In our experimental conditions, 11 bright fringes occupy approximately 3.2 mm, resulting in a fringe spacing of about $h \approx 0.3$ mm.

The geometrical considerations illustrated in Fig.1.8 shows two coherent sources S_1, S_2 to be the virtual images of a point source S. The

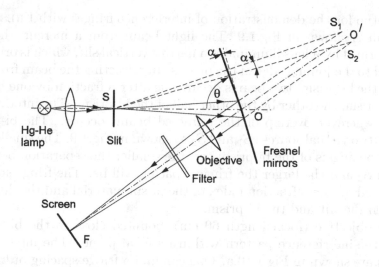

Figure 1.8. Interference scheme with FRESNEL's mirrors. The width of the primary slit is 0.04 mm. The focal length of the objective is 60 cm. Angles of incidence on the mirrors are both approximately 15°. The distance from the slit to the double mirror is about 25 cm.

angular separation of these images is given by the angle α, which is also the angle between the planes of the double mirror.

1.4   Fresnel's bi-prism.

Another way of splitting of a primary light beam into two interfering beams is realized through FRESNEL's *bi-prism*. Such a splitting element is made up of two equal prisms with a very small refraction angle. These prisms are placed together base to base with their refraction edges parallel.

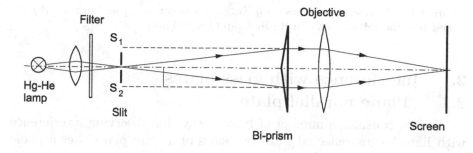

Figure 1.9. Interference with FRESNEL's bi-prism. The width of the primary slit is 0.03 mm, the focal length of the objective is 60 cm. The distance between bi-prism and objective is 25 cm.

A setup for the demonstration of interference fringes with FRESNEL's bi-prism is shown in Fig.1.9. The light beam from a mercury–helium lamp formed by a lens illuminates a narrow vertical slit, which is oriented parallel to the joining edge of the prisms. In refracting the beam from the slit by the bi-prism, two beams occur: one after refraction by one half of the bi-prism, the other by the other half of the bi-prism. The interference occurs wherever overlap of the refracted beams occurs. The bi-prism creates two virtual sources S_1 and S_2 as shown in Fig.1.9. The smaller the refraction angles of the bi-prism are, the smaller the separation between S_1 and S_2 and the larger the fringes spacing will be. The fringe spacing is affected by the refraction index of the prism material and the distance between the slit and the bi-prism.

The objective (focal length 60 cm) mounted close to the bi-prisms forms the interference pattern within its focal plane. The interference fringes are shown in Fig.1.10,a. One can find a fringe spacing of $h \approx 0.7$ mm, which results in about $1.2 \cdot 10^{-3}$ rad for the angular dimension of the spacing.

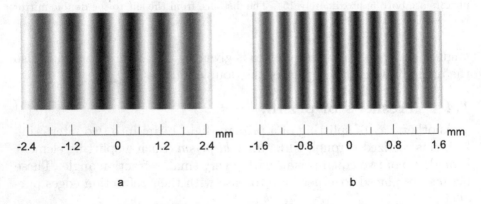

a   b

Figure 1.10. Interference fringes with FRESNEL's bi-prism (a) and mirrors (b) obtained with the yellow lines of a Hg-He lamp ($\bar{\lambda} = 580$ nm).

2.   Interference with glass plates

2.1   Plane parallel plate

We now consider a number of typical ways for observing interference with light beams reflected from two sides of a glass plate. Let a plane parallel glass plate be placed in the light beam from a small circular aperture illuminated by a mercury–helium lamp, as shown in Fig.1.11. Light rays reflected from the two surfaces of the plate are focused by an objective into its focal plane. A portion of the interference pattern, consisting of a system of nearly equidistant bright and dark circular bands, is shown in Fig.1.12. For any point of the aperture S, two virtual coherent sources S_1 and S_2 are created by reflection on the two surfaces of the plate. It is clear that for a given focal length of the objective and a wavelength λ, the width of the fringes is dependent on the thickness of the plate and on the incident angle of the light beam, because these magnitudes determine the mutual position of the virtual sources S_1 and S_2.

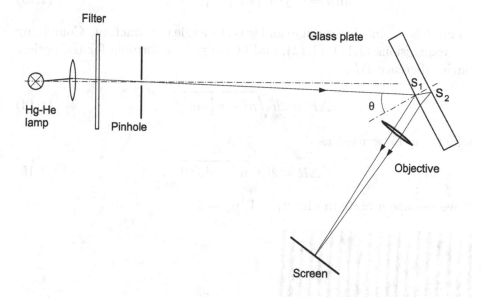

Figure 1.11. Interference scheme with a parallel plane glass plate placed at an angle θ in a light beam. The diameter of the circular aperture is 1.5 mm, the thickness of the glass plate 2 mm. The plate is at a distance of 80 cm from the aperture and inclined at 12° with respect to the axis of the beam. The objective (focal length is 60 cm) is closely mounted to the glass plate.

To derive the formula for the path difference we consider that ray SA falls on point A of the plate surface at angle θ, as shown in Fig.1.13. The interference is caused by two rays: one reflected by the external surface of the plate, ray $SADS_1$, and the other refracted by this surface and then reflected on the inner surface, ray $SABCS_2$. These rays are parallel to each other and maintain the same angle θ with the normal of the plate. The optical path difference between these rays is given by the expression

$$\Delta R = n_2 \overline{AB} + n_2 \overline{BC} - n_1 \overline{AD} \ , \qquad (1.10)$$

where D is the foot of the perpendicular from C to S_1A, n_1 is the refractive index of the medium outside the plate and n_2 is the refractive index of the plate.

If h is the thickness of the plate, we find

$$\overline{AB} = \overline{BC} = \frac{h}{\cos\varphi} \quad , \tag{1.11}$$

$$\overline{AD} = \overline{AC}\sin\theta = 2h\tan\varphi\sin\theta \quad , \tag{1.12}$$

$$\sin\theta = (n_2/n_1)\sin\varphi \quad , \tag{1.13}$$

where θ is the incidence angle and φ is the angle of refraction. Combining the expressions (1.11), (1.12), and (1.13) gives a formula for the optical path difference ΔR,

$$\Delta R = 2h\sqrt{n_2^2 - n_1^2\sin^2\theta} \quad , \tag{1.14}$$

which can be written as

$$\Delta R = 2h\sqrt{n^2 - \sin^2\theta} \quad , \tag{1.15}$$

if we assume a plate in air: $n_1 \approx 1$, $n_2 = n$.

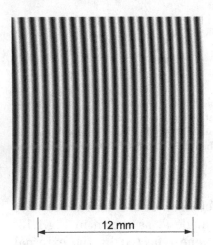

|← 12 mm →|

Figure 1.12 Interference fringes from a parallel plane glass plate. The fringe spacing in the focal plane of the objective is about 0.7 mm. The angular fringe spacing is about $1.2 \cdot 10^{-3}$ rad. The fringes have been obtained for the yellow lines of a Hg-He lamp ($\overline{\lambda} = 580$ nm).

Hence, the corresponding phase difference is

$$\delta = \frac{2\pi}{\lambda}\Delta R = \frac{4\pi h}{\lambda}\sqrt{n^2 - \sin^2\theta} \quad . \tag{1.16}$$

Taking into account the phase change of π which happens at reflection on the upper surface of the plate, the angular positions of the interference maxima of mth order (m is an integer) should be found if $\delta - \pi = 2\pi m$. This condition can be re-written as follows:

$$\frac{4\pi}{\lambda}h\sqrt{n^2 - \sin^2\theta} - \pi = 2\pi m \quad . \tag{1.17}$$

In our case this expression enables the calculation of the fringe spacing in the focal plane of the objective by means of the relationship between the angle of incidence θ and the interference order m. Because the fringes are formed by parallel rays these fringes are said to be *localized at infinity*. One sees from (1.16) that the positions of the maxima depend only on the angle of incidence θ, assuming all others parameters are fixed. For this reason interference fringes of this sort we call *fringes of equal inclination*.

We have seen (Part 1, Chapter 4) that mica can be cleaved into very thin sheets, which are nearly parallel plane plates, transparent for visible light. Such sheets (area of some cm^2) may have a thickness of fractions of a millimeter under conditions of nearly parallel surfaces. Such a sheet should give rise to interference of rays which have a very small path difference due to the small thickness of the mica sheet. Let us discuss the demonstration of such properties of a mica sheet in the interference scheme shown in Fig.1.14,a. A low pressure mercury lamp directly illuminates a mica sheet of the thickness $h = 0.05$ mm and dimensions 2×3 cm. Light rays, falling on the sheet at an angle of incidence $\theta = 45°$, are reflected at the upper surface and at the back side of the sheet and then intersect on a remote screen at a point P.

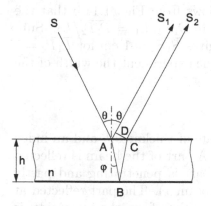

Figure 1.13 Geometrical consideration of the path difference between two rays caused by a plane parallel glass plate.

The mercury lamp ($\bar\lambda = 580$ nm) is at a distance of $L_1 = 1$ m from the mica sheet, and the sheet has a distance of $L_2 = 7$ m from the observation screen. From an illuminating point S, two points of reflection of the light are seen at an angle on the order of $h\cos\theta/L_1 \approx 3 \cdot 10^{-5}$ rad. For this reason we regard the two incident rays as being parallel.

In turn, the same points of reflection are seen from the point P at a smaller angle $\beta = h\cos\theta/L_2 \approx 4 \cdot 10^{-6}$ rad. Due to this fact we also believe that the rays leave the upper surface of the sheet at one point and propagate parallel to each other. In other words, these interfering rays may be represented by one ray reaching the point of observation. Such an approximation gives the same relationship for the path difference as formula (1.15). Really, if assuming light as propagating from the point P back to the sheet, after reflection at both surfaces the two rays created should propagate just in the same directions as the two incident rays before. Let us estimate the linear dimension of the fringe spacing on the screen. Since the angle β can be regarded to be the angle between two interfering rays the fringe spacing may be evaluated to be $\lambda/\beta \approx 14$ cm. A fragment of the interference pattern is shown in Fig.1.15. The extraordinary small angle between the interfering rays allows the use of an extended light source (Fig.1.14,b). Let S_1 and S_2 be located at two opposite borders of the source. The elementary sources positioned at these points are not mutually coherent, therefore they form two independent interference distributions, which are overlapping. Let a certain maximum of one distribution be located at point P_1 and a maximum of the second distribution at point P_2, as shown in Fig. 1.14,b. The resultant interference fringes will form a system of distinct bright and dark bands if the distance $\overline{P_1P_2}$ is much smaller than the fringe spacing x. The resultant fringes will disappear for $\overline{P_1P_2} = x/2$, since minima of the first distribution occur at the positions of maxima of the second distribution. If this is the case we treat the source to be too large to form distinct interference fringes. It follows from Fig. 1.14,b that the condition $\overline{P_1P_2} = x/2$ is just equivalent to $\overline{S_1S_2}/L_1 = \overline{P_1P_2}/L_2$. Substitution of the data of our experiment gives $\overline{S_1S_2} = 1$ cm for $\overline{P_1P_2} = 7$ cm. Hence, for the given parameters of the experiment the width of the source S_1S_2 should not exceed 1 cm.

2.2   Interference patterns

Let a beam fall on a plane parallel plate of thickness h and an index of refraction $n > 1$ as shown in Fig.1.16. A part of the beam is reflected at the first surface (beam $1'$), another part is penetrating and partly reflected, the other part is transmitted (beam 1). The part reflected at the second surface reaches the first surface then from the inner side, is partly transmitted (beam $2'$) and partly reflected. The reflected part then reaches the second surface and so on (beams 2, $3'$, 3 .....).

All transmitted beams are parallel to each other, as well as the beams which origin from the plate in the direction of the first reflected beam. The optical path difference between two neighbouring beams (e.g., 1 and

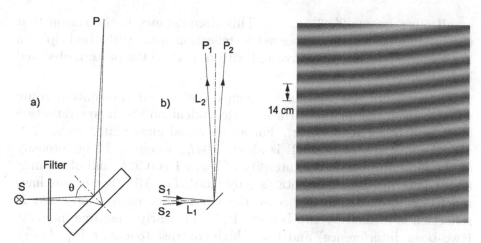

Figure 1.14. Interference with a thin mica sheet. a) A low pressure mercury lamp is spaced by 1 m from a mica sheet. Rays reflected at 45° form fringes at a screen spaced by 7 m from the mica sheet. b) Geometrical considerations when using an extended light source.

Figure 1.15. The interference pattern obtained with a thin sheet of mica with a thickness of 0.05 mm, illuminated by a mercury low pressure lamp. The most bright lines are the yellow ones: $\lambda_1 = 577.9$ nm and $\lambda_2 = 580$ nm.

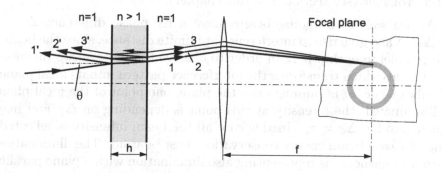

Figure 1.16. Splitting of the incident beam by a plane parallel plate. All neighbouring beams have the same difference in the optical path and interfere at infinity. A lens is used to transform the pattern to its focal plane. The optical axis is given by the normal to the plate through the center of the lens.

2, 2′ and 3′) is the same and, as before, has to satisfy relation (1.15). The only exception is the first reflected beam 1′, which undergoes an additional phase shift of π ($\Delta R = \lambda/2$) due to reflection on a more dense optical medium. Thus this beam 1′ has just the opposite phase

as all other beams $2'$, $3'$, ... This circumstance is the reason that
the interference pattern observed in reflection (e.g. with the help of a
beam splitter) is intensity inverted with respect to the pattern observed
in transmission.

The intensity distribution between reflected and transmitted beam
when passing one of the surfaces is dependent on the power reflection
coefficient $\mathcal{R}$ (or reflectivity). For an uncoated glass plate, where $\mathcal{R} \approx$
4%, the intensity of beam $1'$ is about $0.04I_0$, where I_0 is the intensity
of the incoming beam; the intensity of beam 1 is $0.92I_0$ and of beam $2'$
about $0.037I_0$. Beam 2 contains only $0.0015I_0$. All other beams have
much less intensity. In this case the interference pattern observed in
reflection is formed by two beams $(1', 2')$ of nearly the same intensity
(two-beam interference) and has a high contrast (compare Eq. (1.6)),
while the contrast in the transmitted pattern (formed by the beams 1,
2 with large intensity ratio) is rather low.

If we increase the reflectivity of the glass surfaces by optical layers,
the contrast of the pattern observed in reflection is decreasing, while the
contrast of the pattern ovserved in transmission is enhanced. More and
more outgoing beams have reletively high intensity, and we change from
the case of two-beam-interference to multiple-beam interference. A plane
parallep plate with highly reflecting surfaces is called a FABRY-PEROT
interferometer (see section 4 of this chapter).

As we see, all outgoing beams have a fix phase difference $\Delta \varphi =
2\pi \Delta R / \lambda$ and give rise to interference at infinite distance, since the beams
are parallel to each other. In order to observe the pattern, we therefore
need a lens which transforms the interference pattern into its focal plane.
If only one beam is running onto the plate, one point of the focal plane
is illuminated; the intensity at this point is depending on $\Delta \varphi$ and may
reach zero for $\Delta \varphi = \pi$. In this case all the beam intensity is reflected,
since for each beam energy conservation must be valid. The illumination
with a single beam is representing also illumination with a plane parallel
wave.

When using an extended light source, each point of the plate is pene-
trated by rays which have a large variety of angles θ, thus an interference
pattern is only observed if the surfaces of the plate are really nearly parallel to each other. In this case the lens after the plate is indispensable
to observe the interference pattern, which now is consisting of concentric circles. For observing a bright interference fringe, we need to fulfill
$\Delta R = m\lambda$, where m is an integer number. Eq. (1.15) shows that the
highest order of interference is observed for the most inner circle of the
interference pattern.

If we use a plane parallel monochromatic wave of wavelength λ (e.g. a laser beam) and entrance angle $\theta = 0$ and a plate of certain thickness h and index n, the optical path difference is given by $\Delta R = 2nh = m\lambda$. Since all figures with exception of m are already fixed, in the general case m will not be an integer, and the beam is not transmitted with full intensity. We need to tilt the plate by a small angle θ_0 in order to make m an integer number. Opposite, if we use such a plate inside the cavity of a tunable laser, tilting the plate will change the wavelength of the laser, since the laser will work with a wavelength for which the plate shows most transmission, that means, the condition $2h\sqrt{n^2 - \sin^2\theta} = m\lambda$ is fulfilled with integer m by a small change of λ.

2.3   Air wedge

A space between two plane glass plates can be regarded as an air wedge, provided the surfaces of the plates make a dihedral angle. If this angle is sufficiently small, then reflections and refractions of a primary light ray by the surfaces of the plates will produce two rays, and a noticeable interference pattern should appear wherever (i) a path difference exits between the two rays when they exit the wedge and (ii) this path difference is small enough. In contrast to the previous case of interference the thickness between the two reflecting surfaces is here not invariant. For this reason, an interference pattern can not be formed by a system of parallel rays but rather with inclined rays. Such an interference pattern will most probably be located at some finite distance from the surfaces, rather than at infinity. Nevertheless, for a sufficiently small region of both reflecting surfaces we may regard the thickness h between these surfaces to be nearly invariable (Fig.1.17). Let two rays S_1 and S_2, emitted by a remote monochromatic point source and running nearly parallel to each other, reach the first glass plate at an angle θ. The upper plate in Fig.1.17 we assume to be plane parallel in order to express the angle φ by the incidence angle θ. If this is the case, the path difference between these rays at point C can be represented in the same manner as in previous case shown in Fig.1.13. In the case of non-parallel plates under discussion, interference may occur throughout the inner surface of the first reflecting plate, which depends on the real mutual inclination of the interfering rays. In a particular case, such an interference is located at the inner surface of first plate, and point C belongs to the interference pattern. This kind of interference we call *located on a surface*.

Let us calculate the phase difference of ray S_2ABC with respect to that of ray S_1C at point C, assuming that the upper plate has the index

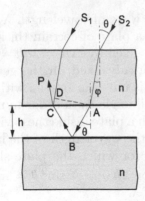

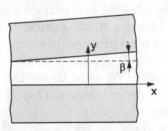

Figure 1.17. Geometrical consideration of the path difference of two rays in the case of interference from an air edge.

Figure 1.18. If two planes make a dihedral angle specified by β and the thickness y in space between them increases along x, the path difference will also change along x. The fringes will be normal to x and parallel to the straight edge of the dihedral angle.

n and the space between the plates is filled with air ($n = 1$):

$$\delta = \frac{2\pi}{\lambda}\Delta R = k(\overline{AB} + \overline{BC} - n\overline{CD} - \lambda/2) \quad , \tag{1.18}$$

where $k = 2\pi/\lambda$. The term $\lambda/2$ appears due to reflection on the inner surface of the second plate as an optically denser medium. Since the geometrical configuration under consideration is very similar to that presented in Fig.1.13, $\overline{AB} + \overline{BC} - n\overline{CD} = 2h\cos\theta$ and the path difference ΔR is given by the expression:

$$\Delta R = 2h\cos\theta - \lambda/2 \quad .$$

Thus, we may write for the phase difference

$$\delta = k\Delta R = 4\pi h\cos\theta/\lambda - \pi \quad .$$

In the particular case where two plane parallel plates make a dihedral angle specified by β, let x be the coordinate along the inner surface of the first plate normal to the edge of the dihedral angle, and let y be the coordinate normal to the plane surface and be orthogonal to x (Fig.1.18). The thickness h is then specified by the y-coordinate. Under such notation for small increments Δx and Δy, the ratio $\Delta y/\Delta x = \tan\beta$ is true, which may be approximated for small angles as $\Delta y \approx \beta\Delta x$. As far as the dihedral angle is considered, the phase difference changes only with the variation in the thickness Δy. Such changes of the phase should depend on the angle β as follows:

$$\Delta\delta = 4\pi\Delta y\cos\theta/\lambda = 4\pi\beta\Delta x\cos\theta/\lambda \quad . \tag{1.19}$$

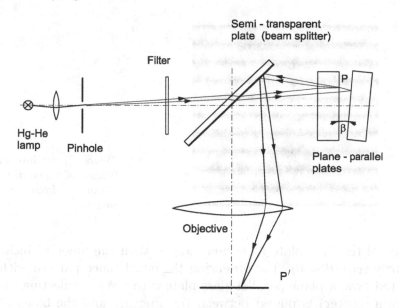

Figure 1.19. Interference scheme for observation of fringes of equal thickness with an air film. Two thick glass plates are illuminated by light from a Hg-He lamp ($\overline{\lambda} = 580$ nm). The thickness of each plate is 3 cm. The inner surfaces of the plates make a small angle. The objective forms an image of the inner surface of the first plate in the plane of observation. The focal length of the objective is 12 cm; the screen is placed 15 cm from the objective; the distance between the objective and the inner surfaces of the plates is about 60 cm.

In lines parallel to the edge of the dihedral angle the thickness is invariable, which results in $\Delta y = 0$. Therefore any interference fringes appear parallel to this edge. These sort of interference fringes is called *fringes of equal thickness*, since every interference fringe appears as a trace over the surface, where the thickness has the same value.

It follows from (1.19) that the fringe spacing $\Delta \widetilde{x}$, following from the requirement $\Delta \delta = 2\pi$, should be constant if the reflecting surfaces are planes:

$$\Delta \widetilde{x} = \frac{\lambda}{2\beta \cos \theta}. \tag{1.20}$$

We see that for a given angle of incidence θ and for a given wavelength λ, the fringe spacing becomes larger for smaller angle β.

In order to illustrate conditions for the appearance of this sort of interference, we consider the experimental scheme shown in Fig.1.19. Two thick parallel plane glass plates are pressed closely to get a small dihedral angle with a horizontal edge. A light beam from a tiny circular aperture illuminated by a mercury–helium lamp falls on the external

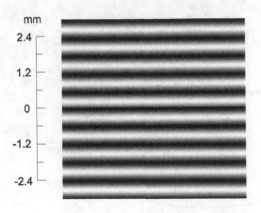

Figure 1.20 Interference fringes of equal thickness obtained from an air wedge.

surface of the first plate at a small angle (then the angle of incidence is nearly zero, $\theta \approx 0$). For observing the interference pattern with the reflected rays, a plane parallel glass plate with a semi–reflecting surface (a beam splitter) is placed between the aperture and the block of the two plates in such a way to permit reflection of the interfering rays approximately at a right angle to the incident light beam.

These interfering rays pass through an objective, which produces an image of the inner surface of the first plate on a screen. Thus interference fringes of equal thickness are formed. The interference pattern consists of equidistant bright and dark parallel bands and is shown in Fig. 1.20. These fringes are horizontal, i.e. parallel to the edge of the dihedral angle.

2.4   Newton's rings

The famous interference experiment suggested for the first time by I.NEWTON (1643-1727) may be demonstrated in the following way. Light from a mercury-helium lamp focused on a pinhole passes through a filter, then a beam-splitting plate positioned at 45° with respect to the direction of propagation, and finally falls on a system composed of a plane parallel plate and a plane convex lens (Fig.1.21).

The air film between the spherical surface of the lens and the plane plate can be regarded to be an air wedge of varying thickness h. By reflections and transmissions of the original rays on the inner surfaces of the lens and the plate two sorts of interfering rays appear. One sort is presented by rays propagating towards the semi-transparent plate. These rays are partially reflected by this plate in the direction of an objective, which images the air wedge on the observation screen. The other sort of interfering rays appear after transmission through the lens and the plate. A second objective (not show in Fig.1.21) images the air

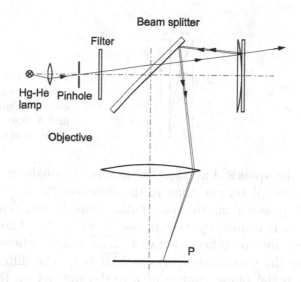

Figure 1.21. Setup for observation of NEWTON's rings. The objective of focal length 12 cm is at a distance of 15 cm from the screen and 60 cm from the system of the lens ($R = 20$ m) and the plane parallel plate. The optical filter provides the green line of $\lambda = 546$ nm.

wedge on a second screen. Both interference patterns are complementary to each other.

In order to derive a formula for the path differences of both sorts of interfering rays, let us consider a ray passing through the lens and then through the air layer at a distance r from the center of the spherical surface (see Fig.1.22). If h is the thickness of the air film for the given r and for the given radius R of the spherical surface of the lens, then, for the relationship between h, r and R, we can write

$$R^2 = r^2 + (R - h)^2 \ .$$

Assuming $R \gg r$ we rewrite the last expression: $R^2 - (R - h)^2 = r^2 \approx 2Rh$, and we get

$$h \approx r^2/(2R) \ .$$

We consider two incident rays S_1 and S_2, regarded as nearly parallel, capable of producing interference of reflected rays as well as transmitted ones (Fig.1.23). Ray S_1, after passing the lens body and being reflected on its spherical surface, falls on point A and is then partially reflected towards point B. The second ray S_2, after passing the lens body up to point B, is also partially reflected by this spherical surface in the same direction as first ray (the direction P_1). Ray S_1AB, after a second

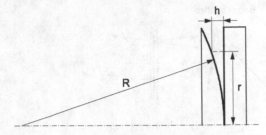

Figure 1.22 Geometrical considerations for an element composed of a lens and A plane parallel plate, being in optical contact.

reflection on the spherical surface, propagates throughout the air film, through the glass plate, and then in the direction P_2. Ray S_2, partially reflected at B, passes along the same path. Thus the interference pattern in transmission is formed by rays in the direction P_2. The geometrical path difference for the reflected rays is $2\overline{AB} + \lambda/2$, whereas the same magnitude for the transmitted rays is $2\overline{AB} + \lambda$. The difference of $\lambda/2$ is evident from the phase change of π of the ray SA on the boundary "air-glass". Ray S_1A, reflected towards point B, changes its phase at A, then the same occurs at point B. Assuming $\overline{AB}$ to be the thickness $h = r^2/(2R)$, we write the phase differences for the reflected δ_1 and the transmitted rays δ_2 as follows:

$$\delta_1 = 2\pi r^2/(\lambda R) + \pi \quad \text{and} \quad \delta_2 = 2\pi r^2/(\lambda R) + 2\pi \quad . \quad (1.21)$$

These relationships show that if δ_1 satisfies conditions to observe intensity minima then δ_2 fulfills the conditions for intensity maxima, and visa versa, since $\delta_2 - \delta_1 = \pi$. This fundamental property of interference is called the *complementary property*.

The interference fringes of both patterns appear as concentric circles with radial distances decreasing successively as r^2 increases. These patterns are called NEWTON's *rings*. If the central circle of the pattern formed by the reflected rays is dark, the transmitted rays provide a pattern with a bright central circle. This particular case always is realized if a so-called "optical contact" exists between the apex area of the lens and the surface of the glass plate. We say "optical contact" if the surfaces touch each other and the thickness between both elements is much smaller than the wavelength of light.

NEWTON's *rings* are shown in Fig.1.24. These patterns are found under the conditions $R = 20$ m, and $\lambda = 546$ nm. The linear dimensions of the sectors of the interference patterns presented in Fig.1.24 are equal to 5 mm at the focal length of the objective $f = 12$ cm; the objective at a distance of 15 cm from the screen and 60 cm from the plane surface of the lens. Early in the development of optics, interference with such a system of lens and plate permitted an estimation of the wavelength of

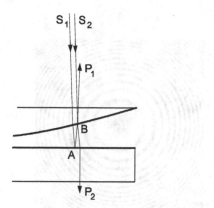

Figure 1.23 Two rays S_1 and S_2 from a remote point source, being nearly parallel, give rise to interfering rays in reflection (in the direction P_1) and transmission (in the direction P_2).

light. We assume the pattern under consideration obtained in reflection to be located in the vicinity of the plane surface of the lens. Thus for the given distances between the positions of the objective, the screen, and this pattern, the diameter of the first bright circle of this pattern should permit an estimation of the first maximum of the interference close to this plane surface. For the diameter we find: $2r = 4.8$ mm. It follows form (1.21) that the radius r associated with this maximum is represented by R and λ should be as $r = \sqrt{\lambda R/2}$. Thus, an estimation of λ is given by $\lambda = 2r^2/R$. In the case under discussion we obtain $\lambda \approx 576$ nm.

We note that fringes composed by transmitted rays appear with a low contrast, because the two interfering rays have unequal intensities (see (1.6)). It follows from the geometrical representation of Fig.1.23 that one transmitted ray S_2BP_2 is affected by one reflection between the inner surfaces, whereas the other transmitted ray S_1ABP_2 is partially reflected at points A and B, and then at the same point of the upper surface of the plate as the first ray. Therefore, the intensity of ray S_1ABP_2 is approximately $1/\mathcal{R}^2$ times smaller than the intensity of ray S_2BP_2, where $\mathcal{R}$ is the reflectivity of the boundary glass-air. Since the beams have nearly normal incidence, $\mathcal{R} \approx (n-1)^2/(n+1)^2 \approx 4\%$ (for uncoated glass surfaces). This value causes the low contrast of the interference pattern.

2.5 A plane parallel plate in a divergent pencil of beams - Testing the quality of a glass surface

We treat a setup for testing the quality (flatness and parallelism) of an optical substrate. A He-Ne laser beam is sent through a short-focus lens or a microscope objective in order to produce a pencil of rays which is practically emitted from a point source (Fig.1.25). The light source

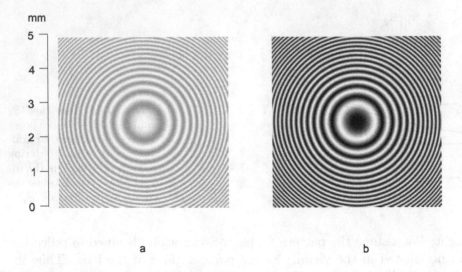

a   b

Figure 1.24. NEWTON's rings in transmitted (a) and reflected rays (b). These patterns are obtained under the conditions $R = 20$ m, $\lambda = 546$ nm. The screen of observation is placed 15 cm from the objective ($f = 12$ cm); the latter is at a distance of 60 cm from the plane surface of the lens.

thus produces a coherent spherical wave. At the focal plane of the lens we place a screen with a hole for observing the interference pattern, arising from the beams reflected on both sides of the glass plate under investigation. The first beam is reflected at the first surface (reflectivity ca. 4%) and interferes with a beam reflected at the second surface. This beam has nearly the same intensity, and the contrast of the fringes is high. We make the following observations:

- If the plate has two plane surfaces which are parallel to each other at distance d, the fringe system consists of circles centered around the optical axis.

- If the plate has still plane surfaces which form now a wedge (of very small angle θ), the fringe system concerns of circles which are centered around a point outside the axis. With increasing value of θ the center of the fringe system moves more and more away from the axis and the fringe system becomes an elliptical one.

- If the plate has schlieren or is uneven on one or both sides, the pattern becomes irregular, and one can find even more than one center or only curved stripes. Examples of observed interference patterns are shown in Fig. 1.26.

In Fig.1.27 the geometry of the problem is shown. Let us consider the phase difference at point C of the plate between rays 1 and 2, both

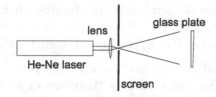

Figure 1.25. Testing the quality of a glass plate by means of a He-Ne laser and a screen. The interference pattern formed by reflected light is observed on the screen.

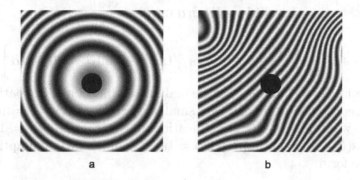

Figure 1.26. Interference fringes obtained with a nearly parallel plane plate (a); and with a plate having non-parallel uneven surfaces (b).

coming from the point source. We assume that the plate with refractive index n_2 is in air of refractive index n_1, and the point source is at distance R from the plate.

Following Fig.1.27, the optical path difference at point C is given by the expression: $\Delta s = n_2 a + n_2 b - n_1 c$, where $a = \overline{AB}$, $b = \overline{BC}$, and $c = \overline{CD}$. Ray 1 receives the plate at point A; here the thickness of the plate is found to be $d_1 = d_0 + x \tan \theta$, where $x = R \tan \alpha$ is the distance between point A and the perpendicular to the first glass surface, and as before θ is the angle associated with the wedge made by the planes of the plate. Ray 1 arrives at the first surface under angle α and runs into the plate, being refracted under angle β, hence the following is valid: $\sin \alpha / \sin \beta = n_2/n_1$. Since the distance a is one side of the triangle ABE, the ratio a/d_1 found from the triangle is

$$a/d_1 = \sin(90 + \theta)/\sin(90 - \beta - \theta) = \cos \theta / \cos(\beta + \theta) \quad .$$

Thus

$$a = d_1 \cos \theta / \cos(\beta + \theta) \quad .$$

If we denote by h the perpendicular to the first surface at point B, we get

$$h = a \cos \beta = d_1 \cos \theta \cos \beta / \cos(\beta + \theta) \quad .$$

The short lines x_1 and x_2 are found to be $x_1 = h \tan \beta$ and $x_2 = h \tan(\beta + 2\theta)$, respectively. The distance c is therefore expressed as $c = (x_1 + x_2) \sin \gamma$, where γ satisfies the relationship $(x + x_1 + x_2) = R \tan \gamma$. Taking into account that b is found from $b = h / \cos(\beta + 2\theta)$, then finally the required optical path difference Δs is given as

$$\Delta s = (d_0 + R \tan \alpha \tan \theta) \frac{\cos \theta}{\cos(\beta + \theta)} \cdot$$

$$\cdot \left\{ n_2 + n_2 \frac{\cos \beta}{\cos(\beta + 2\theta)} - n_1 \cos \beta (\tan \beta + \tan(\beta + 2\theta)) \sin \alpha \right\} \quad .$$

Assuming $n_1 = 1$ and $n_2 = n$ we make an estimation of the path difference under the condition of a very small angle θ, so that $\sin \theta = \tan \theta = \theta$ is valid, what allows the simplifications $\cos \theta / \cos(\beta + \theta) = 1 / \cos \beta$, $\cos \beta / \cos(\beta + \theta) = 1$, $\tan(\beta + 2\theta) = \tan \beta$. Under this assumptions Δs takes the form

$$\Delta s = 2(d_0 + \theta R \tan \alpha) \sqrt{n^2 - \sin^2 \alpha}$$

which is the same path difference as for a plane parallel plate with thickness d_0, but extended by the term $\theta R \tan \alpha$. Further, $d_1 = d_0 + \theta R \tan \alpha$ is the thickness at location x. So one can see, that in this approximation the only difference to a plane parallel plate is a change of the thickness with location x.

For simplicity, we treat further an air plate ($n_2 = 1$), for which we can write $\Delta s = 2(d_0 + \theta R \tan \alpha) \cos \alpha$. Thus we can easily find the condition for α_0: $d(\Delta s)/d\alpha = 0$, where α_0 is the angle under which the highest order of interference is observed. This gives $\tan \alpha_0 = \theta R / d_0$. The order of interference is given by $k = \Delta s / \lambda$ and the phase difference by $\Delta \theta = 2\pi k = 2\pi \Delta s / \lambda$. The highest observed order of interference, which appears in the center of the fringe system, is now not located at the optical axes, but under a certain angle of incidence α_0, and we find the distance between the beam axis and the center of the fringe system, observed in reflection at the observation screen, by $l_0 = 2R \tan \alpha_0 = 2R^2 \theta / d_0$.

We call attention to the fact that in practice the angle θ is very small, therefore beam 1 and 2 runs practically parallel after the plate. With a point source, the interference figure is not only located on the surface of the plate, but is observable even without using a lens in whole space. For example we treat an optical window, 30 mm in diameter, which has at

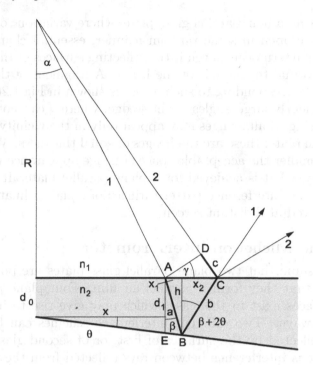

Figure 1.27. Interference at the surfaces of a glass plate forming a wedge.

one place on its circumference a thickness of 3 mm, at the opposite side 2.97 mm, but has flat plane surfaces. We place the plate ($n_2 = 1.5$) at $R = 100$ mm distance from the point source and we find that the highest order of interference (the center of the fringe system) appears under an angle of 4.3° (corresponding to a distance of the fringe center from the point source of $2 \cdot 100$ mm$\cdot \tan 4.3° = 15.04$ mm). One easily can see, that this method is very sensitive to detect small angles θ between the surfaces of optical plates.

We calculate the limit case for a distance of $R = 250$ mm and a distance between the point source and the fringe system center of 1 mm, which we can note without problems with the naked eye. This means that we have an angle $\alpha_0 = 1/500 = 0.002 = 0.11°$. For an air plate with $d_0 = 1$ mm ($n_2 = 1$) we find the highest order under angle α_0 if θ has a value as small as 0.000008. For a plate of 12 mm diameter this corresponds to a change in thickness of 0.0001 mm (or, for $\lambda = 500$ nm, 1/5 of the wavelength)! Due to these considerations, this method can be easily used to adjust the parallelism of air-spaced FABRY-PEROT interferometer plate pairs coated with layers of low reflectivity (for example 20%). Such interferometers are often used in a laser resonator for mode selection. Using a non–parallel glass plate, where variations of thickness of the plate happen in some random manner, essential changes of the interference pattern occur, even if the reflecting surfaces of the plate are positioned normal to the axis of the beam. A fragment of the interference pattern corresponding to such a case is shown in Fig.1.26,b. In the case of sufficiently large angles of the wedge, where θ can not longer be treated as being small, fringes may appear only in the vicinity the upper surface of the plate; these are the fringes of equal thickness. When angle θ becomes smaller the acceptable space of fringe appearance will extend up to infinity, what is achieved for a plane parallel plate. If this is the case we call the interference pattern fringes of equal inclination, which can be observed at a distant screen.

2.6   The Michelson interferometer

Let us assume that two plane parallel glass plates are positioned in such a way that they form a parallel air film. Four plane parallel reflecting surfaces exist in this case, which may give rise to interference in different ways. Two different interference schemes can be realized with rays reflected by the surfaces of first, or of second glass plate. A third scheme is interference between rays reflected from the surfaces of the air film between the plates. This case is now of interest. For this reason we regard the thicknesses of both plates to be much larger than the separation between them. Such a requirement results in a reduction of interference fringes from the glass plates until they nearly disappear. In turn, a sufficiently thin air film will cause highly visible interference fringes of equal inclination, if this air plate is illuminated by parallel rays from a quasi-monochromatic light source. We also assume that the thickness of this air film can be slightly adjusted by turning micrometer screws for tuning the air film from a dihedral wedge to a nearly parallel film. In order to observe changes of the interference pattern caused by such an adjustment of the thickness, the interference fringes formed by parallel rays reflected by the plates are observed by means of a telescope (Fig.1.28).

Light from a mercury lamp illuminates a circular aperture positioned at the first focal plane of an objective. After passing a "yellow" filter ($\overline{\lambda} = 580$ nm), parallel rays fall on a system composed of two thick (thickness 3 cm) plane parallel glass plates. This system inclined at 30° to the incident rays provides an interference pattern formed by reflections on the inner surfaces of the plates, which is observed by means of a telescope.

We assume that the glass plates used for generation of the air plate have very high quality surfaces, provided by fabrication. How parallel

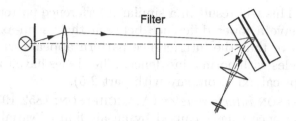

Figure 1.28. Setup for the observation of interference fringes of equal inclination with an air film.

the air plate under consideration is should be adjusted by hand. To perform adjustment, we need a criterion, which can be provided by the shape of the observed interference fringes.

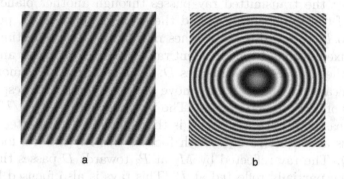

Figure 1.29. Interference fringes of equal inclination with the air film forming a wedge (a) and the air film forming a nearly parallel plate. Light rays falling on the plates at 30° produce fringes of elliptical shape (b).

If the two inner reflecting surfaces first make a dihedral angle then the interference pattern will appear as a system of parallel straight interference fringes (Fig.1.29,a). When slightly turning the adjustment screws, the fringes may curve, tending to the shape of circular arcs, which indicated a decrease of the dihedral angle. If this is the case, we believe that the inner reflecting surfaces are both nearly normal to the incident beam, or their mutual position as being close to be parallel. While performing the adjustment continuously more towards a parallel air film, interference fringes of elliptical shape must appear (Fig.1.29,b). The fringes shown in Fig.1.29,b are obtained under the following conditions: the thickness of the parallel air film is $h = 0.83$ mm, the glass plates make an angle of 30° with respect to the incident rays, and the wavelength is $\lambda = 580$ nm. Now one may repeat the adjustment under a smaller angle of incidence. This will result in a similar interference pattern where the ellipses will have a lower difference between their semi-axes. We can therefore believe that the interference pattern appears as a system of concentric circles with normal incidence. The circles have to be centered around the optical axis (compare with part 2.5).

The MICHELSON *interferometer* (A.MICHELSON, 1852-1931) allows to observe interference fringes caused by an air film of variable thickness at normal incidence. The principal scheme of such a device is shown in Fig.1.30. Light from an extended source S (a mercury-helium lamp in this case) formed by an objective O_1 passes a filter and penetrates to a semi-reflecting surface R of the plane parallel glass plate D positioned at 45° to the incident beam. For this reason, two beams of equal intensity (one is reflected at R, and the other is transmitted), propagating under orthogonal directions, appear after D. The reflected ray passes towards mirror M_1; the transmitted ray passes through another plane parallel plate C of the same thickness and the same material as D, positioned parallel to D. Then this ray reaches mirror M_2. The reflecting surface of M_2 is fixed normal to the incident ray and therefore makes an angle of 45° with the surfaces of both plates D, C. The mirror M_1 is mounted on a mechanical system which may move back and forth with respect to D by means of a micrometer screw. The ray reflected by M_1 at P_2 reaches plate D, passes it once more, and is then partially transmitted through R towards an objective O_2, which focuses this ray into its focal plane (point P). The ray reflected by M_2 at P_3 towards D passes through C and is then partially reflected at R. This rays is also focused by O_2 at the same point P of the focal plane.

Thus both rays under consideration pass the glass material of D and C three times. Since D and C are made from the same glass, the optical path lengths of both rays within the glass plates are the same, whereas the path difference is dependent only on the position of the mirror M_1 relative to the mirror image of M_2, M_2'. The extra glass plate C, called the compensator, permits any effects of dispersion, while light is propagating within D, to be negligible. Since plates D and C are positioned at 45° with respect to incident rays, and since M_1 and M_2 are at 45° with respect to plane surfaces of D and C, the virtual image of M_2 caused by the reflecting surface R will be located parallel to the plane of M_1 (this image M_2' is shown by a dashed line in Fig.1.30).

For this reason, the operating principle of the MICHELSON interferometer is similar to that of a plane parallel air film. The thickness of such an air film is equal to the difference h of the distances of M_1 and M_2 from the reflecting surface R, and can be zero or even have opposite sign. It follows from the fact that for any pair of rays which give rise to

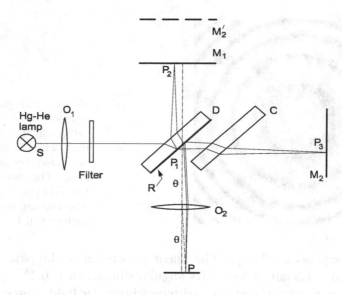

Figure 1.30. The MICHELSON interferometer.

interference, the optical paths within D are completely compensated by C. The arrangement of the mirrors M_1 and M_2 allows inclinations of the planes of the mirrors by means of micrometer screws to adjust the MICHELSON interferometer. Adjustment of the interferometer is performed in a way similar to that considered in the experiment with an air film described before. We also note here the existence of a second interference pattern. This interference pattern is localized in space behind the plate D towards the light source, and can be made visible if an extra beam splitter is used. In fact, every incident ray generates four rays: one pair of them produces the first type of interference considered above, and the other pair may form the second type of interference. Let ray P_1P_2 reach the surface R after reflection by mirror M_1. This ray is partially reflected and propagates toward the light source. The other ray propagating from point P_3 passes through C and D and falls on R, where it is partially transmitted towards the direction of the light source. This is the pair of rays that forms the second interference pattern. It follows from the consideration of the reflections on R that the phase differences associated with each pair of interfering rays are different by π according to the complementary property of interference.

An interference pattern obtained with the adjusted MICHELSON interferometer is shown in Fig.1.31. The geometrical difference between distances from beam splitter R to the mirrors M_1 and M_2 is $h = 0.67$ mm. The interferometer is illuminated by the yellow light ($\overline{\lambda} = 580$ nm)

Figure 1.31 Interference fringes observed with yellow light ($\lambda = 588$ nm) with a MICHELSON interferometer of $h = 0.67$ mm. The focal length of the objective is $f = 12$ cm. The angular width of the pattern is 6.4°.

of a mercury-helium lamp. The linear dimension of the pattern in the focal plane is 1.3 cm, whereas its angular dimension is 6.4°.

In principle, any extended quasi-monochromatic light source placed in the first focal plane of an objective may give rise to interference fringes of equal inclination with the MICHELSON interferometer. In practice, a small circular diaphragm illuminated by a bright source is used (it is needed to permit uniform brightness of the fringes within the area of the interference pattern) to form a system of parallel rays which illuminates the interferometer. For a given inclination θ of an incident ray with respect to the direction of normal incidence, the path difference Δs between interfering rays can be calculated by

$$\Delta s = 2h\cos\theta \quad , \tag{1.22}$$

what follows from the general formula for fringes of equal inclination (1.15) at $n \approx 1$, valid for the case of an air film. The central part of the interference pattern in Fig.1.31 is formed by rays propagating close to normal incidence ($\theta \approx 0$). Therefore, the central interference circle (bright in this case) is associated with the maximal path difference $\Delta s \approx 2h$, as well as the maximal order of interference, whereas fringes at the periphery have lower orders, because Δs decreases with increasing θ. Let us estimate the highest order m of the fringes using the formula:

$$\delta = 4\pi h/\lambda = 2\pi m \quad \text{and} \quad m = 2h/\lambda \quad .$$

Substitution of the numerical values used above gives $m \approx 2280$.

3.   Multiple–beam interference

All the interference schemes considered above can be called *two-beam interference* schemes, because every point in such an interference pattern is formed only by one pair of interfering rays. Another type of interference occurs under superposition of more than two interfering rays. This type of interference is called the *multiple beam interference*.

3.1   The Lummer-Gehrcke interferometer

Let us consider the LUMMER-GEHRCKE *interferometer* (O.LUMMER, 1860-1925, and E.J.GEHRCKE, 1878-1960) as an example of multiple-beam interference. The operating principle of the interferometer consists in successive reflections of one incident ray at two inner surfaces of a plane parallel plate, which is called LUMMER-GEHRCKE plate, and which is usually made from quartz or glass.

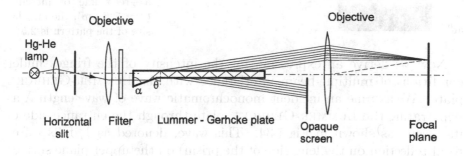

Figure 1.32. Setup for demonstration of multiple-beam interference with a LUMMER-GEHRCKE interferometer.

In the experimental scheme shown in Fig.1.32, a parallel light beam of quasi-monochromatic light is formed from a mercury-helium lamp by means of a horizontal narrow slit, an objective, and an optical filter. This beam penetrates into the parallel plane part of the LUMMER-GEHRCKE plate after total reflection at the inner side of a prism which is mounted at the entrance side of the plate. The reflecting side of the prism has a certain angle α with the reflecting surfaces of the plate, which provides some range of angle of incidence θ_i of the incoming rays for both reflecting surfaces. The angle θ_i is close to the critical angle of total reflection. For this reason a ray, reflected on the inner surface of the parallel plate, is transmitted to the space outside only with small intensity. The intensity of the ray remaining inside the plate decreases slowly and a sufficiently large number of reflections (and transmissions) takes place. Since the reflections occur on both sides of the plate, two systems of rays are leaving the LUMMER-GEHRCKE plate: one system is formed by reflections on the upper side of the plate, and the other one by reflections on the bottom side. If one system is suppressed by a screen and the other one is focused by an objective within its focal plane, then an interference pattern as shown in Fig.1.33 will appear.

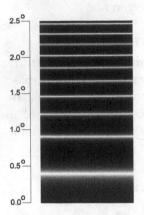

Figure 1.33 Interference fringes with the LUMMER-GEHRCKE interferometer. The thickness of the plate is $h = 4.2$ mm; the angle of the prism $\alpha = 66°$; the plate is illuminated by a green line of mercury ($\lambda = 546$ nm); the angular size of the pattern is $2.5°$.

Now we derive an expression for the intensity of the fringes under conditions of multiple-beam interference with the LUMMER-GEHRCKE plate. We assume an incident monochromatic wave of wavelength λ as penetrating the LUMMER-GEHRCKE plate through the entrance side of its prism, as shown in Fig.1.34. This wave, denoted as 1, falls (after total reflection on the long side of the prism) on the upper plane surface at the angle θ_i, then it is partially reflected (ray 3) and transmitted (ray 2) at the angle θ_t. We introduce the amplitude transmissivity τ and the amplitude reflectivity ρ :

$$\tau = E_2/E_1 \quad \text{and} \quad \rho = E_3/E_1 \ . \tag{1.23}$$

It follows from the FRESNEL formulae (compare Part 1, Chapter 4) that the magnitudes τ and ρ are both real functions of the angles θ_i, θ_t and the refractive index n of the plate, regarding the plate as surrounded by air of the refractive index $n_{air} \approx 1$. Further, ray 3 generates a pair of rays, ray 4 and 5 on the bottom surface, then ray 5 will provide the next pair of rays on the upper surface, and so on. It is clear that each reflected ray of an odd number gives rise to two rays, whose amplitudes have to satisfy (1.23). In other words, the ratios for any transmitted and reflected rays are:

$$\tau = \frac{E_2}{E_1} = \frac{E_4}{E_3} .... \frac{E_{2n}}{E_{2n-1}} .... , \quad \text{and} \quad \rho = \frac{E_3}{E_1} = \frac{E_5}{E_3} .... \frac{E_{2n+1}}{E_{2n-1}} .... \tag{1.24}$$

The rays with even numbers, which propagate outside the plate on its upper or lower side, will contribute to multiple-beam interference.

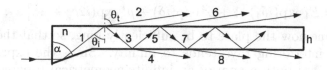

Figure 1.34. The LUMMER-GEHRCKE plate. An incident ray penetrating the plate normal to the entrance side of the prism is reflected and reflected on both sides of the plate to generate a system of parallel rays transmitted out of the plate.

Let us write down the superposition of the fields of all rays transmitted though the upper surface, the field E_{up}, and the lower surface, the field E_{low}, as sums of their complex amplitudes:

$$E_{up} = E_2 \exp(i\delta_2) + E_6 \exp(i\delta_6) + E_{10} \exp(i\delta_{10}) + \ldots \quad ,$$

$$E_{low} = E_4 \exp(i\delta_4) + E_8 \exp(i\delta_8) + E_{12} \exp(i\delta_{12}) + \ldots \quad , \qquad (1.25)$$

where $E_2, E_4, E_6, E_8, E_{10}, E_{12}$ are the real amplitudes of the fields, and $\delta_2, \delta_4, \delta_6, \delta_8, \delta_{10}, \delta_{12}$ are their phases. The phase difference between two neighbouring rays transmitted through each surface is given by

$$\delta = 2khn \cos\theta_i \quad ,$$

since it has to be calculated similarly to the case of fringes of equal inclination (see (1.22)). Thus for the the phase difference δ in (1.25) we can write

$$\delta_6 - \delta_2 = \delta_8 - \delta_4 = \delta_{10} - \delta_6 = \delta_{12} - \delta_8 = \ldots = \delta = 2khn \cos\theta_i \quad .$$

Taking into account the relations above we re-write the right-hand sides in (1.25) in the form

$$E_{up} = E_2 \exp(i\delta_2) \left[1 + \frac{E_6}{E_2} \exp(i\delta) + \frac{E_{10}}{E_2} \exp(i2\delta) + \ldots \right] \quad ,$$

$$E_{low} = E_4 \exp(i\delta_4) \left[1 + \frac{E_8}{E_4} \exp(i\delta) + \frac{E_{12}}{E_4} \exp(i2\delta) + \ldots \right] \quad . \qquad (1.26)$$

The relationships (1.24) allows the representation of all amplitudes in (1.26) in terms of E_1 and the parameters τ and ρ. For example, $E_2 = \tau E_1$, $E_3 = \rho E_1$, $E_4 = \tau\rho E_1$, $E_5 = \rho^2 E_1$, $E_6 = \tau\rho^2 E_1$, $E_7 = \rho^3 E_1$, $E_8 = \tau\rho^3 E_1$, and so on. Substitution of all ratios in (1.26) by the expressions mentioned above gives us for E_{up} and E_{low}

$$E_{up} = E_2 \exp(i\delta_2) \left[1 + \rho^2 \exp(i\delta) + \rho^4 \exp(i2\delta) + \ldots \right] \quad ,$$

$$E_{low} = E_4 \exp(i\delta_4) \left[1 + \rho^2 \exp(i\delta) + \rho^4 \exp(i2\delta) + \; ... \right] \quad . \qquad (1.27)$$

We assume now the plate to be sufficiently long, so that the effective amount of interfering rays tends to infinity. Since the expressions in parentheses are both a sum of an infinite geometrical progression with the factor $\rho^2 \exp(i\delta)$, we find for the complex amplitudes E_{up} and E_{low}

$$E_{up} = E_2 \exp(i\delta_2) \frac{1}{1 - \rho^2 \exp(i\delta)} \quad ,$$

$$E_{low} = E_4 \exp(i\delta_4) \frac{1}{1 - \rho^2 \exp(i\delta)} \quad .$$

The intensities I_{up} and I_{low} of the interference patterns, caused by all transmitted rays, have to be proportional to $E_{up}E_{up}^*$ and $E_{low}E_{low}^*$, respectively:

$$I_{up} \sim E_2^2 \frac{1}{1 + \rho^4 + 2\rho^2 \cos\delta} \quad , \qquad I_{low} \sim E_4^2 \frac{1}{1 + \rho^4 + 2\rho^2 \cos\delta} \quad .$$
$$(1.28)$$

Since $E_2 = \tau E_1$ and $E_4 = \rho\tau E_1$, and $E_1 = \varkappa E_0$, where E_0 specifies the amplitude of the incident ray, and $\varkappa$ is the amplitude transmissivity at normal incidence on the boundary "air-glass", the intensities of rays 2 and 4 are found to be

$$E_2^2 = \varkappa^2\tau^2 E_0^2 \sim \varkappa^2\tau^2 I_0 \quad , \text{ and } \qquad E_4^2 = \varkappa^2\tau^2\rho^2 E_0^2 \sim \varkappa^2\tau^2\rho^2 I_0 \quad .$$

Substitution for E_2^2 and E_4^2 by their forms above gives

$$I_{up} = (\varkappa/\tau)^2 I_0 \frac{\tau^4}{1 + \rho^4 + 2\rho^2 \cos\delta} \quad ,$$

$$I_{low} = (\varkappa\rho/\tau)^2 I_0 \frac{\tau^4}{1 + \rho^4 + 2\rho^2 \cos\delta} \quad . \qquad (1.29)$$

By whatever polarization of the incident beam the LUMMER-GEHRCKE plate is illuminated, the quantity τ^2 is equal to the transmissivity $\mathcal{T}$, and ρ^2 to the reflectivity $\mathcal{R}$ for the boundary "glass-air". Hence, these magnitudes have to satisfy the expression

$$\mathcal{R} + \mathcal{T} = 1 \quad , \qquad (1.30)$$

assuming that any absorption of light energy is absent. Let the transmissivity of the entrance surface (at normal incidence) be $\mathcal{T}_0 = \varkappa^2$. Together with (1.30) this allows the representation of (1.29) in terms of $\mathcal{R}$, $\mathcal{T}$ and $\mathcal{T}_0$:

$$\frac{I_{up}}{I_0}\frac{\mathcal{T}}{\mathcal{T}_0} = \frac{I_{low}}{I_0}\frac{\mathcal{T}}{\mathcal{R}\mathcal{T}_0} = \frac{\mathcal{T}^2}{1 + \mathcal{R}^2 + 2\mathcal{R}\cos\delta} =$$

$$= \frac{(1-\mathcal{R})^2}{(1-\mathcal{R})^2 + 4\mathcal{R}\sin^2(\delta/2)} \quad . \tag{1.31}$$

By means of a quantity

$$\mathcal{F} = \frac{\pi\sqrt{\mathcal{R}}}{1-\mathcal{R}} \quad , \tag{1.32}$$

which is called *reflectivity finesse* of the interference fringes, the relative transmitted intensity I_{up}/I_0 in (1.31) may be written as follows:

$$\frac{I_{up}}{I_0}\frac{\mathcal{T}}{\mathcal{T}_0} = \frac{1}{1 + (2\mathcal{F}/\pi)^2 \sin^2(\delta/2)} \quad . \tag{1.33}$$

The angular positions of the interference maxima result from the following condition:

$$\delta = 2\pi m \quad , \qquad 2hn\cos\theta_i = m\lambda \quad , \tag{1.34}$$

where the order of interference m decreases with increasing angle θ_i. The denominator on the right-hand side of (1.33) equals unity at maxima $(\delta = 2\pi m)$, whereas for minima it takes the values $1 + (2\mathcal{F}/\pi)^2$ under the conditions $\delta = \pi + 2\pi m$.

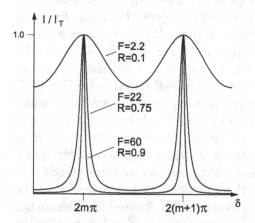

Figure 1.35 Functions $1/[1 + (2F/\pi)^2 \sin^2(\delta/2)]$ calculated at different values of the finesse (which is a function of the reflectivity). In each cases two neighbouring maxima are separated by the phase difference 2π.

It follows from (1.31), that second system of rays, transmitted though the bottom surface of the LUMMER-GEHRCKE plate (rays 4, 8, 12, ... in Fig.1.34), also produces an interference pattern. For normalized intensity I_{low}/I_0, we obtain the same intensity distribution, and the same positions of maxima and minima. The extra factor $(1/\rho)^2 = 1/\mathcal{R}$ in (1.29) will tend to unity at $\mathcal{R} \to 1$.

An important effect of multiple-beam interference results in the concentration of light intensity within narrow bright interference fringes separated by wide dark regions. This effect is enhanced by increasing the reflectivity $\mathcal{R}$, which is equivalent to an increase of the effective number of rays contributing to the generation of the interference pattern. Mathematically, it follows from the increase of terms giving substantial intensity contributions in the sums in (1.25). Intensity distributions

$$I \sim \frac{1}{1 + (2\mathcal{F}/\pi)^2 \sin^2(\delta/2)}$$

as functions of the phase difference δ obtained at $\mathcal{F} = 2.2$ ($\mathcal{R} = 0.1$), 22 ($\mathcal{R} = 0.75$), and 60 ($\mathcal{R} = 0.9$) are shown in Fig.1.35. In the case of the LUMMER-GEHRCKE interferometer, the reflectivity $\mathcal{R}$ is dependent on the angle of incidence θ_i, which is controlled by the angle of the prism α (Fig. 1.33). For a given wavelength λ, and refractive index n, by carefully choosing the angle α one can arrange that the angle of incidence θ_i is only slightly smaller than the critical angle of total reflection. If this is the case, the reflectivity $\mathcal{R}$ is nearly unity (the finesse F becomes sufficient greater than unity), and the interference fringes appear very sharp. Nevertheless, even in this case the number of rays contributing to the interference pattern is limited by the finite length of the glass plate. For a given angle $\alpha = 65.5°$ of the LUMMER-GEHRCKE plate and for a given refractive index $n = 1.52$ (a crown glass) let us estimate the reflectivity $\mathcal{R}$ of rays polarized normal to the plane of incidence. It follows from the FRESNEL formulae that $\mathcal{R}_\perp = \sin^2(\theta_i - \theta_t)/\sin^2(\theta_i + \theta_t)$. A geometrical consideration (Fig.1.33) gives the angle θ_i to be: $\theta_i = 2\alpha - 90° = 41°$. The angle of refraction θ_t is calculated by $\sin\theta_t = 1.52 \cdot \sin 41°$ and we get $\theta_t = 85.7°$. Finally, the reflectivity may be estimated to be $\mathcal{R} \approx 0.77$. The critical angle of incidence θ_i' corresponding to total reflection under the conditions above is estimated to be 41.14°. A small increasing of the angle θ_i from 41° closer to 41.14° would cause the reflectivity $\mathcal{R}$ to tend to unity, and the reflectivity finesse $\mathcal{F}$ becomes much larger. For a plate of 120 mm length and 4.2 mm thickness, one obtains roughly 15 beams which can interfere. Thus the sharpness of the fringes and the spectral resolving power of the interferometer are limited by the number of interfering beams, and the finesse can not exceed substantially the value 15 (compare paragraph 4.1). The interference fringes shown in Fig.1.33 are obtained with the first system of rays leaving the side of the plate opposite to the side containing the prism as shown in Fig.1.32. The LUMMER-GEHRCKE plate used in the experiment has a length of 120 mm and a thickness of $h = 4.2$ mm. The formula (1.31) allows the calculation of the path difference between two neighbouring interfering rays under these conditions ($\theta_i = 41°$; $h = 4.2$ mm; and $n = 1.52$) to 9.64 mm. The interference order, corresponding to the first maximum, follows from (1.34) and is $m = 17655$ for the green mercury line ($\lambda = 546$

nm; a so-called *interference filter* is used for selecting this spectral line to give a high degree of monochromaticity). The interference order m, as well as fringe spacing, decrease with increasing angle θ_t as shown in Fig.1.33.

4.   The Fabry-Perot interferometer

The most important application of the multiple–beam interference phenomenon is the FABRY-PEROT *interferometer* (C.FABRY, 1867-1945, and J.-P.PEROT, 1863-1925), which is widely used and has also great importance in laser physics. This instrument consists of two glass or quartz plates with their plane parallel inner surfaces coated by a layer of high reflectivity. The FABRY-PEROT interferometer has to be adjusted in a way that the inner surfaces of the plates form a plane parallel air film. Alternatively, a parallel quartz or glass plate can be coated on both sides with layers of high reflectivity (then the instrument is called often solid etalon). As with the LUMMER-GEHRCKE interferometer, the splitting of an original incident ray into a system of parallel rays is the operating principle of the FABRY–PEROT interferometer (Fig.1.36). Apart from the system of rays, transmitted through the FABRY-PEROT interferometer, a second system of parallel rays, reflected backwards, is present.

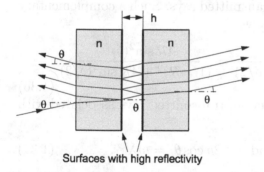

Surfaces with high reflectivity

Figure 1.36 Two plane parallel plates set up a FABRY-PEROT interferometer. An incident ray creates two systems of interfering rays: one is composed by the transmitted rays, the other by reflected rays.

For a given angle of incidence θ_i of the original ray, the transmitted rays will appear as a system of parallel rays leaving the interferometer at the same angle θ_i with respect to the normal to the planes of the plates. For this reason, the intensity obtained due to interference of the transmitted rays has to be described by a formula similar to (1.31), omitting the factor T/T_0. Since it is assumed that the plates are plane parallel, the angle of incidence θ_i remains the same for the first external surface as for the inner surfaces. For this reason the transmissivity for the original ray is equal to T, hence the factor above should be nearly unity. Thus, for the intensity of the interference, obtained by the system of parallel transmitted rays, we may write (see (1.31)):

$$\frac{I_T}{I_0} = \frac{\mathcal{T}^2}{1 + \mathcal{R}^2 + 2\mathcal{R}\cos\delta} = \frac{(1-\mathcal{R})^2}{(1-\mathcal{R})^2 + 4\mathcal{R}\sin^2(\delta/2)} \quad, \qquad (1.35)$$

where, as before, $\mathcal{T}$ is the transmissivity and $\mathcal{R}$ is the reflectivity of the coating layers of the inner surfaces of the plates. The phase difference δ between two neighbouring interference rays is calculated as

$$\delta = 2h\cos\theta_i \quad,$$

where h is the separation of the plates, provided that there an air film exists between the plates and that its refractive index is assumed to be $n_{air} \approx 1$. For a solid etalon (index of refraction n), embedded in air, one gets $\delta = 2h\sqrt{n^2 - \sin\theta_i}$.

We shall now consider the total light flux, which falls on the interferometer. Most of the flux is divided by the inner surfaces into two systems of interfering rays as mentioned above. Reflections on the outer surfaces of both plates may be regarded to be negligible since the reflectivity $\mathcal{R}$ of the inner surfaces is much larger than that of both "air–glass" boundaries under conditions of nearly normal incidence. Under this assumption, for the given original incident ray, the intensity distribution created by the reflected rays may be found as being complementary to the intensity distribution of the transmitted rays. Such a complementary distribution I_R/I_0 is given as follows:

$$\frac{I_R}{I_0} = 1 - \frac{(1-\mathcal{R})^2}{(1-\mathcal{R})^2 + 4\mathcal{R}\sin^2(\delta/2)} = \frac{4\mathcal{R}\sin^2(\delta/2)}{(1-\mathcal{R})^2 + 4\mathcal{R}\sin^2(\delta/2)} \quad.$$
$$(1.36)$$

The angular positions of the system of transmitted rays associated with maxima satisfy the condition

$$\delta = 2\pi m \quad, \qquad \text{and} \qquad 2h\cos\theta_i = m\lambda \quad. \qquad (1.37)$$

At the same time the conditions (1.37) result in minima of the system of reflected rays, and visa versa. This corresponds to the complementary property of interference which appears here in terms of the conservation of the total flux of the incident light (or the total light energy).

Usually the interference pattern generated by the transmitted rays is observed. Since all interfering rays leave the interferometer parallel to each other, a lens is necessary to transform the pattern from infinity to its focal plane. The scheme shown in Fig.1.37 demonstrates the operating principle of a FABRY-PEROT interferometer.

As mentioned before, a FABRY-PEROT interferometer can also consist of one plane parallel glass or quartz plate coated on each side (called *solid*

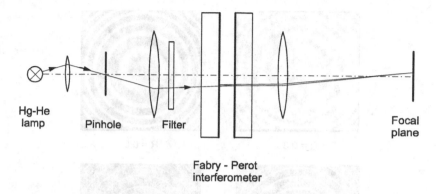

Figure 1.37. Setup for the demonstration of the operating principle of the FABRY-PEROT interferometer.

state etalon). Of course there is no chance for changing the thickness or to adjust the parallelism afterwards; one is dependent on accurate fabrication of the glass plate.

One can coat four areas of a plane parallel plate (thickness $h = 0.8$ mm) on both sides with thin aluminum layers in a way to obtain certain different values of the reflectivity $\mathcal{R}$ of the coated parts. Interference patterns produced by transmitted light, obtained at four values of the reflectivity $\mathcal{R}$: 0.3, 0.5, 0.72 and 0.86, are shown in Fig.1.38. Since the phase difference in (1.37) remains the same with rotating the plates around its surface normal vector, and since the beam is assumed to be axial symmetrical, the interference pattern appears as a system of concentric circles. With an increase in the reflectivity $\mathcal{R}$ (and therefore also of the finesse $\mathcal{F}$) the sharpness of the bright fringes increases.

We call attention to the fact that a high monochromaticity of the incident light is needed here, which will be discussed latter. Such a monochromaticity is provided by an optical interference filter which selects a green line ($\lambda = 546$ nm) of a low-pressure mercury lamp.

4.1 Resolving power and free spectral range

4.1.1 Shape of the interference fringes for high values of $\mathcal{R}$

We have seen that the fringes obtained with a FABRY-PEROT interferometer become sharper with increasing reflectivity $\mathcal{R}$, if a single spectral line (a monochromatic source) is illuminating the instrument. If $\mathcal{R}$ is sufficiently large, the width of the fringes will be much smaller than the separation of two neighbouring fringes (Fig.1.39,a). Let us consider two neighbouring maxima of an intensity distribution formed by the transmitted rays at a high value of $\mathcal{R}$. For the given wavelength λ, separation

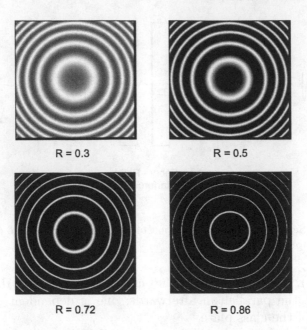

Figure 1.38.   Interference patterns with a FABRY-PEROT interferometer obtained for different reflectivity $\mathcal{R}$ of the reflecting layers. In each case the interferometer has the thickness of 0.8 mm and is illuminated by a green line ($\lambda = 546$ nm) of a mercury lamp.

h, and angle θ_i, we obtain the distribution described by formula (see (1.31) and (1.35)):

$$\frac{I_T}{I_0} = \frac{1}{1 + (2\mathcal{F}/\pi)^2 \sin^2(\delta/2)} \quad , \qquad (1.38)$$

where the finesse is $\mathcal{F} = \pi\sqrt{\mathcal{R}}/(1 - \mathcal{R})$. We introduce a small phase deviation $\Delta\delta$ around the phase amount, corresponding to the mth interference maximum, as follows:

$$\delta = 2\pi m + \Delta\delta \quad ,$$

which allows us to write $\sin^2(\pi m + \Delta\delta/2) = \sin^2(\Delta\delta/2)$. Since $\Delta\delta$ is assumed to be much smaller than 2π, we make the approximation $\sin^2(\Delta\delta/2) \approx (\Delta\delta/2)^2$. Hence the distribution around this maximum takes the form

$$\frac{I_T^{(t)}}{I_0} = \frac{1}{1 + (\mathcal{F}\Delta\delta/\pi)^2} \quad . \qquad (1.39)$$

This function describes the shape of the observed fringes in the vicinity of the maximum. It is a LORENTZian curve. The full width at half maximum (abbreviated very often as FWHM) of the LORENTZian shape, ε, is, in phase terms, twice the deviation $\Delta\delta$, corresponding to one half of the normalized intensity, $I_T^{(t)}/I_0 = 0.5$. For this reason, for the width ε we may write down the condition:

$$\frac{1}{1+(\mathcal{F}\varepsilon/2\pi)^2} = \frac{1}{2} \,,$$

and we get for ε:

$$\varepsilon = 2\pi/\mathcal{F} = 2\frac{1-\mathcal{R}}{\sqrt{\mathcal{R}}} \,. \tag{1.40}$$

Instead of using the finesse $\mathcal{F}$ the LORENTZian curve of the fringe around the maximum may be represented in terms of its width ε in a form:

$$\frac{I^{(t)}}{I_0} = \frac{1}{1+(2\Delta\delta/\varepsilon)^2} \,. \tag{1.41}$$

We can now compare the full width of the transmission peak at half maximum, ε, with the phase difference 2π between two neighbouring

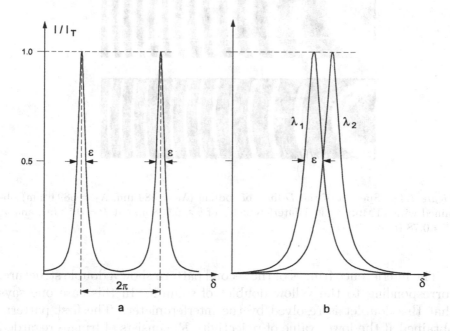

Figure 1.39. Illustrating the width of spectral lines (a); the RAYLEIGH criterion for two spectral lines of wavelengths λ_1, and λ_2, just resolved by means of a FABRY-PEROT interferometer (b).

maxima. We obtain $\mathcal{F} = 2\pi/\varepsilon$, thus, the finesse just tells us the ratio of the width of the bright fringes to their distance. This first definition, well adapted to spectroscopic practice, did lead to the designation "finesse".

4.1.2   The resolving power

Due to the fact that interference patterns obtained with the FABRY-PEROT interferometer consist of fine sharp fringes, if the finesse is sufficiently high, this instrument allows a precise analysis of the wavelength composition emitted by a light source. As examples of such spectra, interference patterns obtained with a FABRY-PEROT interferometer are shown in Fig.1.40. Here the interferometer is illuminated by a low pressure sodium lamp. This lamp emits two bright spectral lines with wavelengths of $\lambda_1 \approx 589$ nm and $\lambda_2 \approx 589.6$ nm (the so-called yellow D-lines of sodium). The first pattern (Fig.1.40,a) is obtained for a reflectivity of the interferometer surfaces of $\mathcal{R} = 0.53$ (provided by aluminium layers), the second pattern (Fig.1.40,b) is obtained for a reflectivity $\mathcal{R} = 0.78$. In both cases the thickness of the interferometer was $h \approx 0.038$ mm.

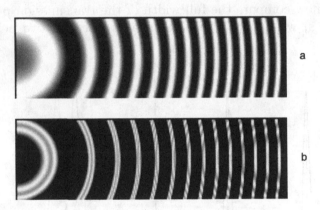

Figure 1.40.   Spectra of the D-lines of sodium ($\lambda_1 = 589$ nm, $\lambda_2 = 589.6$ nm) obtained with a FABRY-PEROT interferometer of $h \approx 0.038$ mm at $R = 0.53$ (a), and at $R = 0.78$ (b).

The interference fringes of the second pattern show a double structure, corresponding to the yellow doublet of sodium. In this case one says that this doublet is resolved by the interferometer. The first pattern, obtained at the lower value of reflectivity $\mathcal{R}$, consists of fringes regarded to be not resolved. The resolving property of an interferometer is usually described by a quantitative measure, which is called the *resolving power*. In terms of the RAYLEIGH criterion (J.W.RAYLEIGH, 1842-1919) two

spectral lines of identical shape and intensity are regarded to be resolved
if their spectral centers are separated by the line width ε (Fig.1.39,b).

Let δ_1 be the phase difference associated with the mth interference
maximum of the wavelength λ_1, and let δ_2 be the phase difference associated with the interference maximum of the same order achieved at λ_2.
According to the RAYLEIGH criterion, the intensity of the first line with
the phase difference $\delta_1 + \varepsilon/2$ should be the same as the intensity of the
second line with $\delta_2 - \varepsilon/2$:

$$\delta_1 + \varepsilon/2 = \delta_2 - \varepsilon/2 \quad . \tag{1.42}$$

Since the spatial separation of the spectral structures achieves a max-
imum at the center of the fringe system, where $\theta_i \approx 0$ and $\cos\theta_i \approx 1$,
let us derive a formula for estimation of the maximal resolving power.
We introduce the wavelength difference $\delta\lambda = \lambda_1 - \lambda_2 > 0$, and then we
re-write (1.42) in terms of wavelengths:

$$\frac{4\pi h}{\lambda_2 + \delta\lambda} + \varepsilon/2 = \frac{4\pi h}{\lambda_2} - \varepsilon/2 \quad ,$$

$$\frac{4\pi h}{\lambda_2}(1 - \delta\lambda/\lambda_2 + \varepsilon) = \frac{4\pi h}{\lambda_2} \quad ,$$

$$\frac{\delta\lambda}{\lambda_2} = \frac{\varepsilon\lambda_2}{4\pi h} \quad .$$

Since $\varepsilon = \pi/\mathcal{F}$, and $m = 4h/\lambda_2$, we get

$$\frac{\lambda}{\delta\lambda} = m\mathcal{F} = m\frac{\pi\sqrt{\mathcal{R}}}{1 - \mathcal{R}} \quad . \tag{1.43}$$

It is the ratio $\lambda/\delta\lambda$ that is the measure of the resolving power. The
resolving power of the FABRY–PEROT interferometer is therefore dependent on the reflectivity $\mathcal{R}$ as well as on the order of interference m (for
the particular case of the center of the pattern). For example, in the case
of the sodium D−lines a FABRY–PEROT interferometer must permit a
ratio $\lambda/\delta\lambda$ to be about 980 in order to resolve such a doublet. One may
estimate the finesse corresponding to the reflectivity $\mathcal{R} = 0.78$ to be
about $\mathcal{F} \approx 12.6$, and the maximal order of interference to be $m \approx 300$
according to $h = 0.038$ mm.

We should mention that it is quite easy to obtain a much higher
resolving power using layers with high reflectivity and a larger spacing
between the interferometer plates.

We will see later (Chapter 2) that we obtain the same formula $\lambda/\delta\lambda = m\mathcal{N}$ as the resolving power of an optical grating, where $\mathcal{N}$ is the number of interfering rays, each coming from one single slit of the grating. Thus $\mathcal{F}$ gives us an estimation of the effective number of rays contributing to the multiple interference pattern. When the reflectivity is high but the number of contributing rays is low (as in the case of the LUMMER-GEHRCKE plate treated at the end of paragraph 3), the effective finesse is given by the number of rays and not by the reflectivity finesse $\mathcal{F} = \pi\sqrt{\mathcal{R}}/(1 - \mathcal{R})$.

4.1.3   Free spectral range

Very high resolving power may be obtained with a "thick" FABRY–PEROT interferometer due to a larger value of the maximal interference order, which is directly proportional to the thickness of the air film: $m_{max} = 2h/\lambda$. On the other hand, the increase of the thickness may result in an overlap of the spectral components belonging to neighbouring interference orders. This may happen if the incident quasi-monochromatic wave covers a wide spectral interval $\Delta\lambda$. Such an overlap of fringes of mth order, corresponding to $\lambda + \Delta\lambda$, with fringes of the next order, $(m + 1)$, corresponding to λ, begins if the condition

$$\lambda(m + 1) = m(\lambda + \Delta\lambda) \quad ,$$

is fulfilled. For central fringes, where θ_i is assumed to be nearly zero, we get $m = 2h/\lambda$, and we find that such an overlap will not occur under the condition $\Delta\lambda \leq \lambda^2/2h$. The quantity

$$\Delta\lambda_{FSR} = \frac{\lambda^2}{2h} = \frac{\lambda}{m} \tag{1.44}$$

is called the *free spectral range* (FSR) of the interferometer. For non-axial incident rays this quantity depends slightly on the angle of incidence θ_i.

In terms of monochromaticity of the quasi-monochromatic wave the inequality may be written in the from

$$m = \frac{2h}{\lambda} \leq \frac{\lambda}{\Delta\lambda} \quad , \tag{1.45}$$

where $\lambda/\Delta\lambda$ is the degree of monochromaticity of the incident quasi-monochromatic light, and m is the maximal interference order of the interferometer which is allowed without obtaining an overlap of the interference fringes. The requirements of a high resolving power can not be met if the incident light possesses a low degree of monochromaticity.

It is of interest to calculate the free spectral range in frequencies using $\lambda\nu = c$, and $|\Delta\nu| = \Delta\lambda c/\lambda^2$. With these formulas one gets

$$\Delta\nu_{FSR} = \frac{c}{2h}$$

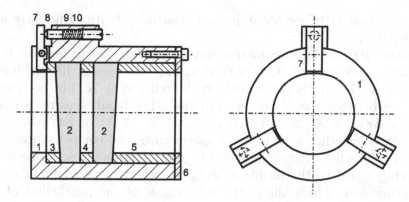

Figure 1.41. A FABRY-PEROT interferometer for spectroscopic use. 1 housing, 2 interferometer plates (the outer surfaces form a wedge to avoid unlike reflections), 3 end ring, 4 spacing ring, 5 compensation ring (fills the remaining length), 6 end cap, 7 clamp, 8 pin, 9 spring, 10 adjustment screw.

which has the advantage of being independent of ν. We may use this formula for the free spectral range when treating the FABRY-PEROT interferometer as the resonator of a laser.

The resolving power of the interferometer can be rewritten using the free spectral range:

$$\frac{\lambda}{\delta\lambda} = \frac{\lambda}{(\Delta\lambda_{FSR}/F)} = F\frac{\lambda}{\Delta\lambda_{FSR}}$$

or $\nu/\Delta\nu = \nu/(\Delta\nu_{FSR}/F)$, where F tells us how many different positions of a fringe can be distinguished within one free spectral range. The smallest resolvable wavelength difference is thus given as

$$\delta\lambda = \Delta\lambda_{FSR}/F \quad .$$

With a spacing of $h = 40$ mm and $\mathcal{R} = 0.97$ (which can be achieved with dielectric multi-layers), we obtain a finesse of $\mathcal{F} = 100$ and a free spectral range (for the light emitted from a red helium-neon laser at 633 nm) of $\Delta\lambda_{FSR} = 0.005$ nm or $\Delta\nu_{FSR} = 3.5$ GHz. The smallest resolvable difference in wavelength is thus 5.10^{-5} nm, or in frequency 35 MHz. Such a high resolution only makes sense if the monochromaticity of the investigated radiation is very high, as in the case of laser radiation. With such an interferometer, we can resolve the spectral composition of the laser radiation (see section 5 of this chapter).

It should be noted that such a high finesse can be only obtained when the surfaces of the interferometer are sufficiently plan and parallel. As a rule, the plate surfaces should not deviate from an ideal plane by more than $\lambda/2\mathcal{F}$ for obtaining a finesse of $\mathcal{F}$ (with sufficiently high reflectivity).

FABRY–PEROT interferometers have been extensively used in high-resolution spectroscopy. One of their main advantages is that the finesse $\mathcal{F} = \pi\sqrt{\mathcal{R}}/(1 - \mathcal{R})$ is dependent on the reflectivity $\mathcal{R}$, but not on the distance between the plates, h. On the other hand, the free spectral range $\Delta\lambda_{FRS} = \lambda^2/2h$ is independent of $\mathcal{R}$. A set of two plane parallel plates together with a set of spacer rings and compensation rings (which always have the same length together) allows the adopting of the resolving power to the problem being investigated. A commonly used mounting unit, which allows the adjustment of the parallelism of the plates by applying forces on the distance ring, is shown in Fig. 1.41.

4.1.4   Example 1: The interference filter

Let us consider the operating principle of a so-called interference filter, which is usually fabricated as a thin solid-state FABRY-PEROT interferometer in which the working plate (refractive index n) is covered by a colored glass plate. For a given reflectivity $\mathcal{R}$ the intensity distribution of the transmitted rays at normal incidence may be treated as having maxima at certain wavelengths, each corresponding to the requirements (1.34). For example, let λ_{-1} correspond to the $(m - 1)^{th}$ maximum, let λ_0 correspond to m^{th} maximum, and let λ_{+1} correspond to $(m + 1)^{th}$ maximum of the transmitted rays (Fig.1.42).

The wavelength difference $\Delta\lambda$ between λ_{-1} and λ_{+1} is two times the free spectral range and may be calculated as follows:

$$\Delta\lambda = \lambda_{-1} - \lambda_{+1} \quad , \quad \text{and} \quad \frac{nh(\lambda_{-1} - \lambda_{+1})}{\lambda_{-1}\lambda_{+1}} = 1 \quad .$$

Thus, $\Delta\lambda_{\pm1} = \lambda_{-1}\lambda_{+1}/nh$. Since, assuming both wavelengths to be only slightly differed from λ_0, we can write

$$\Delta\lambda_{\pm1} = \frac{\lambda_0^2}{nh} \quad . \tag{1.46}$$

The colored glass filter should only be transparent for wavelengths between λ_{-1} and λ_{+1} (Fig.1.42). Thus the combination of glass filter and interferometer will possess the transparency properties of a FABRY-PEROT interferometer but will have only one certain interference maximum. The width ε of the LORENTZian curve, being a measure of the effective transparency region, is determined by the reflectivity $\mathcal{R}$ and the thickness h of the interferometer.

For example, for $h = 0.03$ mm, $n = 1.52$, and $\lambda_0 \approx 579$ nm (a yellow line of mercury) estimation by the formula (1.44) gives: $\Delta\lambda_{\pm1} \approx$

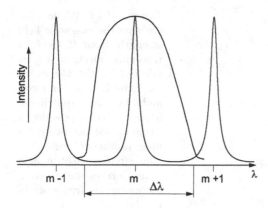

Figure 1.42 Illustrating the operating principle of the interference filter.

7.2 nm. Such a spectral interval may be well selected by a colored glass. In a usual commercial interference filter with aluminium coating layers the reflectivity may achieve 0.85, that permits the width ε of the transparency region to be about 0.32. It means that the effective transparency region is about $(7.2 \text{ nm})/2 \cdot 0.32/2\pi \approx 0.2$ nm wide.

4.1.5 Example 2: Resolution of the Zeeman structure of a spectral line

In order to illustrate how calculations of the resolving power and the spectral range may be performed, we consider an experiment for observation of the ZEEMAN *effect* using a FABRY-PEROT interferometer. The splitting of a spectral line in different frequency components when the emitting lamp is placed in a strong magnetic field was discovered by P.ZEEMAN (1865-1943) in 1896, and its qualitative explanation was given first by G.A.LORENTZ (1853-1928). According to LORENTZ's treatment based on classical electrodynamics, the normal ZEEMAN effect (splitting of a line into three components) may be explained in terms of the LORENTZ force, which acts upon a moving electron. Let an electron follow a circular orbit of the radius r around a positively charged center of e^+. Such a motion of the electron will exist if the actions of the centrifugal force $mr\omega_0^2$ and the COLOUMB force $(1/4\pi\varepsilon_0)e^2/r^2$ are equal:

$$\frac{1}{4\pi\varepsilon_0}\frac{e^2}{r^2} = mr\omega_0^2 \quad , \tag{1.47}$$

where ω_0 is the circular frequency of the electron's orbit when a magnetic field is absent.

With the appearance of a magnetic field a LORENTZ force $F = e(\mathbf{v} \times \mathbf{B})$ will act upon the electron. We should assume that the appearance of the magnetic field will need a finite time interval Δt, in such a way, that the magnitude of this field will increase from zero to the value B.

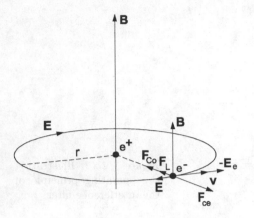

Figure 1.43 While increasing the magnetic field strength from 0 to B a revolving electron e^- is affected by the COLOUMB force, by LORENTZ's force, and by the centrifugal force, which provide a dynamic balance at invariable radius of the orbit. The circular electric field E accelerates the electron, increasing its frequency of revolution.

We also assume the interval Δt to be much longer than the period of the electron's revolution, $T = 2\pi/\omega_0$. For this reason a balance between the COLOUMB force, the LORENTZ force, and the centrifugal force will exist at all times while increasing the magnetic field. However, such a dynamic balance will cause a variation of the frequency of revolution, ω, for an invariable radius r of the electron orbit. Either an increase or a decrease of the frequency occurs due to the existence of a circular electric field, which should accompany the increase of the magnetic field during the period Δt, and which is directed along the electron's trajectory. It is this circular electric field that changes the frequency, because the LORENTZ force, directed normally to the electron's velocity, can not do it. Thus, while increasing the magnetic field from 0 to B the electron is affected by the centrifugal force, by the LORENTZ force and the COLOUMB force as shown in Fig.1.43. The force $-\mathbf{E}e$ is directed parallel to the electron velocity, accelerating the electron and thus increasing its frequency of revolution. The dynamic balance between the three forces as well as the change of the frequency will take place until the magnetic field achieves the value B. We may therefore write a force balance equation as follows:

$$\frac{1}{4\pi\varepsilon_0}\frac{e^2}{r^2} + er\omega B = mr\omega^2 \quad .$$

Equation (1.47) together with the previous one allow a representation of the force balance in the form

$$\omega^2 - 2\omega\frac{eB}{2m} - \omega_0^2 = 0 \quad , \tag{1.48}$$

where the term $eB/(2m)$ is called the LARMOR *frequency* (J.LARMOR, 1857-1942). The equation has two solutions:

$$\omega = \frac{eB}{2m} \pm \sqrt{\omega_0^2 + \left(\frac{eB}{2m}\right)^2} \; .$$

Under laboratory conditions the magnitude of magnetic field is assumed to be smaller than $B \sim 1$ T. Therefore the ratio $(eB/2m)/\omega_0$ is smaller than approximately 10^{-3}, and we can neglect the second term under the square root. Thus we may write down the so-called LORENTZ formula:

$$\omega = \frac{eB}{2m} \pm \omega_0 \; . \tag{1.49}$$

If we look at the orbiting electron from the end of the magnetic field vector **B**, and if its revolution is anti-clockwise, then the circular frequency of the electron is increased by the LARMOR frequency (it is the so-called σ^+- component of the ZEEMAN spectrum). In the opposite case, where ω_0 appears to be negative, the circular frequency of the electron is decreased by the LARMOR frequency (σ^-- component). A careful study of the ZEEMAN effect shows that any motion of an electron along the magnetic field vector will not give rise to extra frequencies, and such motions result in emission of light which undergoes no frequency shift. This light shows linear polarization with the electric field vector directed parallel to the magnetic field and is called π-component of the ZEEMAN spectrum.

When observing the spectrum emitted under the action of a magnetic field in a direction parallel to the magnetic field (e.g. through a small hole in the center of a pole shoe), the π-component has no intensity (this follows from the radiation characteristics of a dipole) and the σ-components of the ZEEMAN spectrum both appear with circular polarization (σ^--component right-hand circularly polarized, σ^+-component left-hand circularly polarized). When observing across the magnetic field, the σ-components will appear to have linear polarization orthogonal to the direction of the magnetic field, whereas the π-component appears to be linearly polarized parallel to this direction.

A setup for a demonstration of the ZEEMAN effect is shown in Fig.1.44. A mercury-cadmium low pressure spectral lamp is mounted between the pole shoes of an electromagnet. The lamp illuminates a circular aperture located in the first focal plane of an objective. A system of parallel rays passes through a filter, and then falls on a FABRY-PEROT interferometer with $h = 4$ mm. The filter selects the red Cd line with wavelength $\lambda = 643.8$ nm. A polarizer allows only the σ-components of the ZEEMAN

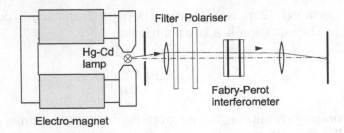

Figure 1.44. Setup for demonstration of the ZEEMAN effect. The ZEEMAN spectrum of the red Cd line ($\lambda = 643.8$ nm) is resolved by a FABRY-PEROT interferometer of the thickness $h = 4$ mm and the reflectivity $R = 0.9$. The magnetic field strength is $B = 0.15$ T. The π component is suppressed by the polarizer used.

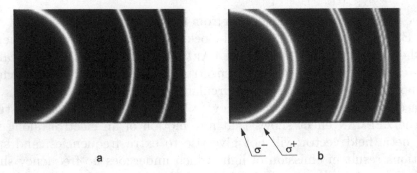

Figure 1.45. Interference pattern of a red Cd-line of $\lambda = 643.8$ nm obtained with a FABRY-PEROT interferometer ($h = 4$ mm; $R = 0.9$) (a); the ZEEMAN spectrum of the σ-components of this line under a magnetic field $B = 0.15$ T with observation normal to the magnetic field, and with the vertically linear polarization.

spectrum to pass, and not the π-component. The spectrum, well resolved by our interferometer, consists of two wavelength components, one with larger and one with smaller wavelength than the wavelength without the field.

The magnitude of the magnetic field applied to the lamp is $B = 0.15$ Tesla. According with (1.49) the deviations of the circular frequencies of the σ-components are $\Delta\omega_Z = \pm eB/2m \approx \pm 1.2 \cdot 10^{10}$ rad/s, which corresponds to a change of the frequency of the emitted light of $\Delta\nu_Z = \Delta\omega_Z/2\pi = 2.10^9$ Hz $= 2$ GHz. In turn, the frequency corresponding to this red line is $\nu_0 = c/\lambda \approx 4.6 \cdot 10^{14}$ Hz, Therefore, the relative changes of the frequencies as well as of the wavelengths of the σ-components are given as

$$\frac{\Delta\lambda_Z}{\lambda} = |\frac{\Delta\nu_Z}{\nu}| \approx \pm 0.43 \cdot 10^{-5} \quad , \tag{1.50}$$

which gives $\Delta\lambda_Z = \pm 2.8 \cdot 10^{-3}$ nm for the change of the wavelength. It implies that the FABRY-PEROT interferometer should provide σ-components, having the wavelengths $\lambda_1 = 643.7972$ nm (σ^+-component) and $\lambda_2 = 643.8028$ nm (σ^--component). This small wavelength difference has to be resolved. It follows from (1.50) that the FABRY-PEROT interferometer has to provide a resolving power larger than $\lambda/(2\delta\lambda_Z) \approx 1.43 \cdot 10^5$. The parameters of the interferometer in use, $h = 4$ mm and $\mathcal{R} = 0.9$, give a resolving power of $4.37 \cdot 10^5$ according with (1.42), that is approximately three times greater than the desired value. A calculation of the spectral range of the interferometer gives $\Delta\lambda_{FSR} = \lambda^2/2h \simeq 3.7 \cdot 10^{-2}$ nm. The separation of the σ-components, $2\Delta\lambda_Z$, has the value $2\Delta\lambda_Z \simeq 3.8 \cdot 10^{-3}$ nm. Thus, under the conditions discussed here, $\Delta\lambda_{FSR} \approx 10\Delta\lambda_Z$. The interference patterns, observed without magnetic field (a), and under action of the magnetic field (b), are shown in Fig.1.45. Each fringe of the original pattern appears as a doublet in the second pattern, with the σ^+-component located further from the center than the σ^--component.

5. Optical resonator of a laser

Aluminium films used for producing reflecting surfaces can not permit a very high magnitude of reflectivity (approximately 95-96% in the visible range of the spectrum). Now we briefly consider a way for fabrication of highly reflecting mirrors (often so-called *laser mirrors*), which possess extremely high magnitudes of reflectivity (up to 99.997 % or even more). Let us assume that a plane parallel glass plate of the refractive index n_0 is covered by a number of dielectric layers as shown in Fig.1.46.

The refractive index of the first layer covering the plate is $n_1 < n_0$, the optical thickness of this layer is $n_1 h_1 = \lambda/4$, where λ is the working

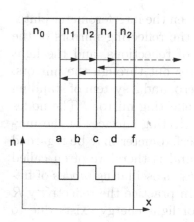

Figure 1.46 A glass plate and the first few dielectric layers. For a given wavelength λ each layer possesses an optical path length of $\lambda/4$.

wavelength of the mirror. The material of the second layer is optically thicker than that of the first layer, $n_2 > n_1$, nevertheless its optical thickness $n_2 h_2$ is also equal to $\lambda/4$. Let a plane monochromatic wave with wavelength λ penetrate the glass surface (a) at normal incidence and let reflect it back, reaching the boundary (b). Since the total optical path difference between boundaries b and a and back is $\lambda/2$, and since the wave undergoes a phase change of π at boundary (a), the total phase difference which occurs at boundary (b) is 2π. In turn, the wave reflected back at boundary (b) has no change in phase, therefore the amplitude of this wave will be increased by the amplitude of the first wave reflected at boundary (a). Further, this wave of increased amplitude reaches boundary (c) with the phase difference of π. The wave reflected back at boundary (c) undergoes a change of the phase of π. Thus, these waves occur in phase, and the amplitude of the wave reflected at boundary (c) is also increased, and so on. This is a sort of multiple-beam interference which allows the achievement of very high magnitudes of reflectivity under conditions of normal incidence for a given wavelength.

Let us assume now that a FABRY-PEROT interferometer is illuminated by quasi-monochromatic light with a certain carrier frequency ν_0. As we have found before, maxima of the transmitted light intensity for nearly normal incidence will appear if the condition $2L/\lambda = m$ is fulfilled, where L is the separation of the mirrors. Since the quasi-monochromatic wave may be regarded as consisting of monochromatic components of frequencies ν within a narrow spectral region around ν_0, the condition caused that the transmitted light is decomposed into a set of frequencies resonant with the interferometer:

$$\nu_m = m\frac{c}{2L} \quad . \tag{1.51}$$

Neighbouring frequencies are distanced by the free spectral range $\nu_m - \nu_{m-1} = \nu_{FSR} = c/2L$.

A light beam, passing into the space between the interferometer plates from outside and then oscillating between the reflecting mirrors of the interferometer, undergoes a large number of reflections, and the light energy of the beam only partially penetrates out through the mirrors. For $\mathcal{R} \rightarrow 1$ any loss of the energy tends to zero, and a system of standing waves is formed in the space between the reflecting mirrors. The nodes of the standing waves are located at the reflecting surfaces of the mirrors. For this reason the FABRY-PEROT interferometer may be regarded as being similar to an oscillating system, and is therefore often called an *optical resonator*. Such a resonator possesses oscillating *modes* of frequencies which satisfy relationship (1.51). In practice the reflectivity $\mathcal{R}$ is smaller than one, and therefore a loss of light energy exists due to

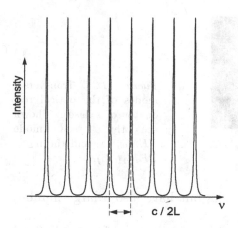

Figure 1.47 Transmission spectrum of a FABRY-PEROT optical resonator.

partial transmission of the light energy through the mirrors. For this reason any resonant frequency of the optical resonator is specified by a certain resonant curve of a finite width ε (see (1.39)).

In reality the light field distribution inside an optical resonator is more complex. Due to diffraction of light at the mirrors, in combination with imperfect surfaces or imperfect adjustment of the mirrors, certain stable distributions of the light field over cross sections of the resonator space appear with different characteristics for every particular kind of optical resonator. Such distributions are called *transversal modes* in contrast to the *axial (or longitudinal) modes* discussed above. With a Cartesian system centered at the axis of the resonator, transverse modes are usually specified as TEM_{mn} modes (transversal electric modes, well known in high frequency electrical engineering), with indexes m and n, which give the number of field strength zeros in the x- and y-direction in a plane perpendicular to the light beam. Two intensity distributions within the cross section of a light beam in an optical resonator are shown in Fig.1.45. The lowest-order mode TEM_{00} has a two-dimensional GAUSSian distribution with increasing radius of the light beam (Fig.1.45,a). The intensity corresponding to TEM_{11} is distributed within four bright spots symmetrically positioned around the center of the beam (Fig.1.45,b).

The idea of maintaining undamped optical oscillations within an optical resonator was realized in the construction of lasers. In these optical devices, the space between both reflecting mirrors contains a light emitting substance, or a so-called *active medium*. This medium is able to amplify a light wave which passes through it (opposite to a normal medium, which always absorbs light energy). One of the most typical gas lasers is the Helium-Neon (He-Ne) laser, where the active medium is a mixture of helium and neon. He-Ne lasers emitting red light at

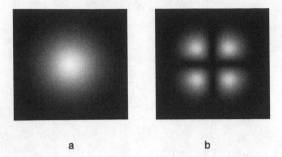

Figure 1.48 Transversal modes of the optical resonator of a laser. Shown are the lowest mode TEM_{00} (a) and the mode TEM_{11} (b).

$\lambda = 632.8$ nm are widely used in physics but have also a large number of technical applications.

A discharge tube, containing a mixture of helium and neon, is fixed along the longitudinal axis of the optical resonator. Usually the resonator consists of two spherical mirrors having dielectric layers which provide extremely high reflectivity $\mathcal{R}$ for a desired wavelength (most frequently the red line of Ne mentioned above). The spectral distribution of intensity is determined by the optical resonator and by the DOPPLER curve which is caused by the movement of the emitting neon atoms in the discharge and which describes the amplification profile of the neon transition. The upper levels of the neon atoms are populated by so-called 'collisions of the second kind' which are characteristic for the helium-neon mixture of the discharge tube. Metastable He atoms transfer their excitation energy to certain energy states of the Ne atoms, which have practically the same excitation energy. These states are much more populated as lower energy states of the Ne atoms. Therefore the Ne atoms can act as an active, amplifying medium for frequencies corresponding to certain optical transitions within the Ne atom. Under typical discharge conditions the effective frequency width $\Delta\nu_D$ of such a DOPPLER curve may reach about 900 MHz at $\lambda = 632.8$ nm (Fig.1.49).

If such an active medium is placed inside an optical resonator, under some specific conditions, which are called the laser generation conditions, standing optical waves are excited, and the discharge tube with the mixture of Ne and He gases emits a narrow, well collimated beam of bright radiation, which propagates strictly in axial direction. The laser radiation is (in the first approximation) considered to be coherent and monochromatic, in contrast to the radiation of all thermal sources.

6.   Scanning Fabry–Perot interferometer

A FABRY-PEROT interferometer regarded as an optical resonant system may be used for the analysis of the frequency spectrum of laser light. A device of this sort should be adjusted in length (mirror distance) in

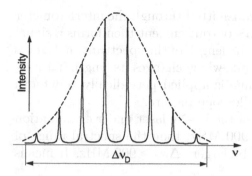

Figure 1.49 A spectrum of a He-Ne laser under the Doppler curve.

order to allow the transmission of a specific spectral component of the incident laser radiation. If the change in length is made periodically, the device is called a *scanning* FABRY-PEROT interferometer or *optical spectrum analyzer*. For unique determination of the frequency spectrum of a light source, the free spectral range of the scanning FABRY-PEROT interferometer has to be larger than the width of the spectrum under investigation. The free spectral range is given by

$$\Delta\nu_{FSR} = \frac{c}{2L_s} \quad , \tag{1.52}$$

where L_s is the length of the scanning interferometer. As before, $\Delta\nu_{FSR}$ is the frequency separation between two neighbouring maxima of same order of interference, and its transmission function is determined by $\Delta\nu_{FSR}$ and the finesse $\mathcal{F}$. In order to examine the spectral intensity distribution of a He-Ne laser at $\lambda = 632.8$ nm, the laser light spectrum with an effective width $\Delta\nu_D$ of the envelope (given by the DOPPLER curve) has to be inside the free spectral range of the scanning interferometer we use for this purpose. Figure 1.50 illustrates the mutual positions of two modes of the scanning interferometer with respect to the laser spectrum, where $\Delta\nu_D < \Delta\nu_{FSR}$ is true. In fact, under the conditions shown in Fig.1.50, no light intensity transmitted through the scanning interferometer will appear, since (for the case illustrated) no component of the laser spectrum is overlapping the transmission curve of the analyzer. Further, by continuously changing the separation L_s (by a small amount, slightly larger than $\lambda/2$), the spectral transmission peak will move over the laser spectrum. If the requirement $\Delta\nu_D < \Delta\nu_{FSR}$ is fulfilled, then only one peak of the analyzer will overlap one spectral component of the laser spectrum, while changing the length of the analyzer. If this overlapping occurs, laser radiation with a certain frequency can pass through the scanning interferometer. A setup for observation of the laser spectrum is shown in Fig.1.51. The light beam of a He-Ne laser ($\lambda = 632.8$ nm) falls on the input mirror of a scanning FABRY-PEROT interferometer. Any light intensity transmitted through the interferometer is received by a photodetector, who's output current then forms a signal on the oscillograph. The change of the length of the spectrum analyzer is performed with help of a piezo ceramic which changes its length if a voltage is applied. The scanning voltage is applied periodically (sawtooth shape) in order to generate an oscilloscope picture.

If we assume that the spectrum of the He-Ne laser under examination has a spectral distribution within 900 MHz, then the spectral range of the scanning interferometer has to be larger: $\Delta\nu_s \geq 900$ MHz. It means

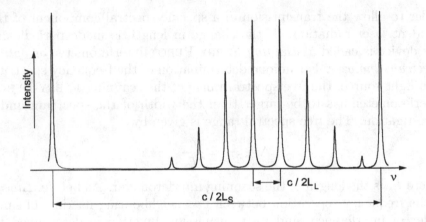

Figure 1.50. The laser spectrum must be located between two transmission maxima of the scanning interferometer.

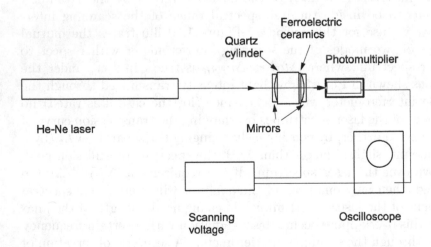

Figure 1.51. Setup for recording the laser spectrum by means of a scanning FABRY-PEROT interferometer.

that the separation of the mirrors of the interferometer (plane plates) should be shorter than $c/(2 \cdot 900 \text{ MHz}) \approx 16.7$ cm. In practice often confocal FABRY-PEROT interferometers are used for scanning interferometers, which consist of spherical mirrors of radius r which are just separated by r. Such interferometers are easier to adjust than interferometers with plane parallel plates, especially if the mirror distance is larger than few cm. The free spectral range is $\Delta\nu_{FSR} = c/4L$ for such interferometers. We use a confocal spectrum analyzer with a free spectral range of 1.5 GHz, corresponding to a separation $L_s = r$ between of the confocal mirrors of $L_s = 5$ cm.

Let us assume the confocal scanning interferometer ($\Delta\nu_{FSR} = c/4L_S$) at L_s to be transparent for a laser mode of λ_m. Thus, the condition

$$n\lambda_m = 4L_s \qquad (1.53)$$

is true where n is an integer. Further, by increasing L_s by ΔL_s, let the interferometer be transparent for a longer laser wavelength λ_{m-1} at the same integer n :

$$n\lambda_{m-1} = 4(L_s + \Delta L_s) \quad . \qquad (1.54)$$

It follows from (1.53) and (1.54) that the displacement ΔL_s satisfies the relationship

$$1 + \frac{\Delta L_s}{L_s} = \frac{\lambda_{m-1}}{\lambda_m} \quad .$$

Since the wavelengths λ_m and λ_{m-1} are separated by the free spectral range of the laser resonator, $\lambda^2/(2L_L)$, we get $\lambda_{m-1} = \lambda_m + \lambda^2/(4L_L)$, and for the displacement ΔL_s we may obtain

$$\Delta L_s \approx \frac{\lambda}{4} \cdot \frac{L_s}{L_L} \quad . \qquad (1.55)$$

For the laser under consideration, $\lambda = 632.8$ nm and $L_L = 1.25$ m (the length of the optical resonator). Because $L_s = 5$ cm, it is seen from (1.54) that a displacement of the mirror spacing of $0.04 \cdot \lambda/4$ is needed for transmitting two axial modes of this laser in sequence. To cover the whole free spectral range of the scanning interferometer, a displacement slightly larger than $\lambda/4$ is necessary.

In order to provide high temperature stability of the primary separation of the mirrors of the scanning interferometer, its base is a quartz tube. One spherical mirror is fixed on one side of the quartz tube. The other mirror is mounted on the plane surface of a thin piecoelectric-ceramic cylinder. The other plane surface of the ceramic cylinder is fixed on the opposite side of the quartz tube (Fig.1.52). With variation of a voltage applied to both sides of the ceramic cylinder, its length changes due to the ferroelectric effect. Linear voltage variation allows all frequencies of the laser spectrum to be scanned. The amplified output signal is shown in Fig.1.53.

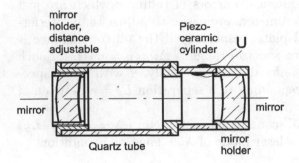

Figure 1.52 Confocal scanning FABRY-PEROT interferometer

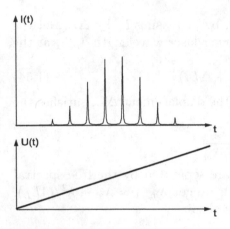

Figure 1.53 The intensity dependency $I(t)$ associated with the observed spectrum obtained by applying a linearly varied voltage $U(t)$ to the scanning interferometer.

SUMMARY

We considered a number of classical interference schemes which illustrate ways of creating interference. One may call attention to the very small size of the region in which the interference fringes are concentrated, as well as to the very small value of the angles between the interfering rays. As we have seen, the angular spacing of the fringes has values which do not exceed 10^{-3} rad at the best conditions.

By selecting some examples of classical interference schemes, we have discussed ways for creating interference of light. In the YOUNG type of interference schemes, two coherent sources are formed by splitting elements such as a double slit or bi-prisms, which give rise to an interference pattern. In practically all the cases we have seen that necessary conditions are small angular dimensions of the light sources and small interference apertures. These two conditions lead to relatively small angular dimensions of obtained interference patterns with an order of magnitude of 10^{-3} rad. These parameters are typical for YOUNG's double slit interferometer, LLOYD mirror, FRESNEL mirrors and so on, where light interference is observed within a pencil of beams diverging from a natural light source.

Observation of light interference in practically parallel beams, as in MICHELSON's interference scheme, is equivalent to the observation of equal inclination interference patterns, and for observation of a clear pattern, the condition of rather high light monochromaticity is essential. Quasi-monochromatic light is created by means of optical filters, extracting a narrow spectral band of the light source radiation (e.g. a single spectral line of a spectral lamp).

We note that the use of the yellow lines of a mercury-helium lamp enables sufficient observation of the interference patterns, because such a light source emits a line spectrum. The yellow spectral lines (mean wavelength $\bar{\lambda} = 580$ nm) are the brightest monochromatic components of this spectrum and this wavelength is close to the spectral sensitivity maximum of the human eye. For these reasons any interference pattern produced with a mercury-helium lamp appears as having distinct bright and dark yellow interference fringes, even without using a filter for quasi-monochromatization of this light source. Optionally, a yellow filter can be used to enhance the contrast of the interference pattern.

Both requirements of light monochromaticity and small angular dimensions of the light source remain valid when treating multiple–beam interference schemes. Optical devices operating with multiple–beam interference have the advantage of extremely narrow interference maxima compared to two-beam interferometers. In order to exploit this advantage, additional monochromaticity of the light source is required. This sensitivity of the patterns of multiple–beam interferometers to the band width and the wavelength composition of of light makes these optical devices predestined for studies of spectral line structures, e.g. for the resolution of the hyperfine structure of atomic spectral lines.

Optical radiation with super narrow band width can be created by means of lasers. The optical resonator of such a laser is nothing more than a modification of the FABRY–PEROT interferometer, and all principles of multiple–beam interference can be applied to describe the resonator.

When using the super narrow band width radiation of a laser for demonstrating optical interference phenomena, one is not restricted to the use of interferometers having optical path differences of just a few

wavelengths. Due to the high monochromaticity of such laser radiation
the coherence length (see Chapter 7) of such light has a range up to
meters, and it is easy to observe interference patterns.

PROBLEMS

1.1 Estimate the distance between the center of the slit and the surface
of the mirror for a LLOYD mirror under experimental conditions. Use
the data given in the headings of Figs.1.5. and 1.6.

1.2 Estimate the small angle α between the two reflecting planes of
the FRESNEL mirror interference scheme shown in Fig.1.8. Use the parameters of the setup and the fringe spacing given in Fig.1.10,b.

1.3 Estimate the order m of the interference fringes obtained with
the experimental setup shown in Fig.1.11. Use the parameters of the
experiment and the interference fringes presented in Fig.1.12.

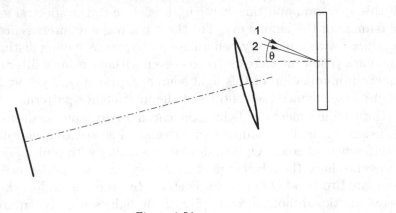

Figure 1.54.

1.4 Let a light ray fall on a plane parallel glass plate at an angle θ, and
let another ray fall on the plate at a smaller angle (Fig.1.54). Two-beam
interference is observed by means of an objective of the focal length f.
Derive the formula for the angular fringe spacing, assuming that λ, n and
the thickness h of the plate are known. Apply your result to estimate
the angular fringe spacing in the experiment shown in Fig.1.12.

1.5 Under the experimental conditions presented in Fig.1.18 and with
the fringes shown in Fig.1.19 estimate the small angle between the plane
surfaces of two glass plates.

1.6 The YOUNG double–slit interferometer is illuminated by white light. Estimate the amount of interference fringes which can be observed, taking into account the wavelength range between $\lambda_r = 690$ nm and $\lambda_v = 420$ nm for white light.

1.7 Let us assume a plane monochromatic wave which falls on a setup for the observation of NEWTON's rings. Estimate the intensity ratio of the transmitted rays between a maximum and a minimum of the pattern, provided that the lens as well as the plate are both made of crown glass with a refractive index $n = 1.52$. Use Fig.1.22.

1.8 An optical interference filter is used for monochromatization of green light with wavelengths around $\lambda_0 = 546$ nm. Aluminium layers of the reflecting surfaces provide a reflectivity of $R = 0.7$ of the active plate ($h = 0.01$ mm, $n = 1.52$). Estimate the degree of monochromaticity permitted by the filter.

SOLUTIONS

1.1 In the setup shown in Fig.1.5 the slit is placed on the focal plane of the objective forming the interference fringes, which are observed on the second focal plane of the objective. Let b be the distance between the center of the slit and its virtual image in the mirror and h be the fringe spacing. Using Fig.1.6 we estimate h as follows: $h \approx 0.5$ mm. For given values $f = 60$ cm and $\overline{\lambda} = 580$ nm we find $b = f.\lambda/h \approx 0.7$ mm. The required separation $b/2$ is estimated to have the value of 0.35 mm.

1.2 Let S be the center of the primary slit. The two virtual images of point S, S_1 and S_2, form two coherent sources, which are causing interference (Fig.1.55). The objective ($f = 60$ cm) is at a distance of 15 cm from the midline of the mirrors. $SO = OO' = 45$ cm, therefore the line $S_1 S_2$ is located at a distance of 60 cm from the objective. The two coherent sources S_1, S_2 are therefore located on the focal plane of the objective. We estimate the fringe spacing from Fig1.10,b to be $h \approx 0.3$ mm. The distance b between S_1 and S_2 is then $b = f.\lambda/h$. In turn, $b = S_1 S_2 \approx SS_2 \cdot \alpha$. Since $SS_2 = 2SO \cos\theta$, the magnitude of α may be found from the expression $\alpha(2SO \cos\theta) = \lambda f/h$, which gives:

$$\alpha = \frac{\lambda f}{2hSO \cos\theta}$$

Substitution of the numerical values gives $\alpha \approx 1.3 \cdot 10^{-3}$ rad.

1.3. The maximum of the m th order of the interference fringes formed by reflection of rays on both sides of the plane parallel plate satisfy

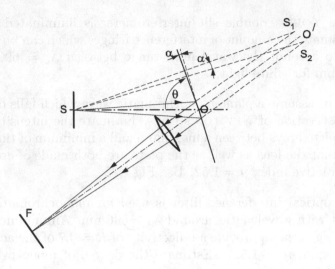

Figure 1.55.

expression (1.17):

$$4\pi nh\cos\varphi/\lambda - \pi = 2\pi m \quad .$$

The unknown angle φ may be found from the law of refraction $\sin\theta = n\sin\varphi$, where $\theta = 12°$ and $n = 1.5$, which gives $\sin\varphi \approx 0.138$ and $\cos\varphi \approx 0.99$. For the given wavelength $\overline{\lambda} = 580$ nm and for $h = 2$ mm the magnitude of $m = 2nh\cos\varphi/\lambda$ is approximately equal to 10^4.

1.4. Let ray 1 fall on the plate at an angle θ and ray 2 at a smaller angle $\theta - \Delta\theta$ (Fig.1.56). Due to reflection and refraction on the plate, ray 1 produces two parallel rays, which are focused by the objective at point P_1 on the focal plane. In the same way ray 2 produces another pair of parallel rays focused at point P_2. We assume that point P_2 corresponds to the mth maximum, and point P_1 to the $(m+1)^{th}$ maximum of interference. According to (1.17) the angular difference $\Delta\varphi$ corresponds to the difference Δm:

$$\Delta\varphi 2hn\sin\varphi = -\Delta m\lambda \quad ,$$

where h is the thickness of the plate and φ is the angle of refraction. Using the law of refraction, $\sin\theta = n\sin\varphi$, we can write $\Delta\varphi = \Delta\theta/n$ and $n\sin\varphi = \sin\theta$. Then we get for $\Delta\theta$

$$\Delta\theta 2h\sin\theta = -\Delta m\lambda n \quad .$$

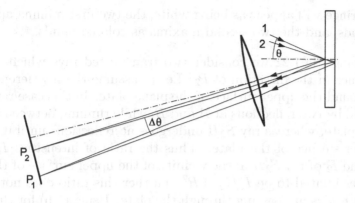

Figure 1.56.

For two neighbouring maxima $|\Delta m| = 1$ is valid, and we obtain

$$\Delta\theta = \frac{n\lambda}{2h\sin\theta} \ .$$

Substitution of numerical values ($\overline{\lambda} = 580$ nm, $n = 1.5$, $h = 2$ mm, $\theta = 12^o$) gives $\Delta\theta \approx 10^{-3}$ rad, which is of the same order of magnitude as the value $1.2 \cdot 10^{-3}$, which is directly found from the pattern in Fig.1.12.

1.5 The interference fringes shown in Fig.1.19 are located close to the inner surfaces of the plates. Let us calculate the linear size of the fringe spacing at this location. In Fig.1.19 we see that 9 bright fringes are distributed within 4.8 mm. The fringe spacing is therefore about $\Delta x \approx 0.53$ mm. For the given distance between the inner surface and the objective, $a = 60$ cm, and the distance from the objective to the screen, $b = 15$ cm, we calculate the fringe spacing at the inner surfaces to be $\widetilde{\Delta x} = \Delta x a/b \approx 2$ mm. According to (1.20) (using $\cos\theta \approx 1$) the required angle β is estimated to be $\beta = \lambda/(2\widetilde{\Delta x}) \approx 1.5 \cdot 10^{-4}$ rad.

1.6 Two adjacent interference fringes of order m for λ_r and $m+1$ for λ_v will be imaged separately if the following inequality is valid: $m\lambda_r \leq (m+1)\lambda_v$. In the limiting case, when $m\lambda_r = (m+1)\lambda_v$, interference fringes of orders higher than m will be smoothed out. Therefore, the highest interference order $m_{\max}$ may be estimated to be $m_{\max} = \lambda_v/(\lambda_r - \lambda_v)$. Substitution of the numerical data gives for $\lambda_v/(\lambda_r - \lambda_v) = 1.5$, which implies $m_{\max} = 1$. In other words, only three interference fringes of orders $m = 0, 1, -1$ can be observed distinctly in the considered case. The central fringe will appear as being white, the two first minima appear as dark bands, and the two second maxima as colored bands.

1.7 Using Fig.1.22 we consider two transmitted rays which produce interference in the direction of P_2. Let us assume the interference is located around the upper surface of the plane plate. In this case ray S_1AB is affected by two reflections at A and B, while running between the lens and the plate, whereas ray S_2B undergoes no reflections until it reaches the upper surface of the plate. Thus the ratio of intensities I_1 of ray S_1AB and I_2 of ray S_2B in the vicinity of the upper surface of the plate may be estimated to be $I_1/I_2 = R^2$. Further this ratio does not change when both rays are passing through the plate. Using (1.6) for the intensity of the transmitted rays, we may write $I = I_2(\mathcal{R} + 1 + 2\mathcal{R}\cos\Delta)$. At normal incidence we apply $\mathcal{R} = (n-1)^2/(n+1)^2$. Taking into account $\mathcal{R} \ll 1$, we approximate $I \approx I_2(1 + 2\mathcal{R}\cos\Delta)$. Then the ratio of the intensities between maxima and minima of the interference pattern is given by

$$V = \frac{1 + 2(n-1)^2/(n+1)^2}{1 - 2(n-1)^2/(n+1)^2} \ .$$

Substitution of the numerical value of n gives $V \approx 27/23 \approx 1.17$.

1.8 By definition the degree of monochromaticity is given as $\lambda/\delta\lambda$, where $\delta\lambda$ is the spectral interval of wavelengths transmitted by the optical filter. In this case $\delta\lambda$ is the width of the transmission curve of one interference maximum (LORENTZian shape). All other transmission maxima have to be suppressed by a colored glass filter. The free spectral range is given by

$$\Delta\lambda_{FSR} = \frac{\lambda^2}{2nh}$$

and the finesse by

$$\mathcal{F} = \frac{\pi\sqrt{\mathcal{R}}}{1 - \mathcal{R}} \ .$$

Thus, the transmitted wavelength range is given by $\delta\lambda = \Delta\lambda_{FSR}/\mathcal{F}$.

The numerical values above give $\Delta\lambda_{FSR} \approx 10$ nm, $\mathcal{F} \approx 9$, $\delta\lambda_{FSR} \approx 1.1$ nm, and the degree of monochromaticity, $\lambda/\delta\lambda \approx 500$.

Chapter 2

DIFFRACTION OF LIGHT

By diffraction usually the phenomenon of deflection of light rays is meant, which occurs when light propagates in the vicinity of the edge of an obstacle, or when light passes through a small aperture in a screen. We have already mentioned (Part 1, Chapter 1) the appearance of fringes around the border of the geometrical shadow of an obstacle, which is caused by diffraction. The well-known HUYGENS principle, which states that secondary spherical waves are emitted by a wavefront, leads to the supposition that the propagation of secondary waves is responsible for this phenomenon. The existence of secondary waves originating from primary wavefronts enables light waves to penetrate the border line of the geometrical shadow.

The extension of HUYGENS treatments by FRESNEL opened a way to a more than qualitative description of diffraction. The principle of HUYGENS–FRESNEL enables us to correctly calculate all problems of diffraction mathematically through the interference of secondary waves. As we have seen in Chapter 1, interference patterns usually have relatively small angular sizes. This is caused by the short wavelength of light and is the reason why the observation of *diffraction fringes* has to be performed under specific geometrical conditions. On the other hand, the small angular sizes of such patterns lead to specifically dimensionless factors which have to be taken into consideration in any diffraction problem. One of these principle factors is the ratio between the wavelength and the linear dimensions of the obstacle. Another one is the ratio between the distance from the obstacle to the region of observation and the wavelength. If this ratio is of moderate size, we have the case of so-called FRESNEL *diffraction* (where rays interfere which include small angles). If this factor is infinity, parallel rays interfere and we speak of FRAUNHOFER *diffraction*. In the latter case, the diffraction pattern appears at infinity, but can be transformed by use of a collecting lens to appear on its focal plane.

1.   Fresnel diffraction

1.1   Circular aperture

The simplest case of FRESNEL diffraction is observed when a monochromatic wave passes a circular aperture in a plane opaque screen. An experimental arrangement, which enables the observation of typical FRESNEL diffraction patterns, is presented in Fig.2.1.

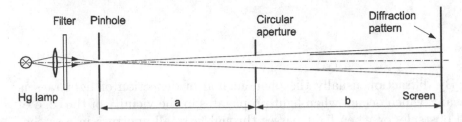

Figure 2.1.   Setup for the observation of FRESNEL diffraction on a circular aperture.

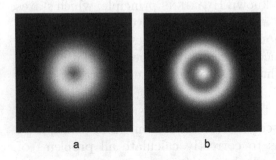

Figure 2.2 Diffraction patterns obtained for a circular aperture of diameter $d = 1.7$ mm at $a = b = 90$ cm (a), and $a = b = 135$ cm (b). The optical filter selects the green mercury line with $\lambda = 535$ nm.

A light beam from a bright source (a mercury lamp can be effectively used), collected by a condenser lens, falls on a pinhole. One can regard the pinhole as a source of diverging spherical waves, which illuminate a circular aperture, e.g. of the diameter $d = 1.7$ mm, at a distance a from the pinhole. The light waves after the aperture, which are usually called diffracted waves, form a diffraction pattern which can be observed on a screen at distance b from the aperture. Figure 2.2 presents two patterns typical for FRESNEL diffraction for the special case of $a = b$ and a circular aperture of the diameter $d = 1.7$ mm. The patterns have an axial symmetry.

The pattern, containing the bright spot at its center, was obtained for $a = b = 90$ cm. The second pattern (Fig.2.2,b) with the dark spot at its center was formed for $a = b = 135$ cm. In the general case, a decrease of the distance $2a$ between the pinhole and the screen results in an increase of the number of bright and dark rings, or diffraction fringes. A dark or a bright spot will appear at the center of the diffraction pattern. Quasi-monochromatic light has to be selected by the optical filter for a clear observation of diffraction, as in any interference experiment, where distinct fringes are formed. All results of the experiment have to be considered for a certain value of wavelength. Due to its large intensity and its good visibility for the human eye, the green line of mercury, $\lambda = 535$ nm, was selected.

1.2 Zone construction of Fresnel

Now we apply the principle of HUYGENS–FRESNEL to an evaluation of the light field caused by diffraction from a circular aperture. Let the point source S, the center O of the circular aperture, and a point of observation P lie on a straight line drawn through O normal to the screen plane (Fig.2.3). We want to calculate the diffraction field at point P.

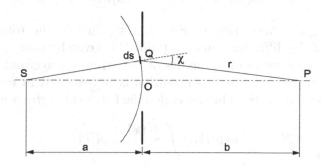

Figure 2.3. Illustrating the derivation of the FRESNEL integral.

We assume that the point source S emits a spherical monochromatic wave with wavelength λ. Hence, light disturbance emitted by S reaches every point of the spherical surface drawn through point O at the same time, the moment t_0. According to HUYGENS' principle, a small element ds on this wavefront around point Q is regarded to be a source of a secondary spherical wave. This wave emitted forward to point P will give a contribution dE to the total diffraction field at point P. Because the time factor $\exp(-i2\pi\nu t_0)$ is common to each contribution given by points Q of the spherical surface, it can be omitted. The total diffraction field amplitude at P should therefore be given by the sum of all terms, varying Q over the total surface of the aperture. Let E_0 be the amplitude

of the spherical monochromatic wave at unit distance from the source
S. The complex amplitude of this wave at point Q is then determined
by the distance a between the point source S and point O :

$$E_Q = \frac{E_0}{a}\exp(ika) \quad , \tag{2.1}$$

where $k = 2\pi/\lambda$. Further, the element dE of the complex amplitude
assigned to one secondary spherical wave produced by the element ds is
considered to have the form

$$dE = K(\chi)E_Q ds \quad , \tag{2.2}$$

where $K(\chi)$ is an inclination factor, which describes the variation of the
amplitude of the secondary waves with direction SQ and is specified by
the angle χ between SQ and QP. Taking into account (2.1) and (2.2),
the contribution dE from the spherical wave emitted by the element ds
can be written as

$$dE = \frac{E_Q K(\chi)}{r}\exp(ikr)ds = \frac{E_0 K(\chi)}{ar}\exp(ika)\exp(ikr)ds \quad , \tag{2.3}$$

where $r = \overline{QP}$. According to FRESNEL's principle the total complex
amplitude of the diffraction field in point P is given by the superposition
of the secondary waves of all points of the primary wavefront inside the
circular aperture. Such a superposition is represented in terms of the
surface integral over dE, which is called the FRESNEL *diffraction integral*:

$$E = \frac{E_0}{a}\exp(ika)\int_S \frac{K(\chi)}{r}\exp(ikr)ds \quad , \tag{2.4}$$

where the subscript S denotes integration over the wavefront within the
limits of the circular aperture.

The integral in (2.4) can be evaluated by means of the method suggested by FRESNEL, called FRESNEL's *zone construction*. Centered on
point P one first draws a sphere with radius $OP + \lambda/2$, then second with
radius $OP + \lambda$, then third with radius $OP + 3\lambda/2$, and so on. These
spheres will intersect the wavefront along circles as shown in Fig.2.4.
Spherical elements of the wavefront between two adjacent circles are
called the FRESNEL *zones*. We mark the zones by an index, beginning
with one at the optical axis.

The surface integral in (2.4) can be seen as a sum of partial integrals,
each calculated within the limits of one appropriate FRESNEL zone:

$$E = \frac{E_0}{a}e^{ika}\int_s \frac{K(\chi)}{r}e^{ikr}ds =$$

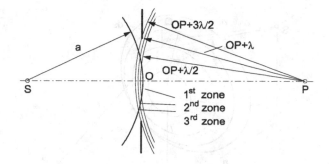

Figure 2.4. FRESNEL's zone construction.

$$= \frac{E_0}{a} e^{ika} \left[\int_{s_1} \frac{K(\chi)}{r} e^{ikr} ds + \dots + \int_{s_n} \frac{K(\chi)}{r} e^{ikr} ds \right] \quad , \qquad (2.5)$$

where s_1, s_2. ..., s_n are designations of the areas of the FRESNEL zones.

The axial symmetry and the circular shape of the FRESNEL zones in the case under consideration (circular aperture) enable us to use the polar angle θ and the azimuth angle ϕ (Fig.2.5) for the calculation of any individual integral term in (2.5). For a given distance a, an element ds of the surface of a FRESNEL zone is given by

$$ds = a^2 \sin \theta d\theta d\phi \quad . \qquad (2.6)$$

It is evident from the triangle SQP that the relationship

$$r^2 = a^2 + (a+b)^2 - 2a(a+b) \cos \theta$$

between $\cos \theta$ and a, b, and r is valid. We differentiate this equation and get

$$r dr = a(a+b) \sin \theta \, d\theta \quad .$$

Finally, using (2.6), we obtain

$$ds = a^2 \sin \theta \, d\theta \, d\phi = \frac{a}{a+b} r dr d\phi \quad .$$

The element ds is now represented in terms of the two distances a and b, which are both experimental parameters.

Further, we suppose that $K(\chi)$ can be approximated by the value K_m, where K_m is constant within the zone with index m, so that we can write K_m in front of the integral. Hence, the integration over ϕ gives

$$E_m = 2\pi \frac{E_0 K_m}{a+b} \exp(ika) \int \exp(ikr) dr \quad .$$

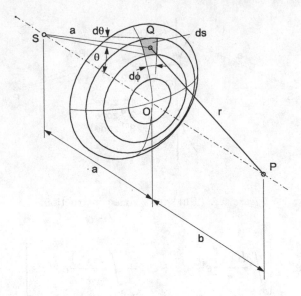

Figure 2.5. Calculation of the FRESNEL integral. The element ds of one FRESNEL zone around point Q is represented in terms of two angles, θ and φ and two distances, a and b.

For the mth zone, the variable r varies from $b + (m-1)\lambda/2$ to $b + m\lambda/2$, hence we get

$$E_m = 2\pi \frac{E_0 K_m}{a+b} \exp(ika) \int_{b+(m-1)\lambda/2}^{b+m\lambda/2} \exp(ikr)dr =$$

$$= -\frac{2\pi i}{k} \frac{E_0 K_m}{a+b} \exp(ik(a+b)) \exp(ikm\lambda/2)\left(1 - \exp(-ik\lambda/2)\right) \quad .$$

Since $k\lambda = 2\pi$, the last two factors may be reduced to

$$\exp(ikm\lambda/2)\left(1 - \exp(-ik\lambda/2)\right) =$$

$$= \exp(i\pi m)\left(1 - \exp(-i\pi)\right) = (-1)^m \cdot 2 \quad ,$$

and we finally get

$$E_m = 2i\lambda \frac{E_0 \exp(ik(a+b))}{a+b}(-1)^{m+1} K_m \quad . \tag{2.7}$$

The total amplitude E at the point of observation P obtained by summarizing over all contributions is given by

$$E = 2i\lambda \frac{E_0}{a+b} \exp(ik(a+b)) \sum_{1}^{n}(-1)^{m+1} K_m \quad . \tag{2.8}$$

We assume now that the distances a and b are very large compared to the wavelength λ. Therefore the variation of the factor K while passing from one zone to the next zone is negligible: $K_m \approx K_{m+1}$. We re-group the terms of the sum in (2.8) as follows:

$$\frac{K_1}{2} + \left(\frac{K_1}{2} + \frac{K_3}{2} - K_2\right) + \left(\frac{K_3}{2} + \frac{K_5}{2} - K_4\right) + \ldots \quad .$$

As the terms in brackets are close to zero, an approximate relation for the sum in (2.8) is found as

$$\sum_{0}^{n} (-1)^{m+1} K_m = \frac{K_1}{2} + \frac{K_n}{2} \quad , \text{when } n \text{ is odd} \quad ,$$

$$\sum_{0}^{n} (-1)^{m+1} K_m = \frac{K_1}{2} - \frac{K_n}{2} \quad , \text{when } n \text{ is even} \quad .$$

Geometrically, the summation in (2.8) can be represented by using a vector diagram method. We suppose that the first zone is divided into nine equal parts. The field produced by this zone at the point of observation is then the sum of nine field contributions. We also assume that each pair of adjacent field contributions is imaged by two small vectors inclined relative to each other by some angle, which is equal to the phase difference between these fields. The vector diagrams of the first and the second open zones are shown in Fig.2.6,a,b. We see that the vectors, starting at point A and summed up, give the total vector of the field at point B. The vector **AB** therefore corresponds to the action of the first zone (Fig.2.6,a). After progressing in a similar way, we obtain the vector **BC**, which shows the value and the direction of the field resulting from the action of the first and the second zone (Fig.2.6,b). A length difference of the vectors **AB** and **BC** exists due to a decrease of the values of K_m with increasing m. This reduction gives rise to a progressive shortening of the lengths of successive vectors, representing the action of the different zones. When the amount of active zones tends to infinity, the end of the resulting vector will approach point O (Fig.2.6,c). Geometrically, the vector **AO** represents the value of the field amplitude, which is created by the completely open wavefront at the observation point P. Let $E(\infty)$ be the field amplitude at point P created by the completely open wavefront, and $E(1)$ be the field amplitude created by first zone, then one sees that the following relationships are valid:

$$E(1) \approx 2E(\infty) \quad , \text{ and } \quad I(1) \approx 4I(\infty) \quad .$$

Therefore, the intensity $I(1)$, arising from first zone, is four times higher, than the intensity $I(\infty)$ from the completely open wavefront. If the geometrical sizes of a, b and diameter d and the wavelength λ lead to an even number of open zones, and if their number is not very big, the action of the zones adds up to a total intensity close to zero. The resulting vector **AC** in Fig.2.6,b shows the field amplitude created by first and second zone, its length is very small. In contrast, the action of an odd number of zones results in an increase of the intensity compared to $I(\infty)$.

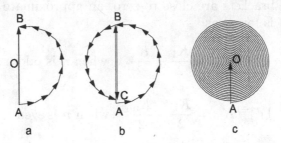

Figure 2.6. The vector diagrams for a circular aperture for the first open zone (a), the first and the second open zones (b), and the completely open wavefront (c).

1.3   A useful relation for Fresnel diffraction

We shall now derive a simple relationship between the geometrical parameters and the wavelength of light which allows estimation of the action of FRESNEL's zones. Let a point source S be at a distance a from an opaque screen with a circular aperture of the diameter d, and let us observe the diffraction pattern at point P at a distance b. The diameter d is chosen so that just n FRESNEL zones are open (Fig.2.7). It means that the radius of the n th zone r_n is equal to the radius of the aperture: $r_n = r = d/2$. According to the method discussed before, the geometrical parameters a, b, and r_n have to be related to the given wavelength λ of the light wave. As we know, the geometrical path difference $SAP - SOP$ must be approximately equal to $n\lambda/2$:

$$\overline{SA} + \overline{AP} - \overline{SP} = n\lambda/2 \quad . \tag{2.9}$$

From the triangles SAP and AOP, one can find the expression

$$\overline{SA} + \overline{AP} - \overline{SP} = \sqrt{a^2 + r_n^2} + \sqrt{b^2 + r_n^2} - (a + b) \quad . \tag{2.10}$$

Since the diameter of the aperture is much less than both distances a and b, the approximations

$$\sqrt{a^2 + r_n^2} \approx a + r_n^2/(2a) \quad \text{and} \quad \sqrt{b^2 + r_n^2} \approx b + r_n^2/(2b)$$

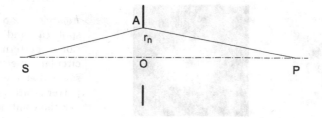

Figure 2.7. Geometrical consideration to obtain useful relationships for FRESNEL diffraction.

are valid, which give, together with (2.9) and (2.10),

$$2(\overline{SA} + \overline{AP} - \overline{SP}) = n\lambda = r_n^2 \left(\frac{1}{a} + \frac{1}{b}\right) .$$

Finally we get the relationship for the desired radius r_n of the aperture in order to open n FRESNEL zones:

$$r_n = \sqrt{n\lambda \frac{ab}{a+b}} . \qquad (2.11)$$

We can use this relationship to calculate the number n of open FRESNEL zones for the given experimental parameters a, b, r_n and λ:

$$n = r_n^2 \frac{a+b}{\lambda ab} . \qquad (2.12)$$

In the general case, for fixed values of a, b, and λ, a given aperture radius $r = d/2$ will lead to a number n' which is not integer. For the experimental parameters of Fig.2.2, we used $\lambda \approx 535$ nm (a green line of mercury) and $d = 1.7$ mm and took values of a and b which led to integer numbers n. For $a = b = 135$ cm (Fig.2.2,a), the dark spot at the center of the pattern results from the action of two open FRESNEL zones ($n = 2$). The second pattern (Fig.2.2,b) was obtained at $a = b = 90$ cm; the action of three open FRESNEL zones ($n = 3$) results in a bright central spot.

1.4 Poisson's spot

The prediction that a bright spot should appear in the center of the shadow behind a small circular disk was deduced from FRESNEL's theory by S.D.POISSON (1781-1840) in 1818. This prediction was in contradiction with common experience and every simple experiment. POISSON inferred from this result that FRESNEL's theory is not correct. However, D.F.J.ARAGO (1786-1843) performed the corresponding experiment with high accuracy and found that this surprising prediction is

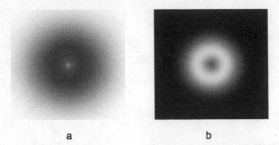

Figure   2.8 POISSON's spot (a) and the diffraction pattern from a circular aperture of the same diameter (b). The patterns are obtained under the conditions $\lambda = 535$ nm, $a = b = 135$ cm. The radii are both 1.7 mm.

a   b

correct. Such a bright spot really exists, later on called POISSON's *spot*. FRESNEL's theory was confirmed in this way.

Let us compare two diffraction patterns, the first obtained from a small metallic disc, and the second from a circular aperture (Fig.2.8). The disk as well the aperture have both the same radius $r = 1.7$ mm. In both cases the diffracting obstacle is placed at equal distance between a point source and the screen of observation ($a = b = 135$ cm). The disk causes a bright spot at the center of the diffraction pattern (Fig.2.8,a), whereas the circular aperture provides appearance of a dark spot (Fig.2.8,b). Such a distinction between the two patterns follows from the different effective number of FRESNEL zones, forming the diffraction pattern in the case of the disk and the aperture. For the parameters above the effective number of FRESNEL zones equals 2 for the aperture. Graphically, the action of the aperture is represented by the vector $\mathbf{AB} + \mathbf{BC} = \mathbf{AC}$ in Fig.2.6,b. In contrast to the aperture, the disk closes the two innermost FRESNEL zones, what results in an increase of the intensity in the center of the disk shadow. Let as before point O be the center of the vector diagram, associated with the action of the disk, point A be the initial point, point B be assigned to one open zone, and point C to both first and second open zones (Fig.2.9,a). If $\mathbf{AO}$ is specifying the completely open wavefront, and $\mathbf{AC}$ the action of first and second open zones, then vector $\mathbf{CO} = \mathbf{AO} - \mathbf{AC}$, directed from C to O, is associated with the action of all zones of the wavefront outside the disk. Evidently an increase of the disk radius causes that point C is shifted towards point O. Thus the length of $\mathbf{CO}$ becomes shorter and the intensity of POISSON's spot decreases. When increasing the radius of the disk without limit, the intensity becomes zero if the wavefront is completely closed. Therefore a more distinct POISSON spot can be achieved for a reasonably larger amount of closed zones, $n \sim 10$. We call attention to the fact that POISSON spot appears whether the number of closed zones is odd or even. Let for example point B_1 be assigned to the action of the first 3 zones, which are closed by the disc. The resulting

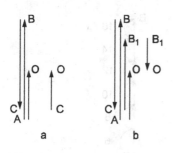

Figure 2.9 Vector diagrams illustrating the appearance of POISSON's spot for two first closed zones (a), and three first closed zones (b).

vector of all open zones is found to be $\mathbf{B_1 O} = \mathbf{AO} - \mathbf{AB_1}$, directing from point B_1 to point O as shown in Fig.2.9,b.

1.5 Fresnel's zone plate

We have seen that diffraction of light by an opaque screen, for example a small metallic disc, can result in a bright spot in the center of the geometrical shadow. We can see this as a focusing effect, or as the amplification of light in the center of a diffraction pattern. Such a focusing effect of FRESNEL's zones can be enhanced by using FRESNEL's *zone plate*.

A zone plate of this sort is a structure of concentric transparent and opaque rings, each having the appropriate radius, drawn, for example, on the plane surface of a thin glass plate (Fig.2.10,a). One can easily produce such a structure, e.g. by means of photography. For the given distances a, and b and the given wavelength λ, each transparent ring of the zone plate is associated with one particular FRESNEL zone; the internal and the external radii r_n and r_{n+1} of the rings should therefore obey formula (2.11). The structure then contains open odd number zones and closed even number zones, or vice versa.

Let E_m and E_{m+2} be the amplitudes of light fields at the point of observation which are caused by the mth zone and the $(m+2)^{th}$ zone, respectively. The phase difference between E_m and E_{m+2} is equal to 2π, whatever the value of m. If the next open zone also has the phase difference 2π, and so on, (the rings of the zone plate follow, for example, in order m, $m+2$, $m+4$, ...), all open zones amplify the field, since the field contributions of these zones are always summarized in phase. Thus the vector-diagram illustrating the action of FRESNEL's zone plate can be represented as in Fig.2.10,b. The vector $\mathbf{AB}$ shows the amplitude $E(n)$ of the resultant field after the action of 8 zones ($m = 2, 4, ..., 16$). This amplitude is estimated to be

$$E(n) \sim nE(1) \approx nE(\infty) \quad ,$$

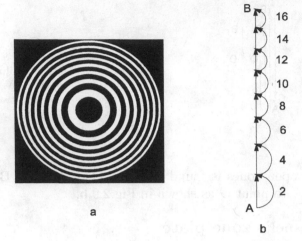

Figure 2.10. A FRESNEL zone plate closing some odd FRESNEL zones (a). The vector diagram associated with this plate (b).

where $E(1)$ is the the amplitude produced by one open zone, and $E(\infty)$ is the amplitude obtained in the case of the completely open wavefront. Therefore, the intensity I at the point of observation P is given by the expression

$$I \sim n^2 I(1) \approx 4n^2 I(\infty) \quad .$$

This simple estimation shows us that the effective intensity of the light field is strongly amplified. A zone plate, containing sixteen even zones, produces an increase of the intensity by a factor 264, compared to the intensity obtained from the first open zone, and an increase by a factor 1024 times compared to the completely open wavefront.

1.6 Fresnel diffraction at a straight edge and at a slit

We consider one of the most important cases of FRESNEL diffraction: diffraction at a straight edge.

A bright light beam from a mercury lamp passes through a filter ($\lambda = 580$ nm, yellow lines of mercury) and is then focused on a narrow vertical slit, which is regarded as a source of a cylindrically divergent light wave (Fig.2.11). A thin metallic plate is mounted at a distance of 62 cm from the slit in such a way that a straight and sharp edge of the plate is parallel to the slit. The plate shields approximately one half of the light beam. The vertical shadow of the plate edge appears on a screen, which is mounted at a distance of 62 cm from the plate. One

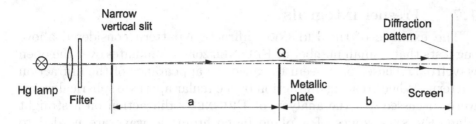

Figure 2.11. Setup for observation FRESNEL diffraction on the straight edge of a metallic plate. A point Q near the plate edge is regarded as a point source. Rays from this source cause the diffraction pattern.

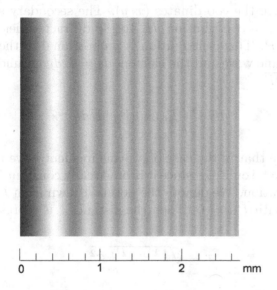

Figure 2.12. Diffraction pattern arising from a straight edge. The left-hand side of the pattern is positioned on the optical axis of Fig.2.11 (specified there by the dash-dot line). It corresponds to the position of the straight edge. The intensity on the left-hand side of the pattern is approximately equal to $I_0/4$, where I_0 is the intensity of the incident light wave.

can observe diffraction fringes, located along the shadow, parallel to the edge of the plate (Fig.2.12).

The dimension of the diffraction pattern is estimated to be about 3 mm; this results in a value of $5 \cdot 10^{-3}$ rad for the appropriate angular size of the region of the fringes.

1.7   Fresnel integrals

The example of the FRESNEL diffraction pattern considered above suggests that a small number of FRESNEL zones dominates over the open wavefront. These zones seem to cause the appearance of the diffraction pattern. When treating diffraction by a circular aperture spherical waves were discussed. In the analysis of FRESNEL's diffraction on a straight edge, FRESNEL zones of a plane monochromatic wave are needed to evaluate the diffraction field.

Let a straight edge be positioned in the (x, y)-semi-plane parallel to the x-axis, and let a monochromatic plane wave of the amplitude E_0 propagate towards the z-axis (Fig.2.13). We consider an arbitrary point Q of the open wavefront, this means Q is a point on the (x, y)-plane. Let the point Q have the coordinates (x, y). The secondary source located at Q produces the contribution dE to the diffracted field at the point of observation P. This contribution is proportional to the amplitude of the incident plane wave and the element $ds = dxdy$ around Q and to the distance $r = \overline{QP}$:

$$dE \sim \frac{1}{r} E_0 \exp(ikr) dxdy \quad . \tag{2.13}$$

Here we assume that in the case of a plane incident wave the inclination factor $K(\chi)$ can be omitted, since it is practically constant for any remote point of observation. We denote the normal drawn from P to the (x, y)-plane by PO with $\overline{PO} = L$. Then the distance r is represented by L as follows:

$$r = \sqrt{L^2 + y^2 + x^2} \quad .$$

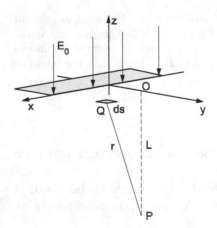

Figure 2.13 Calculation of the contribution of an element ds of the wavefront to the FRESNEL diffraction on a straight edge.

The diffracted field E at the point of observation P can therefore be expressed in terms of an integral:

$$E \sim \int_{-\infty}^{\infty} \int_0^y \frac{1}{\sqrt{L^2 + y^2 + x^2}} e^{ik\sqrt{L^2 + y^2 + x^2}} dx\, dy \quad,$$

where the variable x runs from $-\infty$ to $+\infty$, whereas the variable y varies from 0 to some finite value. Since the plate is assumed to be infinite with respect to the x-coordinate, the field E should not depend on the variable x, which enables us to replace the right-hand part of the integral by a linear integral over y:

$$E \sim \int_0^y \frac{1}{\sqrt{L^2 + y^2}} e^{ik\sqrt{L^2 + y^2}} dy \quad.$$

We also think that the diffracted field should be affected much more by the phase factor $\exp(ik\sqrt{L^2 + y^2})$ than by the amplitude factor $1/\sqrt{L^2 + y^2}$. For this reason the expression for the diffracted field can be re-written as

$$E \sim \int_0^y \exp(ik\sqrt{L^2 + y^2})dy \quad. \tag{2.14}$$

We have esimated that only a limited number of FRESNEL zones with small order give a sufficient contribution to the diffracted field. This fact allows us to take into account only a narrow band of the wavefront parallel to the straight edge. In other words, we can assume that the inequality $y \ll L$ is valid. This permits the use of the approximation $\sqrt{L^2 + y^2} \approx L + y^2/(2L)$. Hence, the expression (2.14) for the diffracted field can be re-written as

$$E \sim \int_0^y \exp(iky^2/2L)dy \quad,$$

where the invariable term $\exp(ikL)$ is omitted. Using $k = 2\pi/\lambda$ and introducing the new variable $\eta = y\sqrt{2/(\lambda L)}$, we obtain

$$E \sim \int_0^\eta e^{i\pi\eta^2/2}d\eta = \int_0^\eta \cos(\pi\eta^2/2)\, d\eta + i \int_0^\eta \sin(\pi\eta^2/2)\, d\eta \quad. \tag{2.15}$$

The calculation of the diffracted field in (2.15) is usually performed by means of two FRESNEL *integrals*:

$$X(\eta) = \int_0^\eta \cos(\pi\eta^2/2)\, d\eta \quad \text{and} \quad Y(\eta) = \int_0^\eta \sin(\pi\eta^2/2)\, d\eta \quad. \tag{2.16}$$

Geometrically, for the analysis of the diffracted field, a vector diagram called the CORNU *spiral* is used, which is a parametric plot representing the functions $X(\eta)$ and $Y(\eta)$. The CORNU spiral shown in Fig.2.14 contains two symmetric branches: one is centered on the focus F and the other on the focus F'.

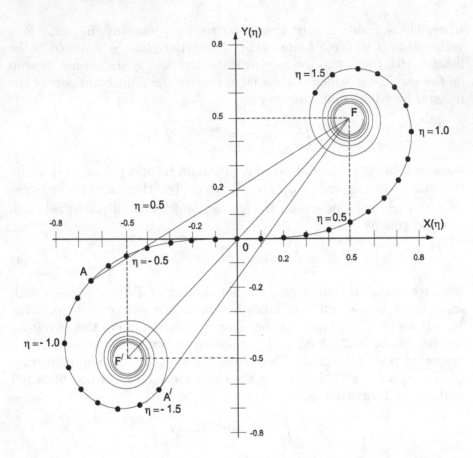

Figure 2.14. The CORNU spiral. Whatever the position of the point of observation P, the distance AF represents the amplitude of the diffracted field at P. Certain values of parameter η are specified by small dark circles on the spiral.

The left part of the CORNU spiral (A.CORNU, 1841-1902) describes the action of the secondary sources of the plane wavefront at $y < 0$. The action of all secondary sources, corresponding to the plane wavefront at $y > 0$, is shown by the vector OF. The field amplitude at the point of observation P, resulting from the completely open front, is given by the vector FF' which connects the focii of the spiral.

The CORNU spiral enables the estimation of the field amplitude at P. For all points P, this amplitude is represented by the vector AF (Fig.2.14), ending at point F. The position of A is determined by the position of P relative to the edge of the geometrical shadow. For instance, if P is located on the boundary of the shadow, then A coincides with O, and the field amplitude at P is given by the vector $OF = FF'/2$. Hence, the resulting amplitude is one half of the amplitude of the incident wave; the intensity at point P is seen to be $I = I_0/4$ (Fig.2.15). A set of values of parameter $\eta = y\sqrt{2/(\lambda L)}$, specified by small dark circles on the spiral allows us the estimation of distance y for a given wavelength λ and for the length L. For example, the first maximum of the intensity observed near the edge is specified by $\eta \approx \sqrt{2}$, which gives for the distance $y \approx \sqrt{\lambda L}$.

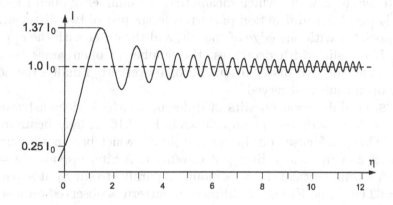

Figure 2.15. The intensity distribution within the diffraction fringes from a straight edge. The distribution is given by the function $\left[(X(\eta) - 0.5)^2 + (Y(\eta) - 0.5)^2\right]/2$, where $\eta = y\sqrt{2/(\lambda L)}$.

When moving the point of observation P out of the geometrical shadow, the point A will move to the left part of the CORNU spiral towards point F'. While moving along the CORNU spiral from point A to A' and finally to point F' the intensity will achive maxima and minima (Fig.2.15). With increasing distance from the geometrical shadow the amplitude of these oscillations decreases and the value of the intensity tends to I_0. The intensity distribution for the FRESNEL diffraction pattern at a straight edge is shown in Fig.2.15. The intensity of the diffraction pattern, calculated by means of the CORNU spiral, is proportional to the sum

$$I(\eta) = I_0 \frac{(X(\eta) - 0.5)^2 + (Y(\eta) - 0.5)^2}{2},$$

where $X(\eta)$, and $Y(\eta)$ vary from 0.5 ($y = -\infty$) to -0.5 ($y = \infty$). At the point corresponding to the boundary of the geometrical shadow, where $X(\eta) = Y(\eta) = 0$, we have $I(\eta) = I_0/4$ as it has been found above. The completely open wavefront is associated with $X(\eta) = Y(\eta) = -0.5$, or $y = \infty$; the intensity is seen to be the constant: $I(\eta) = I_0$. For $y = -\infty$ the point of observation is distanced far from the straight edge inside the shadow region. Here the wavefront is completely closed, and the intensity is equal to zero, because the functions $X(\eta)$ and $Y(\eta)$ are both equal to 0.5.

1.8   Fresnel diffraction by a slit

When we consider diffraction of a monochromatic wave with a cylindrically divergent wavefront by a slit we can build on the previous section: If the slit is wide (which means that the number of open FRESNEL zones is large), the diffraction pattern consists just of two distributions, one associated with one edge of the slit, and the other with the opposite edge. If the slit width decreases, the number of open zones becomes smaller, and the two distributions become overlaid. Finally, the action of one open zone is observed.

FRESNEL diffraction on slits of different widths is demonstrated by means of the experimental setup shown in Fig.2.16. A light beam from a mercury lamp is focused on the vertical slit S_1, which becomes a source of light with a cylindrically divergent wavefront. A filter provides ($\overline{\lambda} = 580$ mn) . A second vertical slit S_2 is mounted parallel to slit S_1 at a distance of $a = 62$ cm. The FRESNEL diffraction pattern is observed on a screen at distance of $b = a = 62$ cm from S_2.

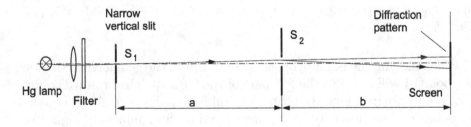

Figure 2.16.   Setup for observation of FRESNEL diffraction by a slit.

The width of S_2 can be varied. A decrease in this width from 3 mm to 0.8 mm is accompanied by a decrease in the number of the observed diffraction fringes (each being a bright vertical stripe) and a decrease of the total width of the pattern. This variation is caused by decreasing the number of FRESNEL zones of the cylindrical wavefront opened by

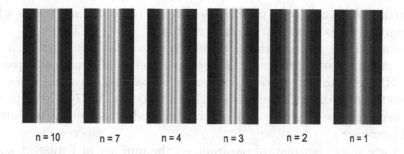

$n = 10$ $n = 7$ $n = 4$ $n = 3$ $n = 2$ $n = 1$

Figure 2.17. Diffraction patterns obtained for different slit widths. The number of open FRESNEL zones is indicated.

S_2. Fig.2.17 shows six diffraction patterns obtained at different widths of S_2. The number of open FRESNEL zones is varied from 10 to 1.

When the slit S_2 is narrow it can easily be seen in Fig.2.17 that a bright or a dark diffraction fringe occurs in the center of the pattern, depending on the number n of the open FRESNEL zones. For odd numbers one observes a bright stripe and for even numbers a dark stripe. For the used wavelength $\lambda = 580$ nm and for $a = b = 62$ cm, the width d of S_2 assigned to $n = 7$ is 2.2 mm, for $n = 4$, 1.7 mm, for $n = 3$, 1.4 mm, for $n = 2$, 1.2 mm, and 0.85 mm for $n = 1$. It is easy to verify that in the case under consideration the following relationship between the number n of open FRESNEL zones, the wavelength λ, and the width d of slit S_2 is approximately valid:

$$\sqrt{n\lambda a/2} \approx d/2 \quad , \tag{2.17}$$

which is linked to the relation (2.11) for $a = b$.

The diffraction fringes are parallel to the edges of a slit, but their features are crucially dependant on the slit width. We see that the fringes corresponding to the widely open slit are grouped near to the geometric shadow; the fringes resemble to two systems of fringes, created by two independent straight edges (cf. Fig.2.12). By decreasing the slit width, the center of the diffraction pattern becomes filled by diffraction fringes. The total number of the fringes gets smaller, but the amplitudes of their intensity increase. For the given number n of open FRESNEL zones, a maximum or a minimum will occur at the center of the diffraction pattern.

2. Fraunhofer diffraction

We have seen that the zone construction of FRESNEL works well as a method for the evaluation of the diffracted field. This method uses approximate calculations of the interference of secondary waves which are emitted by points of an open wavefront.

We also have seen in the previous chapter that one particular and important case of interference is the interference of practically parallel rays. The diffraction pattern composed by parallel rays and called the FRAUNHOFER diffraction pattern (J.FRAUNHOFER, 1787-1826) is formed in a very remote plane or can be transformed by an objective into its focal plane.

If, for given experimental parameters, the number of FRESNEL zones, estimated from Eqs. (2.12) or (2.18), is larger than one ($n > 1$), we have the case of FRESNEL diffraction, and the wavefront at the place of the diffracting obstacle has a spherical or elliptical (or any curved) shape. When n is found to be one or smaller, the wavefront can be regarded as being nearly planar. This is the case of the FRAUNHOFER diffraction.

2.1   Fraunhofer diffraction by a slit

The experimental arrangement used for the demonstration of FRES-NEL diffraction by a slit (Fig.2.16) can also be used to show FRAUN-HOFER diffraction, provided that the width d of the slit and the distances a and b satisfy the new parameters of the FRAUNHOFER diffraction. In this case the effective number of FRESNEL zones n should be smaller than one. We consider such an effective number for the particular case $b = a$. Then, according to (2.18), the relationship

$$n = \frac{d^2}{2a\lambda} < 1 \tag{2.18}$$

between λ, d, and a has to be fulfilled. The FRAUNHOFER diffraction pattern from a vertical slit obtained for the parameters $\lambda = 580$ nm, $d = 0.3$ mm, and $a = 80$ cm is shown in Fig.2.18. Using (2.18), the effective number n is estimated to be about 0.1, therefore the conditions for FRAUNHOFER diffraction are satisfied.

There always exists a bright maximum in the center of a FRAUNHOFER diffraction pattern. This is the *principal maximum*, or the maximum of zero order. The following maxima of higher orders are located symmetrically and equidistant around the principal maximum. The width of the principle maximum of the pattern shown in Fig.7.18 is about 1.5 mm, whereas its angular width is $\theta \approx 3.8 \times 10^{-3}$ rad. The pattern is stretched along the horizontal direction when the slit is vertical. The central or principal maximum is approximately twice as wide as each maximum of higher order.

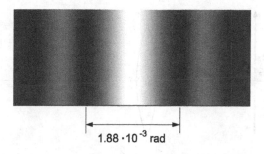

1.88 ·10^{-3} rad

Figure 2.18. FRAUNHOFER diffraction pattern from a vertical slit of 0.3 mm width. The wavelength is $\lambda = 580$ nm, the slit is equally spaced 80 cm from the light source and the screen.

2.2   The Fraunhofer diffraction integral

The fact that the approximation of plane waves is used when FRAUNHOFER diffraction is considered allows a simple way for the calculation of the diffraction field in terms of the so-called FRAUNHOFER *diffraction integral*. Let us illustrate such an integral for the example of FRAUNHOFER diffraction by a vertical slit.

Let η be the vertical coordinate parallel to the slit, ξ be the coordinate normal to the slit, and z be oriented along the wave propagation (Fig.2.19). Also let the two edges of the slit have the coordinates $\xi = -d/2$ and $\xi = d/2$, respectively. In accord with the general conditions for FRAUNHOFER diffraction, the slit is illuminated by a plane monochromatic wave of wavelength λ. This assumption provides the same initial phase for every secondary wave emitted by the wavefront within the slit space at a moment t_0. For this reason the factor $\exp(-i2\pi\nu t_0)$ is the same for each secondary wave and can be omitted. Let us assume that the FRAUNHOFER diffraction pattern is located on the second focal plane of a positive lens (Fig.2.19). For a given angle of diffraction θ, all plane waves which leave the surface of the slit at the angle θ will interfere at one point on the second focal plane. This point has the coordinate $x = f \sin\theta$, where f is the focal length of the lens. Let two points of the luminous surface of the slit, which emit plane waves under angle θ, be separated by a small distance ξ. Then the path difference between these waves is $\xi \sin\theta$. Therefore, such a pair of plane waves contributes to the desired complex amplitude with an amount proportional to the expression

$$dE \sim E_0 \exp(ik\xi \sin\theta)d\xi \quad ,$$

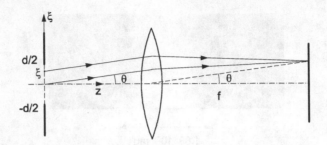

Figure 2.19. FRAUNHOFER diffraction by a slit. Two parallel rays leaving the luminous surface of the slit at angle θ are focused by a lens into one point on its focal plane. The x-coordinate of this point is $f \sin \theta$, where f is the focal length of the lens.

where E_0 is the amplitude of the incident plane wave. The total complex amplitude $E(\theta)$ is caused by the action of all points, which are assumed as equally distributed between the coordinates $\xi = -d/2$ and $\xi = d/2$. The total contribution then is represented by the FRAUNHOFER diffraction integral in the form

$$E(\theta) \sim E_0 \int_{-d/2}^{d/2} \exp(ik\xi \sin \theta) d\xi \quad . \qquad (2.19)$$

For the angular distribution $E(\theta)$ of the complex amplitude, we obtain

$$E(\theta) \sim E_0 \int_{-d/2}^{d/2} \exp(ik\xi \sin \theta) d\xi = E_0 d \frac{\sin(u)}{u} \quad , \qquad (2.20)$$

where

$$u = \frac{\pi d \sin \theta}{\lambda} \quad . \qquad (2.21)$$

For the angular distribution of the intensity within the focal plane we obtain the expression

$$I(\theta) = I_0 \left(\frac{\sin(u)}{u} \right)^2 \quad , \qquad (2.22)$$

where I_0 is the intensity at the center of the distribution (at $\theta = 0$, as the intensity of non–diffracted waves). The functions $E(\theta)$ and $I(\theta)$ are shown in Fig.2.20. The zeros of both distributions $E(\theta)$ and $I(\theta)$ occur under the following conditions:

$$\sin \theta_m = m \frac{\lambda}{d} \quad , \text{ with } \quad m = \pm 1, \pm 2, \pm 3, \ldots \quad . \qquad (2.23)$$

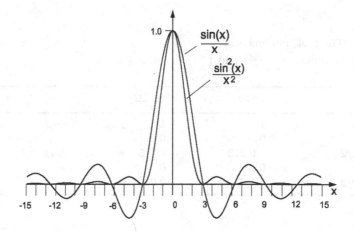

Figure 2.20. The fuctions $\frac{sin(x)}{x}$ and $\left(\frac{sin(x)}{x}\right)^2$

The principal maximum is located between the first zeros which appear for $m = 1$ and at $m = -1$. The angular width of the principal maximum $\Delta\theta$ is given by the angles θ_{+1} and θ_{-1}, both assigned to the first zeros of the distribution:

$$\Delta\theta = \theta_{+1} - \theta_{-1} \quad , \qquad \sin\theta_{+1} = \frac{\lambda}{d} \quad , \qquad \sin\theta_{-1} = -\frac{\lambda}{d} \quad .$$

Since $d \gg \lambda$ is valid in most cases, we can use the approximations

$$\theta_{+1} \approx \frac{\lambda}{d} \quad , \qquad \theta_{-1} \approx -\frac{\lambda}{d} \quad , \qquad \text{and} \qquad \Delta\theta \approx 2\frac{\lambda}{d} \quad .$$

Calculations show that the main part of the intensity, more than 80%, is concentrated within the principal maximum. The first few maxima and minima of the functions $(\sin x)/x$ and $(\sin^2 x)/x^2$ are given in Table 2.1.

At a remote distance between the plane of the diffracting aperture and the plane of the diffraction image, the observation of the FRAUNHOFER diffraction pattern is possible without the use of any lens. Let the aperture of arbitrary shape be illuminated by a plane monochromatic wave of the wavelength λ, propagating normal to the plane of the aperture, let the z-axis of a Cartesian system be directed along the direction of the wave propagation, and let the plane of the aperture be the (ξ, η)-plane of this system (Fig.2.21). Let a remote plane normal to the z-axis be the plane of observation, and let (x, y) be the Cartesian coordinates within this plane.

Table 2.1. The first maxima and minima of the functions $\frac{sin(x)}{x}$ and $\left(\frac{sin(x)}{x}\right)^2$ at $x \geq 0$

x	$\frac{sin(x)}{x}$	$\frac{sin^2(x)}{x^2}$	
0	1	1	*max*
π	0	0	*min*
$3\pi/2$	-0.212	0.045	*max*
2π	0	0	*min*
$5\pi/2$	0.127	0.016	*max*

The action of spherical waves associated with secondary sources of the wavefront at the position of the aperture is represented by a FRESNEL integral similar to that in (2.4):

$$E \sim E_0 \exp(-i2\pi\nu t_0) \int_S \frac{1}{r} K(\chi) \exp(ikr) ds \quad , \qquad (2.24)$$

where all spherical waves are assumed to be emitted at the time t_0, $K(\chi)$ is the inclination factor, and r is the distance between a point on the (ξ, η)-plane and another point on the (x, y)-plane. Both factors, $\exp(-i2\pi\nu t_0)$ and $K(\chi)$, can be omitted as before. We introduce the distance L between the (ξ, η)-plane and the (x, y)-plane, and we get r as follows:

$$r = \sqrt{L^2 + (x - \xi)^2 + (y - \eta)^2} \quad .$$

Since a remote region of observation is of interest, we expand r with respect two small parameters, $(x - \xi)^2/L^2 \ll 1$ and $(y - \eta)^2/L^2 \ll 1$,

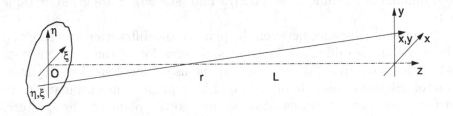

Figure 2.21. For the calculation of the FRAUNHOFER diffraction integral. An aperture of arbitrary shape is placed on the (ξ, η)-plane, and the diffraction pattern is observed in the remote (x, y)-plane. This plane is parallel to the (ξ, η)-plane and located at a distance L.

and use the approximation

$$r \approx L \left[1 + \frac{(x - \xi)^2}{2L^2} + \frac{(y - \eta)^2}{2L^2} \right] .$$

We re-write the last expression as follows:

$$r \approx L - \frac{x\xi + y\eta}{L} + \frac{x^2 + y^2}{2L} + \frac{\xi^2 + \eta^2}{2L} . \qquad (2.25)$$

We now assume that the distance L is large enough to allow us to neglect the term $(\xi^2 + \eta^2)/(2L)$:

$$\frac{\xi^2 + \eta^2}{2L} \ll \frac{|x\xi + y\eta|}{L} . \qquad (2.26)$$

If this is the case, then for a given (x, y)-point of observation the phase in (2.26) depends linearly on the coordinates ξ and η:

$$\exp \left[ik \left(L - \frac{x\xi + y\eta}{L} + \frac{x^2 + y^2}{2L} \right) \right] . \qquad (2.27)$$

With fixed (x, y)- coordinates, such a phase dependency, being linear with regard to the (ξ, η)-coordinates, should result in a FRAUNHOFER integral. As before, the amplitude factor in the integral can be regarded as invariable, this means $r \approx L$. By approximations as above the FRAUN-HOFER integral can finally be represented in the form

$$E(x, y) \sim \frac{E_0}{L} \exp(ikr_0 + ik(x^2 + y^2)/(2L)) \cdot$$

$$\cdot \int_\eta \int_\xi \exp\left[-ik(x\xi + y\eta)/L \right] d\xi \, d\eta \quad, \qquad (2.28)$$

where the integration is performed over the area of the aperture.

2.2.1 Rectangular aperture

Let us use the same experimental arrangement as before. We now place a rectangular aperture with sides $a = 0.1$ mm and $b = 0.2$ mm into the light beam instead of the slit. Let a pinhole be now the light source. Let the longer side b be oriented along the vertical direction. The FRAUNHOFER diffraction pattern formed by the rectangular aperture is shown in Fig.2.22. It is seen that the principal maximum of the diffraction pattern is surrounded by first minima.

The distance Δx between the two first zeros of the intensity distribution in the horizontal direction is measured on the screen to be $\Delta x = 9.3$

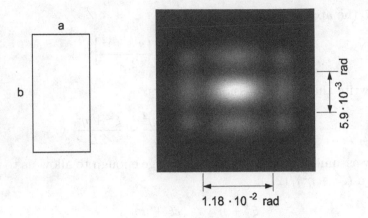

Figure 2.22. The mutual orientation of a rectangular aperture (a) and the FRAUN-
HOFER diffraction pattern (b) associated with the aperture. This pattern is obtained
under the conditions $a = 0.1$ mm, $b = 0.2$ mm, and $\lambda = 580$ nm. The rectangular
aperture is equally spaced by 80 cm from the light source and from the screen.

mm, whereas in vertical direction the two first zeros are separated by
$\Delta y = 4.7$ mm. Since the distance between the aperture and the plane
of observation is $L = 80$ cm, the angular dimensions associated with Δx
and Δy can be estimated to be $\Delta\theta \approx 1.18 \times 10^{-2}$ rad and $\Delta\zeta \approx 5.9 \times 10^{-3}$
rad, respectively.

Let us consider the evaluation of such a diffraction pattern in terms
of the FRAUNHOFER diffraction integral. Since the luminous aperture
is a rectangle, the surface integral in (2.28) is represented by two linear
integrals as follows:

$$E(x,y) \sim E_0 \int_\eta \exp\left[-iky\eta/L\right] d\eta \int_\xi \exp\left[-ikx\xi/L\right] d\xi \quad , \qquad (2.29)$$

where the constant phase factor is omitted. It is obvious that each
integral in (2.29) is a function similar to (2.20). Therefore, for the distribution $E(x,y)$ we can write

$$E = E_0 ab \frac{\sin(u)}{u} \frac{\sin(v)}{v} \quad , \qquad (2.30)$$

where

$$u = \frac{kax}{2L} \quad , \qquad v = \frac{kby}{2L} \quad , \qquad (2.31)$$

and $k = 2\pi/\lambda$. By introducing a pair of angular variables θ and ζ, called
the *angles of diffraction*, in such a way that

$$\sin\theta = x/L \quad , \text{ and } \qquad \sin\zeta = y/L \quad ,$$

the phase variables u and v take the form

$$u = \pi a \sin \theta / \lambda \quad , \text{and} \quad v = \pi b \sin \zeta / \lambda \quad . \tag{2.32}$$

The angular width $\Delta \theta$ in the horizontal direction can be approximated in the case of small angles to be $\Delta \theta = 2\lambda/a$, whereas the same magnitude assigned to the vertical direction is $\Delta \zeta = 2\lambda/b$. Assuming the wavelength $\lambda = 580$ nm for the quasi-monochromatic light in the experiment above, the angular widths of the diffraction pattern obtained from the rectangle with the sides $a = 0.1$ mm and $b = 0.2$ mm is $\Delta\theta \approx (2 \cdot 5.8/0.01) \cdot 10^{-5} = 1.16 \cdot 10^{-2}$ rad and $\Delta\zeta \approx (2 \cdot 5.8/0.02) \cdot 10^{-5} = 5.8 \cdot 10^{-3}$ rad, in good agreement with the data found from the measurements.

2.3   Circular aperture

FRAUNHOFER diffraction on a circular aperture is one of the most important cases of this type of diffraction, since most of the optical devices used in practice are built with a circular aperture.

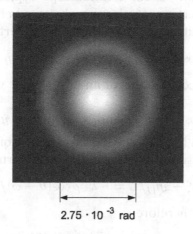

2.75 $\cdot$ 10^{-3} rad

Figure 2.23. FRAUNHOFER diffraction pattern of a circular aperture. This pattern is obtained under the following conditions: Radius of the aperture $a = 0.25$ mm, and wavelength $\lambda = 580$ nm. The aperture is equally spaced by 80 cm from the light source and the screen for observation.

A small circular aperture of radius $a = 0.25$ mm is now placed in the light path ($\lambda = 580$ nm) instead of a rectangular one and results in the FRAUNHOFER diffraction pattern shown in Fig.2.23. A bright circle exists in the center of the pattern, which is called the AIRY *circle* (G.B.AIRY, 1801-1892). The diameter of first dark circle of the diffraction pattern is found to be about $D = 2.2$ mm. This value corresponds

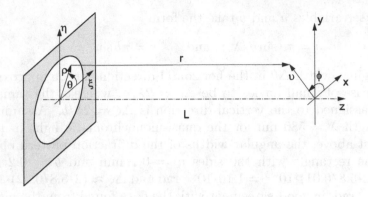

Figure 2.24.　Two polar coordinate systems are used for the calculation of the FRAUN-HOFER diffraction integral in the case of a circular aperture.

to an angular diameter of $\Delta\theta \approx 2.75 \times 10^{-3}$ rad for the given distance of $L = 80$ cm between the aperture and the screen for observation.

Let us derive a few useful relationships typical for FRAUNHOFER diffraction by an aperture of circular shape. Let (ρ, θ) be the polar coordinates of an arbitrary (ξ, η)-point of the aperture (Fig.2.24)

$$\rho \cos\theta = \xi \quad, \text{ and } \quad \rho \sin\theta = \eta \quad, \tag{2.33}$$

and let (v, ϕ) be the coordinates of an (x, y)-point within the diffraction pattern,

$$v \cos\phi = x \quad, \text{ and } \quad v \sin\phi = y \quad. \tag{2.34}$$

In accordance with (2.33) and (2.34) the phase of the integrand of the FRAUNHOFER diffraction integral will take the form

$$-k(x\xi + y\eta)/L = -kv\rho\cos(\theta - \phi)/L \quad,$$

and the integrand is therefore

$$\exp\left[-ikv\rho\cos(\theta - \phi)/L\right] \quad.$$

On substituting $d\xi\,d\eta$ by $\rho\,d\rho\,d\theta$ in (2.30), we obtain the expression

$$E \sim E_0 \int_0^a \int_0^{2\pi} \exp\left[-ikv\rho\cos(\theta - \phi)/L\right] \rho\,d\rho\,d\theta$$

for the complex amplitude E, where a is the radius of the aperture. The two-dimensional integral in the right-hand part of the last expression can be represented by means of the BESSEL function $J_1(x)$:

$$E \sim E_0 \int_0^a \int_0^{2\pi} \exp\left[-ikv\rho\cos(\theta - \phi)/L\right] \rho\,d\rho\,d\theta =$$

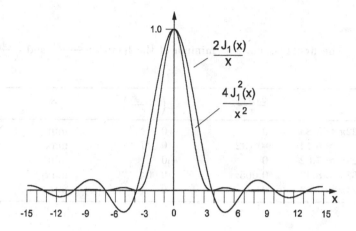

Figure 2.25. The functions $\frac{2J_1(x)}{x}$ and $\left(\frac{2J_1(x)}{x}\right)^2$.

$$= E_0 \pi a^2 \frac{2J_1(u)}{u} \quad , \text{ where } \quad u = \frac{2\pi a v}{\lambda L} \quad . \tag{2.35}$$

The function $2J_1(x)/x$, assigned to the angular distribution of the field amplitude, and the function $4J_1^2(x)/x^2$, associated with the angular distribution of the intensity, are shown in Fig.2.25. The first few maxima and minima of the functions $2J_1(x)/x$ and $4J_1^2(x)/x^2$ are presented in Table 2.2. The first zeros of $2J_1(x)/x$ and $4J_1^2(x)/x^2$ occur at $x = \pm 0.61\pi$, where

$$x = 2\pi a v/(\lambda L) \quad .$$

Thus the diameter D of the center of the first dark ring in Fig.2.23 is given by

$$D = 1.22 \frac{\lambda}{a} L \quad . \tag{2.36}$$

The angular diameter of the principal maximum is found to be

$$\Delta\theta = \frac{D}{L} = 1.22 \frac{\lambda}{a} \quad . \tag{2.37}$$

Calculations show that more than 90% of the light intensity is concentrated within the principal maximum. Assuming the wavelength of the quasi-monochromatic light in the experiment is again $\lambda = 580$ nm, and the radius of the aperture is $a = 0.25$ mm, we estimate the angular diameter of the principal maximum to be $\Delta\theta = 1.22\lambda/a \approx 2.8 \cdot 10^{-3}$ rad, which is in good agreement with the measurements.

Table 2.2. The first maxima and minima of the fuctions $\frac{2J_1(x)}{x}$ and $\left(\frac{2J_1(x)}{x}\right)^2$ at $x \geq 0$.

x	$\frac{2J_1(x)}{x}$	$\left(\frac{2J_1(x)}{x}\right)^2$	
$1.22\pi \approx 3.83$	0	0	min
$1.64\pi \approx 5.14$	-0.132	0.017	max
$2.23\pi \approx 7.02$	0	0	min
$2.68\pi \approx 8.42$	0.065	0.004	max
$3.24\pi \approx 10.2$	0	0	min

2.4   Geometrical optics and diffraction

The concept of light rays is based on the assumption that the wavelength considered is nearly zero with respect to the size of any aperture or obstacle, which is of interest in the geometrical optics. Now we can state that it is the number of the FRESNEL zones which allows us to distinguish between either the geometrical optics, or the case of FRESNEL diffraction, or the case of FRAUNHOFER diffraction.

Let an aperture of the size d at a distance a from the screen of observation be illuminated by a monochromatic plane wave of the wavelength λ. The amount n of the open FRESNEL zones observed at the central point of the screen,

$$n = \frac{d^2}{\lambda a} \quad , \tag{2.38}$$

can be regarded as the parameter which allows us to estimate which regime is valid. We hold the *geometrical optics approximation* to be true if

$$n \gg 1 \quad .$$

In the vicinity of the edges of the aperture, light waves are always deflected at angles of the order of λ/d due to diffraction. Nevertheless, violations of this sort can be regarded as negligible when the angular size of the aperture, given approximately by d/a, is still much greater than the diffraction angle λ/d: $d/a \gg \lambda/d$. We get the inequality $d^2 \gg \lambda a$ which is similar to $n \gg 1$.

We say that FRESNEL diffraction is taking place when the diffraction angle λ/d has the same order as the angular dimension of the aperture. This condition can be expressed by the inequality $\lambda/d \leq d/a$ which shows that a moderate number n of FRESNEL zones is open: $n = d^2/(\lambda a) \geq 1$.

The requirement for the FRAUNHOFER diffraction, represented by the inequality (2.28) can be written as

$$\frac{d^2}{2L} \ll \frac{xd}{L} \; .$$

Since $x/L = \sin\theta$, where θ is the angle of diffraction, we can write the equivalent inequality $d^2/(2L) \ll d\sin\theta$. In order to estimate the left-hand side of this inequality we substitute $d\sin\theta = \lambda$ for the right-hand side, which is true in the case of FRAUNHOFER diffraction, by regarding θ as corresponding to the first minima of the diffraction pattern. This gives $d^2/(2L) \ll \lambda$, which is similar to the inequality

$$n = d^2/(2\lambda L) \ll 1 \quad ,$$

which means that the effective number of FRESNEL zones is less or much less than one. The inequality $d^2/(2L) \ll d\sin\theta$ is similar to $d/(2L) \ll \theta$, if the angle of diffraction θ is assumed to be nearly zero ($\sin\theta \approx \theta$). Since d/L can be regarded as the angular size of the aperture, then the inequality $d/(2L) \ll \theta$ tells us that any angle of diffraction is much larger than the angular size of the aperture in the case of the FRAUNHOFER diffraction. The inequality $d^2/(2L) \ll d\sin\theta$ allows us further a treatment in phase terms. If the angle θ is approximately one half of the angular width of the principal maximum, then $kd\sin\theta/2 \simeq \pi$ (as before $k = 2\pi/\lambda$). Therefore, the approximation of FRAUNHOFER diffraction is true if $kd^2/(2L) \ll \pi$. In turn, contributions to the phase caused by the term $k(\xi^2 + \eta^2)/(2L)$ can be neglected. If this is the case, we say the diffraction is formed by practically plane waves.

3. Fraunhofer diffraction in optical instruments

3.1 Amplitude diffraction grating

A diffraction grating is an optical instrument, which can be considered as a system which causes a periodic variation of the amplitude or the phase, or both, of the incident wave. In the simplest case of the *amplitude diffraction grating*, a periodic amplitude variation of the incident wave is caused by means of the periodic transmissivity of this grating.

Let us consider an amplitude diffraction grating consisting of parallel grooves ruled on a surface of a plane parallel glass plate, so that N transparent stripes of width a arise. Let the (ξ, η)-plane coincide with the plane of the grating, ξ be the direction of the grooves, and d be the period of the grooves in η–direction (Fig.2.26). We also assume that a plane monochromatic wave falls on the grating, in the positive direction of the z-axis. Since all grooves and transparent stripes are identical, the diffracted field can be represented in terms of a sum of partial contributions, each being the contribution from a single slit of width a. The same phase difference exists between every pair of waves from two neighbouring slits separated by the period d of the grating. In such a way, there arises a multiple beam interference (which has been considered in the previous chapter) due to the superposition of the partially diffracted waves.

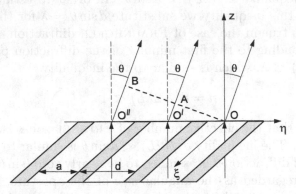

Figure 2.26. Illustrating the operating principle of the amplitude diffraction grating. The phase difference between interfering rays, diffracted at the same angle θ, increases linearly from the first groove ($m = 0$) to the other ones as follows: $mkd\sin\theta$, where m is the number of the groove at a distance md from the first groove, hence $\overline{O'A} = d\sin\theta$, $\overline{O''B} = 2d\sin\theta$, and so on.

Let η be the coordinate of an arbitrary point within the first slit with respect to the center of the slit so that

$$-a/2 \leq \eta \leq a/2 \quad .$$

The η–coordinate of a point positioned in the same manner within the mth slit is given by the expression

$$\eta_m = \eta + md \quad .$$

The complex amplitude of the diffracted field from the diffraction grating is regarded as an angular distribution $E(\theta)$, and represented in terms of the FRAUNHOFER diffraction integral. We write it in the form

$$E(\theta) \sim E_0 \sum_{m=0}^{N-1} \int \exp(-ik\eta_m)d\eta =$$

$$= E_0 \sum_{m=0}^{N-1} \int \exp\left[-ik(\eta\sin\theta + md\sin\theta)\right] d\eta \qquad (2.39)$$

and perform the integration over all transparent parts of the slits. The integral factor

$$\int \exp\left[-ik(\eta \sin \theta)\right] d\eta$$

is independent of m, whereas the integral factor

$$\int \exp\left[-ik(md \sin \theta)\right] d\eta$$

is independent of η. Thereby, the sum of integrals in (2.39) can be written as follows:

$$E(\theta) \sim E_0 \int_{-a/2}^{a/2} \exp(-ik\eta \sin \theta) d\eta \cdot \sum_{m=0}^{N-1} \exp(-ikmd \sin \theta) \quad , \qquad (2.40)$$

where the action of a single slit shows up as the integral over the transparent part a (cf.(2.19)):

$$E_0 \int_{-a/2}^{a/2} \exp(-ik\eta \sin \theta) d\eta = E_0 a \frac{\sin(u)}{u} \quad , \text{ with } \quad u = \pi \frac{a \sin \theta}{\lambda} \quad .$$

The sum in (2.40) allows the representation in terms of the multiple interference of $(N-1)$ beams. This sum is a sum of the geometric progression of the factor $\exp(-ikd \sin \theta)$. Its magnitude is known to be the fraction

$$\frac{1 - \exp(-ikNd \sin \theta)}{1 - \exp(-ikd \sin \theta)} \quad .$$

Thus for the field $E(\theta)$, we obtain

$$E(\theta) \sim E_0 a \frac{\sin(u)}{u} \frac{1 - \exp(-ikNd \sin \theta)}{1 - \exp(-ikd \sin \theta)} \quad .$$

This complex amplitude allows us to find the angular distribution of the intensity $I \sim E(\theta)E^*(\theta)$. We note that

$$\frac{1 - \exp(-ikNd \sin \theta)}{1 - \exp(-ikd \sin \theta)} \cdot \frac{1 - \exp(ikNd \sin \theta)}{1 - \exp(ikd \sin \theta)} = \frac{\sin^2(Nkd(\sin \theta)/2)}{\sin^2(kd(\sin \theta)/2)} \quad ,$$

hence

$$E(\theta)E^*(\theta) \sim E_0^2 \frac{\sin^2(u)}{u^2} \frac{\sin^2(Nkd(\sin \theta)/2)}{\sin^2(kd(\sin \theta)/2)} \quad ,$$

and finally

$$I(\theta) = I_0 \frac{\sin^2(\pi a \sin\theta/\lambda)}{(\pi a \sin\theta/\lambda)^2} \frac{\sin^2(Nd\pi(\sin\theta)/\lambda)}{\sin^2(d\pi(\sin\theta)/\lambda)} \quad , \tag{2.41}$$

where I_0 is the intensity of the incident wave of amplitude E_0.

In order to analyze multiple beam interference from N grooves we denote the phase difference between any two neighbouring beams with δ:

$$\delta = \frac{2\pi d \sin\theta}{\lambda} \ . \tag{2.42}$$

We now rewrite the second fraction in (2.41) as

$$f(\delta) = \frac{\sin^2(N\delta/2)}{\sin^2(\delta/2)} \ . \tag{2.43}$$

If $\delta/2 = \pi m$, where m is an integer, the terms $\sin^2(N\delta/2)$ and $\sin^2(\delta/2)$ are both equal to zero, and $f(\delta)$ has maxima

$$f(2\pi m) = N^2 \ .$$

Substituting $\delta/2 = \pi m$ for $2\pi d \sin\theta/\lambda$ we find the diffraction angles, corresponding to the maxima to be satisfied by the requirement

$$\sin\theta = m\lambda/d \quad , \qquad m = 0, \pm 1, \pm 2, \ldots \quad . \tag{2.44}$$

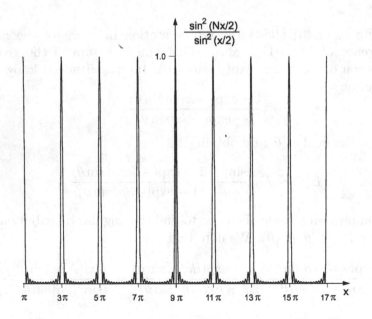

Figure 2.27. The function $\left(\frac{\sin(Nx/2)}{\sin(x/2)}\right)^2$ for $N = 16$.

Here m is called the order of the diffraction pattern. These maxima are called the principal maxima of diffraction. The function $f(\delta)$ is shown in Fig.2.27 (for $N = 16$). The principal maxima are separated by points of zero intensity, where $\sin^2(N\delta/2) = 0$. The angular positions of zero intensity (arising between two adjacent principal maxima, one of order m and the other of order $m + 1$) obey the following relationship:

$$\sin\theta = \frac{m\lambda}{d} + \frac{n\lambda}{Nd} \quad , \tag{2.45}$$

where $n = 1, ... N - 1$. Thus $N - 1$ zeros exist between two neighbouring principal maxima. A few principal and secondary maxima are shown in Fig.2.28.

The effect of the finite width of the transparent parts (single slit width a) on the angular intensity distribution is represented by the ratio

$$\frac{\sin^2(\delta a/(2d))}{(\delta a/(2d))^2} \quad .$$

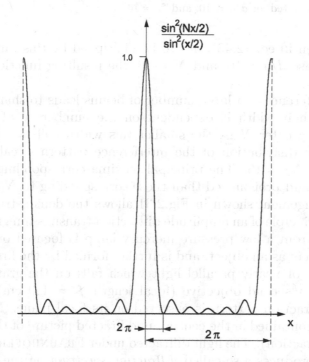

Figure 2.28. Principal and secondary maxima of the function $\sin^2(\delta N/2)/\sin^2(\delta/2)$, calculated for $N = 7$.

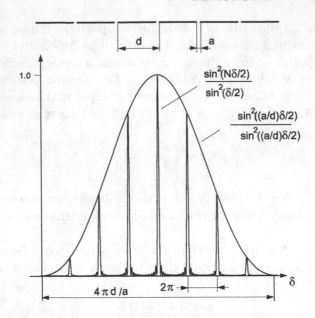

$$\frac{\sin^2(N\delta/2)}{\sin^2(\delta/2)}$$

$$\frac{\sin^2((a/d)\delta/2)}{\sin^2((a/d)\delta/2)}$$

Figure 2.29. The effect of the finite width a of the transparent parts of the periodic structure. Calculated for $d/a = 10$, and $N = 16$.

$f(\delta)$, as given in eq. (2.43), has to be multiplied by this ratio. For the particular case $d/a = 10$, and $N = 16$, the resulting function is shown in Fig.2.29.

The interference of a large number of beams leads to sharp principal maxima. Their width is dependent on the number N of interfering beams: the greater N is, the smaller this width will be. In Fig.2.30 the intensity distribution of the interference pattern is calculated for $N_1 = 8$ and $N_2 = 16$. The principal maxima corresponding to N_2 are more sharp and pronounced than those corresponding to N_1.

The arrangement shown in Fig.2.31 allows the demonstration of the operating principle of an amplitude diffraction transmission grating. The light beam from a low pressure mercury lamp is focused on a vertical slit, which acts as an object, and is further formed by the first objective into a beam of nearly parallel light which falls on the grating. After the grating a second objective (focal length $f = 400$ cm) is placed, and the diffraction pattern is observed on its focal plane. Most of the intensity is contained in the central, undiffracted picture of the slit (zero order of diffraction). The light diffracted under FRAUNHOFER diffraction conditions produces a so-called *diffraction spectrum* on the focal plane of second objective, since, following eq. (2.44), each discrete wavelength of the light source is diffracted at a certain angle.

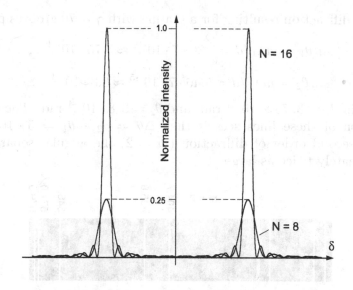

Figure 2.30. The effect of multiple-beam interference leads to sharper maxima when the number of beams N increases.

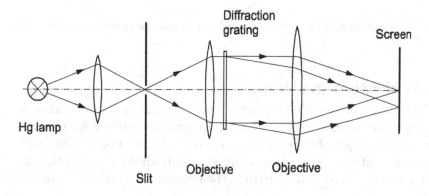

Figure 2.31. Experiment to demonstrate the operating principle of an amplitude diffraction transmission grating. A grating with $\gamma = 70$ grooves per mm is used.

Several bright lines, corresponding to the first two orders of diffraction ($m = 1$ and $m = 2$), are shown in Fig.2.32 and form the spectrum of the light source. These lines show up in order of increasing wavelength: a violet line ($\lambda = 404.7$ nm), a blue line ($\lambda = 435.8$ nm), a green line ($\lambda = 546.1$ nm) and the yellow doublet ($\lambda = 577$ and $\lambda = 579$ nm).

An amplitude diffraction grating used to generate the spectrum of a light source is often characterized as having γ grooves per mm, where γ is the inverse of the grating constant d: $\gamma = 1/d$. For the lines $\lambda_1 = 535.8$ nm and $\lambda_2 = 546.1$ nm, and for the order $m = 1$, let us estimate the angles of diffraction resulting for a grating with $\gamma = 70$ grooves per mm:

$$\sin\theta_1 = m\lambda_1/d = 5.36 \cdot 7 \cdot 10^{-5} \approx 3.75 \cdot 10^{-4} \quad,$$

$$\sin\theta_2 = m\lambda_2/d = 5.46 \cdot 7 \cdot 10^{-5} \approx 3.8 \cdot 10^{-4} \quad.$$

We obtain $\theta_1 = 3.75 \times 10^{-4}$ rad, and $\theta_2 = 3.8 \cdot 10^{-4}$ rad. The angular separation of these lines sees is then $\Delta\theta = \theta_2 - \theta_1 = 5 \cdot 10^{-6}$ rad. For the second order of diffraction ($m = 2$) the angular separation is approximately twice as large.

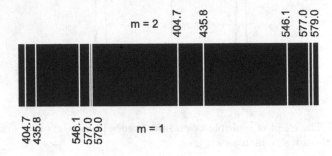

Figure 2.32. Bright lines in the spectrum of a low pressure mercury lamp in the first and second orders.

3.2   Reflection grating

The fabrication of so-called *reflection gratings* allows one to achieve a very high number of grooves per mm (up to approximately 2000 grooves per mm; with holographic techniques up to 3600 per mm). This sort of diffraction gratings is used to achieve a high angular separation in diffraction spectra. We now illustrate the operating principle of such a device.

Let us discuss the production of the diffraction maximum of first order by means of a reflection grating, as schematically represented in Fig.2.33. Let two beams S_1A and S_2O fall on the grating surface at an angle φ. Principally, waves reflected by the surface can leave this surface at any angle; let us take into account reflected waves, those leaving the grating surface at the particular angle θ, propagating along directions AP_1 and OP_2. Let AC be normal to S_2O and BO be normal to AP_1. Then the path difference between the waves reflected at the angle φ will be $\overline{AB} - \overline{CO}$. It is seen from Fig.2.33 that this path difference can be represented in terms of the angles θ and φ, and the period d:

$$\overline{AB} - \overline{CO} = d\sin\theta - d\sin\varphi \quad.$$

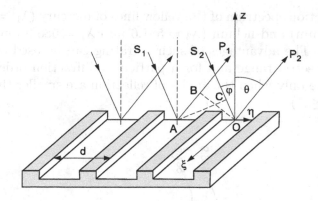

Figure 2.33. Geometrical considerations for a reflecting diffraction grating.

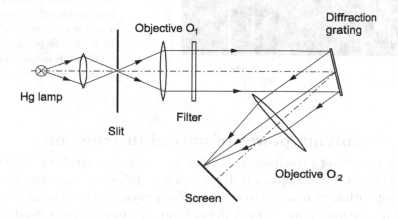

Figure 2.34. Setup for the demonstration of the operation principle of a reflecting grating

Hence, the condition for observing a principal maximum of m^{tn} order takes the form

$$m\frac{\lambda}{d} = \sin\theta - \sin\varphi. \qquad (2.46)$$

In the particular case $m = 1$ it follows from (2.46), that

$$\sin\theta = \lambda/d + \sin\varphi. \qquad (2.47)$$

In order to demonstrate the operating principle of a reflection grating with a high value of grating constant γ, light beam from a He-Hg lamp passes through a "yellow" filter and falls on the grating surface at the angle $\varphi = 12^o$ (Fig.2.34). γ is specified by 1200 grooves per mm ($d = 1/1200 = 8.3 \times 10^{-4}$ mm), which gives a value of 0.7 for $\lambda/d = \lambda\gamma$. A first order diffraction spectrum of the yellow lines of mercury ($\lambda_1 = 577.0$ nm, $\lambda_2 = 579.0$ nm) and helium ($\lambda_3 = 587.6$ nm, $\lambda_4 = 588.9$ nm) is shown in Fig.2.36. The advantages of such a grating can be used only within a narrow spectral range and for a particular diffraction order, because (2.47) is true only when the angles of reflection are smaller than 90^0 (or while $\sin\varphi \leq 1$).

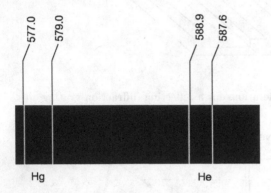

Figure 2.35 A first order spectrum of the yellow lines emitted from a Hg-He lamp obtained by means of a reflecting grating of $\gamma = 1200$ grooves per mm. The spectrum is obtained at the angle of incidence $\varphi = 12^o$.

4.   Resolving power of optical instruments

The operating principles of optical instruments have been treated in terms of geometrical optics in Part 1. These principles are based on the transformation of homocentrical pencils of rays. Such a transformation provides a point image from a point source (if one neglects geometrical errors of the imaging systems). This concept of a point in the frame of geometrical optics will still be true when the smallest dimension of the object is much larger than the wavelength. On the other hand, when considering an image in detail, even the picture of a ideal point will have a finite dimension, owing to diffraction.

The finite value of the wavelength of light determines a lower limit value of the diffraction angle, which provides a finite image size of a distant point source on the focal plane of a lens or an objective. This diffraction gives rise to a principal restriction for the minimal dimensions within optical images, and it has to be taken into account when discussing the resolving power of optical instruments. The problem of the spatial resolution of two elements of an optical image can be solved by applying the RAYLEIGH criterion. It has already been illustrated in the example of the resolving power of the FABRY–PEROT interferometer (see Chapter 1). Now we consider the resolving power of the diffraction grating and the telescope.

4.1   Resolving power of a diffraction grating

According to the RAYLEIGH criterion two adjacent principal maxima of λ_1 and λ_2 of the same order m will be observed separately if the first diffraction minimum of the first spectral line is positioned at the same angle as the principal maximum of the second spectral line (Fig.2.36). According to (2.45) the angle of the first minimum after the primary maximum of mth order of λ_1 (for $n = 1$) is given by

$$\sin \theta = m\frac{\lambda_1}{d} + \frac{\lambda_1}{Nd} \quad .$$

The RAYLEIGH criterion is just fulfilled if the mth maximum of λ_2 is positioned at the same angle θ, which gives the condition

$$m\frac{\lambda_2}{d} = m\frac{\lambda_1}{d} + \frac{\lambda_1}{Nd} \quad . \tag{2.48}$$

Let the wavelength λ_2 be different from λ_1 by an amount $\Delta\lambda$: $\lambda_2 = \lambda_1 + \Delta\lambda$. Thus, we get

$$m\frac{\lambda_1 + \Delta\lambda}{d} = m\frac{\lambda_1}{d} + \frac{\lambda_1}{Nd}$$

or

$$m\frac{\Delta\lambda}{d} = \frac{\lambda_1}{Nd} \quad ,$$

and finally, omitting the subscript of λ_1 for calculating the resolving power of the diffraction grating, we find the formula

$$\frac{\lambda}{\Delta\lambda} = mN \quad . \tag{2.49}$$

The resolving power of the diffraction grating is equal to the product of the spectral order m and the number of used grooves N, which is at the same time the number of interfering beams. It should be emphasized, that this relation holds for every type of interferometer, when N is treated as the number of interfering beams.

When discussing the FABRY–PEROT interferometer (Chapter 1), we have seen that any spectral device can be characterized by its free spectral range. In the case of the diffraction grating, the larger the wavelength λ is, the larger the diffraction angle will be. The angle also increases with the order of diffraction m. In the case of a line spectrum formed by a set of spectral lines within a spectral interval $\delta\lambda$, these lines form groups for each diffraction order. Lines of shorter wavelengths can be diffracted for a given order $m + 1$ at the same angles as lines of longer wavelengths of mth order. If this is the case, the line groups of

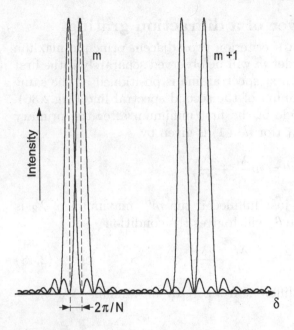

Figure 2.36 Illustrating the resolving power of the diffraction grating. The RAYLEIGH criterion in the case of the diffraction spectrum of an amplitude diffraction grating.

$(m + 1)^{th}$ and m^{th} orders would overlap if the spectral interval $\delta\lambda$ is rather wide. Let λ be the shortest and $\lambda + \delta\lambda$ the longest wavelength of such a spectrum. This overlap begins if

$$(\lambda + \delta\lambda)m = (m + 1)\lambda$$

is valid. From this relationship we get $\delta\lambda = \lambda/m$ for the free spectral range. In order to avoid this sort of an overlap (e.g. for identifying spectral lines without doubts) we must provide the condition

$$\delta\lambda \leq \lambda/m \quad , \tag{2.50}$$

where λ is the shortest wavelength of the spectrum. The requirement (2.50) holds usually if the diffraction order m is not high, but can contradict the requirement (2.49). We see that we have obtained the same results as for multiple-beam interference, Eq. (1.44).

In case of a diffraction grating the order of interference is small, for example $m = 2$, but the number of interfering rays is large (e.g. for a grating with 1200 grooves per mm and a dimension 60 mm, we get $N = 72000$). This gives $\lambda/\Delta\lambda = 144000$, and, assuming $\lambda = 600$ nm, a free spectral range of 300 nm. In contrary, for a FABRY-PEROT-interferometer, usually the order of interference is high (e.g., for $\lambda = 600$ nm and a spacing of 30 mm, we obtain $m = 10^5$), while the number of interfering beams is not too large (approximately given by the finesse $\mathcal{F}$,

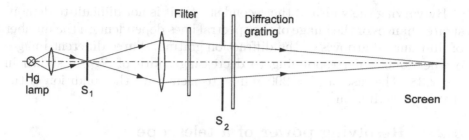

Figure 2.37. Demonstration of the effect of a finite illuminated part of the diffraction grating. An amplitude diffraction grating of $\gamma = 100$ grooves per mm is used.

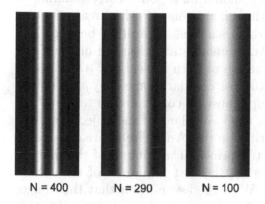

N = 400   N = 290   N = 100

Figure 2.38 Images of the yellow doublet of mercury after an amplitude grating at different numbers of illuminated grooves.

e.g. $\mathcal{F} = 50$). Thus the resolving power may be higher (in our example, $\lambda/\Delta\lambda = 5.10^{6}$), but the free spectral range $\delta\lambda_{FSR}$ is much smaller (in our example, $\delta\lambda = 6.10^{-3}$ nm).

Figure 2.37 shows a simple arrangement for the demonstration of the resolving power of an amplitude diffraction grating. The bright beam of a mercury lamp is focused on a narrow slit S_1, then its image is formed by an objective on a remote screen. A second widely open slit S_2 and an amplitude diffraction grating with $\gamma = 100$ grooves per mm are placed in the beam one after the other. An optical filter enables the selection of the yellow doublet of the mercury spectrum of $\lambda_1 = 577$ nm and $\lambda_2 = 579$ nm. To resolve these lines, a power $\lambda/\Delta\lambda$ of about 290 is necessary. To observe separated images of these spectral lines at the first diffraction order $m = 1$, the amount of open grooves of the grating should be larger than 290 to permit a good resolution of the yellow doublet ($\lambda/\Delta\lambda \leq mN$). For the given value of $\gamma = 100$, the illuminated part of the grating surface has to be wider than 3 mm in order to exceed the RAYLEIGH criterion.

By varying the width of the second slit S_2, it is not difficult to demonstrate changes of the image of the spectral lines dependent on the number of illuminated grooves of the diffraction grating. Three different images of the doublet corresponding to decreasing width of S_2 are shown in Fig.2.38. The case of $N = 290$ is in agreement with the condition of the RAYLEIGH criterion.

4.2   Resolving power of a telescope

Let us assume that the light beam from a star passes along the optical axis and through the objective of a refractive telescope. Geometrical optics treats the image of the star as located in the center of the focal plane of the objective. This image should be a point, corresponding to the focus of the objective (neglecting aberrations). Such a treatment still holds until the question arises which smallest separation between two point-images can be achieved. Any remote star is imaged on the focal plane by an AIRY circle, the diffraction pattern typical of FRAUNHOFER diffraction on an objective aperture of circular shape. By this reason the resolving power of the telescope must be considered in terms of the separation of two AIRY circles on the focal plane of the objective.

According to the RAYLEIGH criterion two AIRY circles are considered to be resolved if the position of the principal maximum of the first intensity distribution is just located at the first minimum of the second intensity distribution (Fig.2.39). With the assumption that the images are formed by light waves with the same intensities, both diffraction patterns have to be of the same brightness; if this is the case we can apply formula (2.37) to both intensity distributions to analyze the mutual positions of these patterns.

For a given radius a of the objective, and wavelength λ, the angular separation $\Delta\theta$ between the minimum of the first pattern and the maximum of the second is given by the expression $\Delta\theta = 0.61\lambda/a$. Assuming that the diffraction angles are sufficiently small ($\sin\theta \approx \theta$), the angular separation of two stars consistent with the RAYLEIGH criterion, is represented by

$$\theta = 0.61\frac{\lambda}{a} = 1.22\frac{\lambda}{D} \ , \tag{2.51}$$

where D is the diameter of the objective.

Fig.2.40 shows a simple setup for the demonstration of the resolving power of a telescopic system. A diaphragm, containing two pinholes separated in the horizontal direction by $d = 2$ mm, is illuminated by the bright light of a mercury lamp ($\lambda = 546$ nm). The pinholes act as a model for a double star. The distance L between the objective of the telescope and the aperture of the two pinholes is $L = 14$ m. The diam-

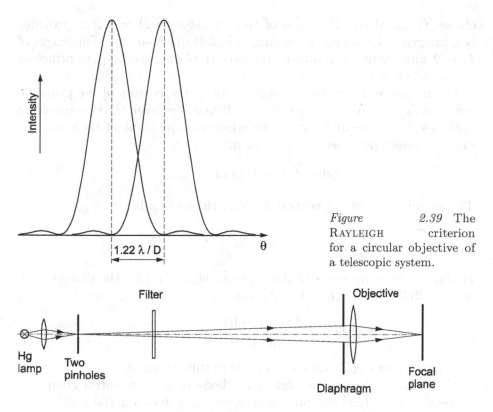

Figure   *2.39* The RAYLEIGH criterion for a circular objective of a telescopic system.

Figure 2.40. Setup for the observation of a double-star image via a telescopic system. The focal length of the telescope objective is $f = 30$ cm, its diameter 30 mm; the two pinholes, separated by 2 mm from each other, are positioned 14 m from the telescope. The green light ($\lambda = 546$ nm) from a mercury lamp is used.

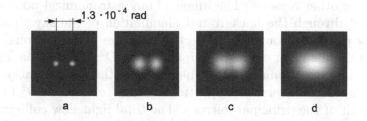

Figure 2.41. Images of a "double star" obtained at different diameters of a diaphragm in front of the telescopic objective.

eter of the objective is 30 mm. Under these conditions one can observe two distinct small bright spots (AIRY circles) in the focal plane of the objective (Fig.2.41,a). A circular diaphragm, diameter $D = 9$ mm, placed in front the objective, gives rise to an increase of the size of both AIRY circles (Fig.2.41,b). Decrease of the diameter to $D = 5$ mm provides two images which starts overlaping (Fig.2.41,c). A smaller diaphragm of $D = 2$ mm results in a diffraction pattern of two overlapping principal maxima (Fig.2.41,d).

Let us estimate the angular separation of the centers of the principle maxima $\Delta\varphi$, taking into account the distance between the centers of the pinholes $d = 2$ mm and the distance between the plane of these sources and the telescope objective, $L = 14$ m:

$$\Delta\theta = d/L \approx 1.43 \cdot 10^{-4} \text{ rad} \quad .$$

The angular size of the central AIRY circle is given by

$$\theta = 1.22\frac{\lambda}{D} \quad .$$

Based on the parameters of the experiment we find for the diameter D, which fulfills the RAYLEIGH criterion:

$$D = \frac{1.22 \cdot 5.46 \cdot 10^{-4}}{1.43 \cdot 10^{-4}} \approx 4.7 \text{ mm}$$

The pattern in Fig.2.41,c corresponds to this value.

Nowadays elaborate numerical methods allow a powerful computer correction of optical astronomical images obtained on the focal plane of the giant mirrors of reflecting telescopic systems, which permit the RAYLEIGH criterion to be exceeded. The use of telescopes with large diameter objectives or mirrors, and even the construction of giant reflecting telescopes is stimulated not only by increasing the resolving power but also for other reasons. The image of any astronomical point source is received through the background illumination of the sky. The total light flow passing through the objective aperture is proportional to the square of the diameter of the principal mirror, D^2; thus the background light flow is proportional to D^2. The area σ of the optical image (AIRY circle) of a star is inversely proportional to D^2: $\sigma \sim f^2\lambda^2/D^2$ (f is the focal length of the principal mirror). The total light flow collimated by the principal mirror within the area σ is also directly proportional to D^2, therefore the image intensity, or its brightness, is dependent on D^4 as the ratio $D^2/\sigma \sim D^4/(f^2\lambda^2)$. Hence, the use of a telescope with a giant reflected mirror provides a considerable gain in the relative brightness of such a point source against the background brightness of the sky. The ratio of brightnesses as a measure of this advantage is proportional to D^2.

SUMMARY

Diffraction as an ordinary phenomenon for all wave processes in optics possesses a number of specific features. These features are determined by the small value of the wavelength of light. In fact, when diffraction of light is considered, the geometrical parameters of an optical experiment are linked to the wavelength by particular relations (see (2.12)).

In view of the small light wavelength λ, the diffraction angle for an obstacle of a geometrical dimension d is on the order of λ/d. Thus, in order to observe diffraction phenomena for typical macroscopic obstacles $d \gg \lambda$, a long distance between an obstacle and the observation region is required.

Another facet of diffraction as a violation of the straight path of light propagation is the principally unavoidable effect, that always an image of limited size is formed in every optical instrument from a point light source.

Special methods of computations for the diffraction fields based on the FRESNEL and the FRAUNHOFER integral approximations give good results for most of the diffraction problems. One of the practical applications of the mathematical apparatus of diffraction theory may be found in FOURIER optics.

PROBLEMS

2.1 A plane monochromatic wave of intensity I_0 falls normally on a semi–infinite straight edge limited at point A (Fig.2.42). Using the HUYGENS–FRESNEL principle, find the intensity of the diffracted wave at point P located on the boundary of the shadow.

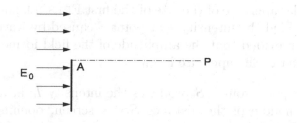

Figure 2.42.

2.2 A plane monochromatic wave of intensity I_0 falls normally on a semi–infinite plane parallel glass plate of thickness d with refractive index n (Fig.2.43). What is the intensity of transmitted wave at point P located on the perpendicular to the edge of the plate? What are the maxima and minima of the intensity? One may assume that reflection on both sides of the plate is negligible.

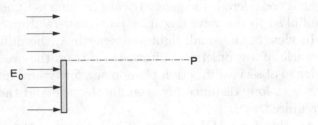

Figure 2.43.

2.3 A plane monochromatic wave of wavelength λ falls normally on a square aperture with sides a. An image of the aperture is observed on a screen at a distance L from the aperture. Estimate the distance L where the size of the diffraction pattern is approximately the same as the size of the original aperture.

2.4 Green light of $\lambda = 530$ nm falls on a diffraction grating of length $l = 12$ mm and period $d = 1.5$ μm. Calculate the angular dimension of the primary maximum and the resolving power of the diffraction grating.

2.5 Estimate the size of sensitive cells of the human eye, provided that the entrance pupil of the eye is about 1 mm in diameter under usual sunlight conditions.

2.6 A point source S is placed in front of an infinite reflecting screen. A circular aperture of the size of the first FRESNEL zone is cut out of the screen. Find the intensity I at point S caused by waves reflected by the screen, provided that the amplitude of the field formed by the source at the center of the aperture is E_0.

2.7 A point source S produces the intensity I_0 at a distant point $S^{/}$. At the middle of the distance $SS^{/}$ a screen, opening only two thirds of the first FRESNEL zone, is inserted normal to $SS^{/}$. Estimate the intensity at point $S^{/}$.

SOLUTIONS

2.1 If the edge is absent, every plane normal to the propagation direction will be a phase plane of the plane monochromatic wave. Hence, in this case the field as well as the intensity of the wave at the point P will be the same as at at point A. According to the HUYGENS-FRESNEL principle, the field $\mathbf{E}$ at point P may by regarded as a result of radiation of all elementary sources positioned within the phase plane at a distance of AP from the point P. The edge blocks one half of all these elementary sources, so the field at the point P becomes $\mathbf{E}/2$ and the intensity $I_p = I_0/4$.

2.2 According to the HUYGENS-FRESNEL principle, the field at the point in question is formed by two waves: one passes directly to the point, the other through the glass plate. These waves both have the same amplitude $E_0/2$. The wave passing through the plate is retarded by $d(n-1)/c$ with respect to the wave passing directly to the point. The complex amplitude of the resulting field E at the point is represented by the following expression:

$$E = \frac{E_0}{2} + \frac{E_0}{2}\exp(i\omega dn/c) = \frac{E_0}{2}[1 + \exp(ikd(n-1))] \quad .$$

For the intensity EE^* at point P, one gets

$$\frac{E_0^2}{4}[1 + \exp(ikdn)][1 + \exp(-ikdn)] = E_0^2\cos^2(kd(n-1)/2) \quad .$$

For the case $d = 0$ the intensity is equal to that in the case where the plate is absent, therefore, to the value of E_0^2. An analogous result follows for the assumption $n = 1$. The intensity has maxima E_0^2 if

$$kd(n-1) = 2\pi m \quad ,$$

where $m = 0, \pm 1, \pm 2, \dots$. Minima of the intensity, $EE^* = 0$, will occur if

$$kd(n-1) = \pi(2m+1) \quad .$$

2.3 The diffraction pattern from the aperture will look like a square with sides $2L\lambda/a$, for the distance L, providing a visible FRESNEL diffraction effect. From the condition $a = 2L\lambda/a$ we find, for L, the following estimation $L \simeq a^2/(2\lambda)$. We see, that, apart from detailed diffraction manifestations, the size of the diffraction pattern will have approximately the same value as the original, if the condition for one or two open FRESNEL zones is given.

2.4 The angular dimension of the primary maximum is equal to $2\lambda/d = 0.7$ rad $\simeq 40^o$. The resolving power of the grating is expressed by the formula $\lambda/\Delta\lambda = Nm$ and is dependent on the total number of the grating grooves N and on the order of diffraction m. The total number of the grating grooves is equal to $N = l/d = 8000$. The highest usable order of diffraction m_{max} will give rise to the highest value of the resolving power $\lambda/\Delta\lambda$. The value for m_{max} may be found from the condition $\lambda m_{max} = d \sin 90^o$, which gives the following estimation for m_{max}: $m_{max} \leq d/\lambda = 2.830$. Hence, $m_{max} = 2$, and one gets for the highest achievable resolving power $\lambda/\Delta\lambda = 16000$.

2.5 The resolving power of the human eye illuminated by bright sunlight is caused by its entrance pupil d as well as the wavelength of light, λ. According to the general formula for the resolving power of an optical device with circular entrance aperture, one can estimate the minimum angular size θ of an image occuring on the sensitive area of the eye: $\theta \sim \lambda/d \approx 5.5 \cdot 10^{-4}$ rad, for $\lambda = 550$ nm. For the focal length of normal eye $f \approx 17$ mm the size of the minimum area of a sensitive cell is estimated to be $\theta f \sim 10^{-2}$ mm $= 10$ μm.

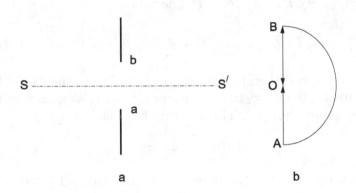

Figure 2.44.

2.6 The reflecting action of the full screen at point S is the same as caused by a point source S', which is a virtual image of S caused by this screen (Fig.2.44,a). Since S' and S are at the same distance from the plane of the screen, the amplitude of the reflected field at point S is equal to $E_0/2$, resulting in the intensity $E_0^2/4$. The action of the screen without the first FRESNEL zone is that of a point source $S^{/}$ in the case, where the first zone is closed. In the vector diagram AOB (Fig.2.44,b) this action is presented by the vector **BO** which has equal magnitude as **AO**. In turn $|\textbf{AO}| = E_0/2$. Hence, the intensity in question is $E_0^2/4$.

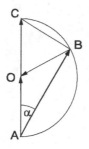

Figure 2.45.

2.7 We assume the vector diagram for the first FRESNEL zone to be presented by the semi-circle drawn from A to C as shown in Fig.2.45. If **AC** is associated with the amplitude of the fully open zone, then **AB** will specify the resulting amplitude for a partially open zone. Since the triangle ABC fits into the semi-circle, its angle ABC is a right angle, therefore $|\textbf{AB}| = |\textbf{AC}| \cos \alpha$. According to the principles for the construction of the vector diagram, $|\textbf{AC}|$ is directly proportional to the radius of the opened portion of the aperture, hence $|\textbf{AC}|/|\textbf{AB}| = r_1/r_2$, where r_1 and r_2 are the radii of the apertures for the 1st and 2/3 of the 1st FRESNEL zones, respectively. Since $r_2/r_1 = \sqrt{2/3}$, one finds $\cos \alpha = r_2/r_1 = \sqrt{2/3}$. Further, for the intensity I in question one can write $I = |\textbf{AB}|^2$. Let $|\textbf{AC}| = 2E_0$, where $I_0 = E_0^2$. Then for I we get $I = 4E_0^2(\cos \alpha)^2 = 8E_0^2/3 = 8I_0/3$.

Chapter 3

FOURIER OPTICS

FOURIER *optics* is a special part of optics connected to transformation and processing of optical images. The large amount of information provided in the form of two-dimensional images requires substantial effort in order to finally obtain an optical image of high quality. Such problems gain foremost significance when considering photographs of the earth's surface from aircraft and for images of the earth and other objects taken from spacecraft. Such images are taken at long distances, sometimes even of astronomical scale. Therefore, many factors exist which distort an image and are represented as noise of the final optical image. In this respect, the nature of image processing problems is quite similar to well known radio-technique problems - extraction of the useful signal from a background of noises. The subject of FOURIER optics is to study and to develop methods for optical image processing using optical systems [1]. Thus, this area of optics includes a number of integral transformations of an image, such as FOURIER transform, spatial filtration, convolution operation and many others.

1. Properties of Fourier spectra in optics

1.1 Spatial frequencies

When considering diffraction of a plane monochromatic wave caused by a planar screen we have discussed the special case of FRAUNHOFER diffraction (Chapter 2). Under the conditions of FRAUNHOFER diffraction the optical field on a remote plane of observation can be considered to be formed by a system of plane monochromatic waves, each propagating in a certain direction with respect to the normal drawn from the center of the screen to the plane of observation. The normal itself corresponds to propagation of undiffracted waves, those with an incident angle on this plane of zero.

Let, as before, (ξ, η) and (x, y) be the coordinates of the screen of diffraction and the plane of observation, respectively, and let L be the distance between these planes along the normal mentioned above. Then, for a given wavelength λ of a monochromatic wave, the magnitude of the wave vector is $k = 2\pi/\lambda$. The complex amplitude $E(x, y)$ at point (x,y) is represented in terms of the complex amplitude E_0 of the plane monochromatic wave, incident on the screen, by means of the following expression (see (2.28)):

$$E(x, y) \sim \frac{E_0}{L} \exp(ikL + ik(x^2 + y^2)/(2L)) \cdot$$

$$\cdot \int_\eta \int_\xi \exp\left[-ik(x\xi + y\eta)/L\right] d\xi \, d\eta \quad , \tag{3.1}$$

with integration over the area of the aperture of the screen.

It can be seen from (3.1) that the right-hand side of the expression contains one exponential item, which has quadratic form in the phase:

$$ik(x^2 + y^2)/(2L) \quad . \tag{3.2}$$

Such a quadratic dependency is caused by the propagation of spherical secondary waves from the screen to the plane of observation. We should keep in mind that, according to the HUYGENS–FRESNEL principle, these secondary waves are induced by the initial incident plane wave over the aperture, and they must be regarded as being spherical waves. One can expect that for a very remote plane of observation (at infinity) the term in the form (3.2) may be neglected with respect to other exponential terms of the integral in (3.1), being linear in the coordinates x, and y. Fortunately, the focal planes of every spherical positive thin lens fulfil the requirements of planes at a distance of infinity. Thus such a lens should realize a phase transformation which completely compensates the quadratic phase term, provided the screen which causes diffraction is positioned on the first focal plane of the lens, and the plane of observation is the second focal plane of the lens. A mathematical proof of the statement above will be done subsequently. Here, we represent an explicit form of the relationship between two distributions of the complex amplitude: one over the first focal plane (ξ, η), and the other over the second focal plane (x, y) of a thin lens:

$$E(x, y) \sim \int_\eta \int_\xi E(\xi, \eta) \exp\left[-ik(x\xi + y\eta)/f\right] d\xi \, d\eta \quad , \tag{3.3}$$

where $E(\xi, \eta)$ is the distribution of the complex amplitude within the limits of the aperture of the screen (at the first focal plane), and f is the focal length of the lens. The sign "∼" emphasizes the fact that a constant factor in front of the integral is omitted. Let us re-write the phase factor of the integrand (3.3) in the form

$$\exp\left(-ik\frac{\xi x + \eta y}{f}\right) = \exp\left[-2\pi i\left(\frac{x}{\lambda f}\xi + \frac{y}{\lambda f}\eta\right)\right] =$$

$$= \exp[-2\pi i(f_x\xi + f_y\eta)] \quad ,$$

where f_x and f_y are so-called *spatial frequencies*, which have the form

$$f_x = \frac{x}{\lambda f} \quad , \quad \text{and} \quad f_y = \frac{y}{\lambda f} \quad . \tag{3.4}$$

Using these new variables, we represent the left-hand side of (3.3) as being a complex function of the spatial frequencies in the form $E(x, y)$ and the right-hand side of (3.3) in a form of a *two-dimensional* FOURIER *spectrum* of the complex function $E(\xi, \eta)$:

$$E(f_x, f_y) \sim \int\limits_{\xi}\int\limits_{\eta} E(\xi, \eta) \exp\left[-2\pi i(f_x\xi + f_y\eta)\right] d\xi d\eta \quad . \tag{3.5}$$

The function $E(f_x, f_y)$ is called the *spatial spectrum* of the complex function $E(\xi, \eta)$.

We should direct our attention to the fact that any distribution over the aperture (the first focal plane) is assumed always to be limited. For example: In the case of a plane monochromatic wave incident on the screen we can regard $E(\xi, \eta)$ as having an invariable magnitude throughout the aperture of the screen, and to have a value of zero beyond the limits of the aperture. If this is the case, the spatial spectrum is caused by the shape of the aperture, which is usually described in terms of a so-called *pupil function* $P(\xi, \eta)$. Simple apertures such as a narrow slit or rectangular and round holes considered in the previous chapter are typical examples of pupil functions. Since in every experiment the incident light beam is limited, a special pupil function has to be applied which effects the spatial resolution of the observed spatial spectrum. We have seen that the resolving power of any telescopic system is limited by the entrance pupil of that system. The pupil function has the same meaning for problems of FOURIER optics concerned with the formation of spatial spectra.

Even if the original distribution $E(\xi, \eta)$ diverges from an uniform field of the monochromatic wave, the pupil function will effect the spatial spectrum in general. Thus, in any case the limits of the integral in (3.5) may be considered to range from $-\infty$ to $+\infty$. Then, the integral will take the explicit form of a two-dimensional FOURIER spectrum:

$$E(f_x, f_y) \sim \int\limits_{-\infty}^{+\infty} \int\limits_{-\infty}^{+\infty} E(\xi, \eta) \exp\left[-2\pi i(f_x\xi + f_y\eta)\right] d\xi d\eta \quad , \qquad (3.6)$$

where $E(\xi, \eta)$ has to be restricted to the aperture.

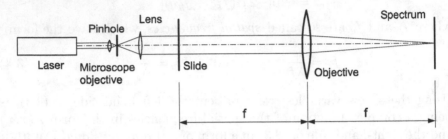

Figure 3.1. Setup for demonstration spatial spectra from slides. An uniform illumination is formed by the beam of a He-Ne laser ($\lambda = 632$ nm) by means of the microscope objective, a pinhole and a composed lens. A slide is positioned on the first focal plane and its Fourier spectrum is formed on the second focal plane of the objective ($f = 180$ cm).

The setup shown in Fig.3.1 allows the demonstration of spatial spectra arising from slides of different transparency forms. For example, we use the slide shown in Fig.3.2,a. Its transparency is varying with the coordinates (ξ, η). Thus, the complex amplitude $E(\xi, \eta)$ of the light coming from a He-Ne laser ($\lambda = 632.8$ nm) is found to vary with the transparency of the slide. The slide is placed on the first focal plane of an objective ($f = 180$ cm), and the spatial spectrum is formed on the second focal plane of the objective. In order to get a nearly uniform illumination of the slide, a microscope objective and a pinhole are placed into the laser beam, and a lens is used to form a plane light wave with a nearly uniform distribution of light energy over the slide. The intensity distribution in the second focal plane, corresponding to the spatial spectrum, is shown in Fig.3.2,b. In order to enhance weak intensity parts, this intensity distribution is shown again, but overexposed, in Fig.3.3,c. We notice in Figs.3.2,b, c that the intensity distribution shows an axial symmetry, even if the initial image is asymmetrical as in Fig.3.2,a.

The axial symmetry of the intensity distribution results from the method for the preparation of the amplitude distribution $E(\xi, \eta)$ of the original object. The *amplitude transparency* of the slide is described by

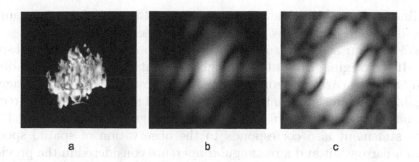

a   b   c

Figure 3.2. Transparent object (a); the intensity distribution associated with its spatial spectrum (b); an overexposured picture of the intensity distribution (c). The size of the slide is 15×15 mm, the size of the spectrum 6×6 mm. $\lambda = 632.8$ nm; the focal length of the objective is $f = 180$ cm.

a real function, usually denoted as $t(\xi, \eta)$. Assuming, that the amplitude E_0 of the incident plane monochromatic wave has a real constant value over the slide's area, the amplitude $E(\xi, \eta)$ directly behind the slide has a real magnitude

$$E(\xi, \eta) = E_0 t(\xi, \eta) \quad . \tag{3.7}$$

Hence the integral in (3.6) presents the spatial spectrum as a complex function, which should satisfy the requirements

$$E(-f_x, -f_y) = E^*(f_x, f_y) \quad , \text{ and } \quad E^*(-f_x, -f_y) = E(f_x, f_y) \quad . \tag{3.8}$$

These requirements are fulfilled by every FOURIER spectrum from a real function.

The average intensity of the spectral image has to be calculated from the expression

$$\overline{I(f_x, f_y)} \sim \overline{E(f_x, f_y)E^*(f_x, f_y)} \quad .$$

Due to (3.8), the right-hand side of this expression is equivalent to

$$\overline{E(f_x, f_y)E^*(f_x, f_y)} = \overline{E^*(-f_x, -f_y)E(-f_x, -f_y)} \quad ,$$

and the average intensity can therefore be represented as

$$\overline{I(f_x, f_y)} = \overline{E^*(-f_x, -f_y)E(-f_x, -f_y)} = \overline{I(-f_x, -f_y)} \quad .$$

Since $\overline{I(f_x, f_y)} = \overline{I(-f_x, -f_y)}$, and since the axial symmetry of the intensity distribution is expressed just by the transformations

$$f_x \rightarrow -f_x \quad , \text{ and } \quad f_y \rightarrow -f_y ,$$

we find confirmed that the axial symmetry of the spatial spectrum under consideration is caused by the reasons discussed above.

We turn our attention to an another common property of spatial spectra: If the original image has a bright element along a particular direction, as the nearly horizontal strip in the lower part of the image in Fig.3.2,a, then a significant part of the spatial spectrum will be directed normal to the bright part of the original image, as is seen in Fig.3.2,b. This statement also corresponds to the observation of spatial spectra from a narrow slit and a rectangular aperture considered in the previous chapter.

1.2   Image construction with parallel rays

The problem of improving an image received by an optical system is widely encountered in practice. Such a problem may often be solved by means of taking influence on the spatial spectrum of the original image, which results in the required transformations of the image. In Fig.3.3 a simple setup is shown which fulfills the requirements for performing such an optical transformation.

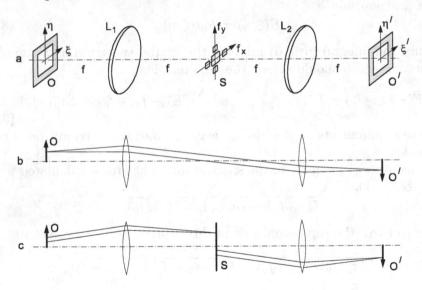

Figure 3.3.   Image construction with parallel rays. (a) perspective drawing, (b) imaging beams, (c) beams forming the spatial spectra

Two identical positive lenses are arranged in such a way, as to transform an incoming parallel beam into an outgoing parallel beam. In order to provide this property, the lenses L_1 and L_2 have to be separated by $2f$, where f is the focal length of both lenses. Let a transparent object, the object $O(\xi, \eta)$, be placed on the first focal plane of lens L_1. Then an image of this object, $O'(\xi', \eta')$, should appear on the second focal plane of the second lens, L_2. Simultaneously, light rays from the original image form a spatial spectrum S on the second focal plane of lens L_1. This spatial spectrum S is the object for lens L_2 which forms the spatial spectrum of the object on its second focal plane. Image $O'(\xi', \eta')$ is thus the spatial spectrum of S. Thus, we can treat the image formation $O \rightarrow O'$ as performed in two steps: the first consists in the formation of the spatial spectrum $S(f_x, f_y)$ by the first lens, the second consists in the formation of image $O'(\xi', \eta')$ from this spatial spectrum.

If we change the direction of the light from right to left, O' is the object and O the image, and S is the spatial spectrum of O'. Let us denote by $E'(\xi', \eta')$ a distribution of the complex amplitude of the image $O'(\xi', \eta')$, then the optical transformation performed by the second lens can be represented in terms of the *two-dimensional* FOURIER *integral*, or the inverse FOURIER transform of function $E(f_x, f_y)$:

$$E'(\xi', \eta') \sim \int_{-\infty}^{+\infty} \int_{-\infty}^{+\infty} E(f_x, f_y) \exp\left[2\pi i (f_x \xi + f_y \eta)\right] df_x df_y \quad . \qquad (3.9)$$

Here, as before, the requirement of a limitation of the integrand over the first focal plane of lens L_2 is assumed to be true.

Since a thin positive lens has the property of the FOURIER transform presented by the integrals (3.6) and (3.9), the formation of images of different kinds can be carried out by using two-dimensional FOURIER analysis. This method is based on a set of theorems and general requirements connected with the properties of functions which are subjected to FOURIER transform. Additionally, a set of helpful ideal mathematical objects of two-dimensional FOURIER analysis, for example the *two-dimensional* DIRAC δ-function, is used. The theorems mentioned above will be formulated later.

Now we discuss a method of image construction associated with the two-dimensional DIRAC δ-function as an important object in FOURIER optics. Let us introduce the symbol $F\{..\}$ specifying the integral of the FOURIER spectrum, and $F^{-1}\{..\}$ as the integral represented by the FOURIER integral. By using such short notations the spatial spectrum S of an object O and the image O' associated with S can be written as

$$S(f_x, f_y) = F\{O(\xi, \eta)\} \quad , \qquad O'(\xi', \eta') = F^{-1}\{S(f_x, f_y)\} \quad . \qquad (3.10)$$

For example, if the object function $O(\xi, \eta)$ has a quadratic shape with side length a, the spatial spectrum found in the previous chapter, see

(2.30), takes now the form

$$S(f_x, f_y) = F\{O(\xi, \eta)\} = E_0 \frac{a^2}{f^2} \frac{\sin(\pi f_x a)}{\pi f_x a} \frac{\sin(\pi f_y a)}{\pi f_y a} \quad .$$

Let us assume that the original image becomes progressively larger. Then the area of the spectrum between first zero values of the function $S(f_x, f_y)$ becomes smaller and smaller, whereas the peak amplitude increases. For $a \to \infty$, the set of functions $S(f_x, f_y)$ converge monotonically to the two-dimensional DIRAC δ-function as follows:

$$\delta(f_x, f_y) = \lim_{a \to \infty} S(f_x, f_y) \quad .$$

In the particular case under discussion, the function $\delta(f_x, f_y)$ has the following explicit form:

$$\delta(f_x, f_y) = E_0 \lim_{a \to \infty} \frac{a^2}{f^2} \frac{\sin(\pi f_x a)}{\pi f_x a} \frac{\sin(\pi f_y a)}{\pi f_y a} \quad . \tag{3.11}$$

We should pay attention to the fact that a similar form of δ-function can be achieved with other shapes of the original aperture, for example with a circular one. Thus, with an entirely open wave front, we can expect that the FOURIER spectrum of the incident monochromatic wave converges to a point. In conformation with the optical scheme shown in Fig.3.3 such a point source would produce a new entirely open wave front originating at the second lens. Although transformations of this kind are not possible under real experimental conditions due to the finite dimensions of every pupil function, the concept of the two-dimensional δ-function is very productive to analyze practical situations in FOURIER optics.

1.3   Theorems of two-dimensional Fourier analysis

Common requirements assumed to be valid with respect to any integrals, as well as to two-dimensional FOURIER integrals, are that the modulus of the integrand has to be restricted throughout the area of integration. Because the integrand functions have the form of the instantaneous intensity over the area of any object as well as over the spatial frequencies, we will consider this common requirement to be true.

We can also assume that there is no light energy loss while the light rays are propagating through the optical system, so the flux after the last optical element is the same as the incoming flux. The considerations above are formalized in from of the following theorems:

1 The PARSEVAL **theorem.** Let be $S = F\{O\}$, then the following is valid:

$$\int\limits_{-\infty}^{\infty} \int\limits_{-\infty}^{\infty} |O(\xi,\eta)|^2 d\xi d\eta = \int\limits_{-\infty}^{\infty} \int\limits_{-\infty}^{\infty} |S(f_x, f_y)|^2 df_x df_y \quad . \tag{3.12}$$

In fact this theorem represents the energy conservation law.

2 The **integral** FOURIER **theorem.** For every point of the function $O(\xi,\eta)$

$$F\{F^{-1}\{O(\xi,\eta)\}\} = FF^{-1}\{\{O(\xi,\eta)\}\} = O(\xi,\eta)$$

is valid.

3 The **linearity property.** Let $S(f_x, f_y) = F\{O(\xi,\eta)\}$, and $S^{/}(f_x, f_y) = F\{O^{/}(\xi^{/}, \eta^{/})\}$, then, with constants a and b, the following is valid:

$$F\{a \cdot O + b \cdot O^{/}\} = a \cdot S + b \cdot S^{/} \quad .$$

4 The **displacement property.** Let $S(f_x, f_y) = F\{O(\xi,\eta)\}$, then, with real magnitudes a, and b, for $O(\xi - a, \eta - b)$ the expression

$$F\{O(\xi - a, \eta - b)\} = \exp[2\pi i(f_x a + f_y b)]S(f_x, f_y) \tag{3.13}$$

is valid.

5 The **similarity property.** Let $S(f_x, f_y) = F\{O(\xi,\eta)\}$, then, with real magnitudes $a \neq 0$, and $b \neq 0$, for $O(\xi/a, \eta/b)$ the expression

$$F\{O(\xi/a, \eta/b)\} = |ab|S(f_x, f_y) \tag{3.14}$$

is valid.

6 The **convolution theorem.** Let $S = F\{O\}$ and $S^{/} = F\{O^{/}\}$, then the following integral relationship is valid:

$$F\left\{\int\limits_{-\infty}^{\infty} \int\limits_{-\infty}^{\infty} O(\xi^{/}, \eta^{/}) \cdot O^{/}(\xi^{/} - \xi, \eta^{/} - \eta) d\xi^{/} d\eta^{/}\right\} = S(f_x, f_y)S^{/}(f_x, f_y) \quad .$$

$$\tag{3.15}$$

The two dimensional integral on the left-hand side of (3.15) is called the two dimensional *convolution integral*. The convolution integral is usually denoted by the symbol "$\otimes$". Using this notation the spectrum of the convolution takes the form

$$F\{O \otimes O^/\} = S \cdot S^/ \tag{3.16}$$

where, by definition,

$$O \otimes O^/ = \int\limits_{-\infty}^{\infty} \int\limits_{-\infty}^{\infty} O(\xi^/, \eta^/) \cdot O^/(\xi^/ - \xi, \eta^/ - \eta) d\xi^/ d\eta^/ \quad . \tag{3.17}$$

Along with (3.15) a similar relationship between the product $OO^/$ and the convolution of the spectra exists in the form

$$F\{S \otimes S^/\} = O \cdot O^/ \quad . \tag{3.18}$$

7 The **autocorrelation theorem.** By definition, for a given object function O, the autocorrelation function is given by the convolution integral in the form

$$O \otimes O^* = \int\limits_{-\infty}^{\infty} \int\limits_{-\infty}^{\infty} O(\xi^/, \eta^/) \cdot O^*(\xi^/ - \xi, \eta^/ - \eta) d\xi^/ d\eta^/ \quad . \tag{3.19}$$

The spectrum of the autocorrelation function is given by

$$F\{O \otimes O^*\} = |S(f_x, f_y)|^2 \quad , \tag{3.20}$$

where, as before, $S(f_x, f_y) = F\{O\}$.

In conformity with the experimentally realized arrangement of a nearly ideal optical system, shown in Fig.3.3, one can treat the integral theorem as connected with the situation where a plane monochromatic wave passes through both lenses with entirely open wave fronts. The linearity property states that two parts of one original image give a spatial spectrum as the sum of two particular spectra, provided that these parts of the original image do not overlap. The matter of the displacement property is that, with any displacement of the original image over the first focal plane by magnitudes a and b, the spatial spectrum remains at the center of the optical axis, but it has to have a phase factor $\exp(2\pi i(f_x a + f_y b))$. According to the similarity property, with any change of the scale of the original image by means of the two factors $1/a$ and $1/b$, the scale of the spatial spectrum will be changed as $|ab|$; that means, any decrease of the image leads to an increase of the spectrum, and vice versa. The convolution theorem establishes a relationship between the product of two spectra and two images associated with these spectra. The autocorrelation theorem is an explicit form of the important case of the previous theorem where $O^{/} = O^*$.

The theorems and properties considered above allow the derivation of new properties of spatial spectra and relationships between objects and their spectra. Let us consider that a substantial part O of an original image is limited by a pupil of quadratic shape as function P. Then, one can represent the action of the pupil aperture by the product PO. In accordance with the second part of the convolution theorem the spatial spectrum of the effective image should be calculated to be

$$F\{PO\} = F\{P\} \otimes F\{O\} \quad . \tag{3.21}$$

Let us now regard the pupil as increasing without limit, up to infinity, which gives $F\{P\}$ as tending to the two dimensional δ function. In turn, in accordance with the operating principle of the optical system shown in Fig.3.3, the image constructed by the second lens should become a copy of O, and we get the spectrum $F\{P\} \otimes F\{O\}$ as a copy of $F\{O\}$. Hence, we state another definition of the δ function in the form of a convolution integral:

$$F\{O\} \otimes \delta(f_x, f_y) = F\{O\} \quad . \tag{3.22}$$

The same treatment may be applied to the problem of the action of a pupil function on the spatial spectrum within the second focal plane of the first lens. It is clear that an analogous relationship should exist between the two dimensional $\delta(\xi, \eta)$ function and the object function in the form

$$O(\xi, \eta) \otimes \delta(\xi, \eta) = O(\xi, \eta) \quad . \tag{3.23}$$

The image shown in Fig.3.4,a illustrates another form of the convolution operation, where a small rectangle is repeated at the nodes of a quadratic spatial grid. Let d be the side of this grid, and $R(\xi, \eta)$ be the transparency function, describing one elementary rectangle, then the object function which would represent such an image over the whole plane should take the form

$$O(\xi, \eta) = \sum_{m,n} R(\xi, \eta)\delta(\xi - md, \eta - nd) \quad , \tag{3.24}$$

where m and n are integer numbers ranging from $-\infty$ to $+\infty$. Every term of the sum represents one particular element of the grid; and the coordinates (ξ, η) of the center of this element satisfy $\xi = md$, and $\eta = nd$. For the same reason as with the convolution integral, the spatial

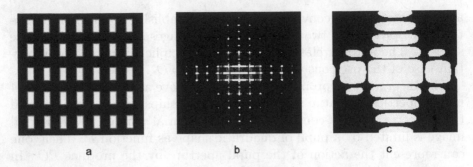

$$a \qquad\qquad\qquad b \qquad\qquad\qquad c$$

Figure 3.4. The image of a grid of rectangles (a); its spatial spectrum (b), and the spatial spectrum of one element of the grid (c). The size of the slide is 15×15 mm, the size of the spectrum 6×6 mm. $\lambda = 632.8$ nm; the focal length of the objective is $f = 180$ cm.

spectrum of the function $O(\xi, \eta)$ has to be represented in terms of the product of two spatial spectra. One is the spatial spectrum of the spatial grid of period d, which is denoted as $G(f_x, f_y)$, and the other is the spatial spectrum of the rectangle, $F\{R(\xi, \eta)\}$. The quadratic grid can be considered to be composed from two crossed linear diffraction gratings of the same period d. Since these diffraction gratings should simulate a two dimensional δ function, the width of each elementary stripe of these gratings is assumed to be nearly zero. Hence, the spatial spectrum, which would arise from one grating, for example in x−direction, should have the form

$$\sum_m \delta(f_x - mf_{x0}) \quad ,$$

where $f_{x0} = \lambda f/d$, and f is the focal length of the lens, which follows from the expression found in the case of the amplitude diffraction grating with $a \to 0$ in formula (2.22). In the same way, the spatial spectrum in the y−direction has the similar form

$$\sum_n \delta(f_y - nf_{y0}) \quad ,$$

where $f_{y0} = \lambda f/d$. The total spatial spectrum of the grid is therefore given by the expression

$$G(f_x, f_y) = \sum_{m,n} \delta(f_x - mf_{x0})\delta(f_y - mf_{y0}) \quad . \tag{3.25}$$

Thus, in the ideal case of an unlimited pupil, the spatial spectrum from the object should take the form of the product

$$F\{O\} = F\{R\} \cdot G = F\{R\} \cdot \sum_{m,n} \delta(f_x - mf_{x0})\delta(f_y - mf_{y0}) \quad . \quad (3.26)$$

It follows from (3.26) that the function $F\{O\}$ will have a non-zero value at every point f_x, f_y, where a peak of G coincides with a non-zero value of function $F\{R\}$. Because the original grid of rectangles is framed by a quadratic pupil, the total transparency function has to take the form $P(\xi, \eta) \cdot O(\xi, \eta)$, where $P(\xi, \eta)$ specifies this pupil. According to the convolution theorem, the spectrum of this product gives rise to the following spatial spectrum:

$$F\{P\} \oplus F\{O\} = F\{P\} \oplus (F\{R\} \cdot G) \quad . \quad (3.27)$$

Due to the finite dimensions of the pupil, every elementary bright area of the intensity distribution of this spatial spectrum should take the form of an intensity distribution around the central peak associated with the intensity distribution of function $F\{P\}$. The effect of a final dimension of the pupil is demonstrated in Fig.3.4,b, where each bright area consists of a few first maxima of the intensity distribution typical for FRAUNHOFER diffraction on a quadratic aperture. It is also seen that the locations of all elementary bright areas follow the intensity distribution in the spatial spectrum of one element of the original image, which is shown in Fig.3.4,c. We stress that the spatial spectra under discussion were found using the experimental setup shown in Fig.3.1.

2. Isoplanatic linear systems

The operating principle of optical systems can be seen as modifications of the solid angle of the propagation of the incoming light beams. Mathematically, such modifications can be treated in terms of linear operators, with the exception of violations caused by aberrations of optical systems. The mathematical tools connected with linear operators in FOURIER optics allow simple solutions for such tasks through a set of general principles established from properties of optical elements. Such an approach based on linear operators applied to FOURIER optics is closely related to the same task in radio-techniques, where many methods of linear filtration were developed in order to treat the action of a linear radio circuit on its input signal. FOURIER analysis, using linear operators, has also a significant place in the description of linear electric circuits. Because of this fact, a set of important principles of FOURIER optics were adopted into the theory of linear radio circuits.

In optics, the basic ideas related to linear operators may be found from the operating principles of various optical devices. However, the mathematical treatment of the HUYGENS–FRESNEL principle already gives an example of such a linear operator in the form of the FRESNEL diffraction integral. We now treat the FRESNEL diffraction in terms of linear operators in order to establish a set of basic properties, valid for light propagation in general.

Let us discuss the propagation of monochromatic light waves between two planes: the input plane (ξ, η) and the output plane (x, y). We assume that a distribution of the complex amplitude $E(\xi, \eta)$ exists over the aperture of the input plane, in contrast to the case considered in the previous section, where the incident wave was a plane monochromatic wave. Nevertheless, the complex amplitude $E(x, y)$ at point (x, y) is represented, as before, by the superposition of the complex amplitudes of all spherical waves emitted within the aperture in an integral form:

$$E(x, y) = \frac{1}{i\lambda} \int_{\xi, \eta} E(\xi, \eta) \frac{1}{r} K(\chi) \exp(ikr) d\xi d\eta \ , \qquad (3.28)$$

where $K(\chi)$ is the inclination factor, r is the distance between points (ξ, η) and (x, y). It is clear that the integral in (3.28) has the form of a linear operator, which allows the calculation of the complex amplitude over the output plane via the amplitude distributed over the input plane. Under the approximation of FRESNEL diffraction, where

$$K(\chi) \approx 1, \text{ and } r \approx L + (\xi - x)^2/(2L) + (\eta - y)^2/(2L) \ ,$$

this integral takes the form of a convolution integral,

$$E(x, y) = \frac{\exp(ikL)}{i\lambda L} \cdot$$

$$\cdot \int_{\xi, \eta} E(\xi, \eta) \exp\left\{ ik\left[(\xi - x)^2/(2L) + (\eta - y)^2/(2L) \right] \right\} d\xi d\eta \ , \qquad (3.29)$$

because the phase term of the integrand includes the coordinate differences. By definition, this linear operator (which is dependent on the coordinate differences between two points, one belonging to the input plane and the other to the output plane) describes the light propagation in a so-called *isoplanatic*, or *space invariant linear system*. A convolution integral of the linear operator exists, which allows the deduction of the properties of the isoplanatic linear system by means of two-dimensional FOURIER analysis. In terms of an isoplanatic linear systems every "point signal" from the input plane should be transferred into a signal on the output plane by a function, called the *impulse response*, in analogy to radio circuits. In the particular case considered above, the impulse function is given by

$$h(\xi, \eta) = \frac{\exp(ikL)}{i\lambda L} \exp\left\{ik\left[(\xi)^2/(2L) + (\eta)^2/(2L)\right]\right\} \quad . \tag{3.30}$$

We say "point signal" assuming that each point source on the input plane is represented by the two-dimensional $\delta(\xi, \eta)$ function; thus a "point signal" of unity amplitude causes a response on the output plane in the form (3.30). For a given "input signal" in the form $E(\xi, \eta)$, the isoplanar property establishes a rule for calculating the output signal $E(x, y)$ via the convolution in operator form:

$$E(x, y) = E(\xi, \eta) \otimes h(\xi, \eta) \quad . \tag{3.31}$$

We direct our attention to the fact that light propagation between two planes under conditions of FRAUNHOFER diffraction should also satisfy the isoplanar linear requirement, since FRAUNHOFER diffraction is considered as being a limit case of FRESNEL diffraction. Assuming that a thin positive lens provides FOURIER transforms to images on its focal plane, we will now provide the isoplanar properties of thin positive lenses as well. The important property of the δ function (3.23) deduced by empirical conclusions from the properties of FOURIER spectra now takes an interpretation in terms of the isoplanar property.

In general, let $\mathcal{P}$ be a linear operator implementing the image transformation of an input object $O(\xi, \eta)$ into an output image in an isoplanar optical system. Then the output image is represented by

$$O_{out} = \mathcal{P}\{O(\xi, \eta)\} \quad .$$

We apply the operator $\mathcal{P}$ to both sides of the equality (3.23):

$$\mathcal{P}\{O(\xi, \eta) \otimes \delta(\xi, \eta)\} = \mathcal{P}\{O(\xi, \eta)\} \quad . \tag{3.32}$$

Because the convolution integral is a linear operator, one can change the order of the two operators on the left side:

$$\mathcal{P}\{O(\xi, \eta) \otimes \delta(\xi, \eta)\} = O(\xi, \eta) \otimes \mathcal{P}\{\delta(\xi, \eta)\} \quad .$$

Since the system is isoplanar, the operation $\mathcal{P}\{\delta(\xi, \eta)\}$ results in the impulse response of the system: $\mathcal{P}\{\delta(\xi, \eta)\} = h(\xi, \eta)$. Substitutions of $\mathcal{P}\{\delta(\xi, \eta)\}$ by $h(\xi, \eta)$ on the left-hand side and of $\mathcal{P}\{O(\xi, \eta)\}$ by O_{out} gives the required relationship between the input and output images in the general form:

$$O_{out} = O(\xi, \eta) \otimes h(\xi, \eta) \quad . \tag{3.33}$$

Thus, the output image of the isoplanar linear system is found to be a convolution integral of two integrands: one of the input image and the other of the impulse response of the system. This important relationship can be represented in terms of spatial spectra by applying the linear operator $F\{\}$ to both sides of (3.22). According to the second part of the convolution theorem, the function $F\{O(\xi, \eta) \otimes h(\xi, \eta)\}$ takes the form

$$F\{O(\xi, \eta)\} \cdot F\{h(\xi, \eta)\} = F\{O(\xi, \eta)\} \cdot H(f_x, f_y)$$

where

$$H(f_x, f_y) = F\{h(\xi, \eta)\} \tag{3.34}$$

is the spatial spectrum of the impulse response; the function $H(f_x, f_y)$ is called the *transfer function* of the optical system. The function $F\{O_{out}\}$, obtained after applying the operator $F\{\}$ to the left-hand side of (3.33), is the spatial spectrum of the output image, which now has the form

$$F\{O_{out}\} = F\{O(\xi, \eta)\} \cdot H(f_x, f_y) \quad . \tag{3.35}$$

Hence the spatial spectrum of the output image in the isoplanatic linear optical system can be calculated as the product of the spatial spectrum of the input image with the transfer function of the optical system. In this way, image construction by an isoplanatic linear optical system becomes similar to the propagation of a signal through a linear radio system, or through a linear filter. In other words, the operating principal of an optical system having the isoplanar linear property is similar to linear filtration by a passive linear radio filter with invariable parameters.

2.1   Transfer function for Fresnel diffraction

Let us calculate the transfer function $H(f_x, f_y)$, associated with the impulse response in the form (3.30), for FRESNEL diffraction. According to the definition (3.34), the transfer function calculated by means of the two-dimensional FOURIER spectrum of the function $h(\xi, \eta)$ must take the form

$$H(f_x, f_y) = \frac{\exp(ikL)}{i\lambda L} \int\limits_{-\infty}^{\infty} \int\limits_{-\infty}^{\infty} d\xi \, d\eta \cdot \tag{3.36}$$

$$\cdot \exp\left[ik\left(\frac{\xi^2}{2L} + \frac{\eta^2}{2L}\right)\right] \exp\left[-ik\left(\frac{\xi x}{L} + \frac{\eta y}{L}\right)\right] \quad .$$

Using the new variables

$$\tau_1 = (\xi - x)\sqrt{\frac{2}{\lambda L}} \quad , \quad \text{and} \quad \tau_2 = (\eta - y)\sqrt{\frac{2}{\lambda L}}$$

we re-write the function $H(f_x, f_y)$ as follows:

$$H(f_x, f_y) = \frac{\exp(ikL)}{2i} \exp\left(-i\pi\lambda L(f_x^2 + f_y^2)\right) \cdot$$

$$\cdot \int\limits_{-\infty}^{\infty}\int\limits_{-\infty}^{\infty} \exp\left[i\frac{\pi}{2}(\tau_1^2 + \tau_2^2)\right] d\tau_1 d\tau_2 \quad . \tag{3.37}$$

Since

$$\exp\left[i\frac{\pi}{2}(\tau_1^2 + \tau_2^2)\right] = \cos\left[\frac{\pi}{2}(\tau_1^2 + \tau_2^2)\right] + i\sin\left[\frac{\pi}{2}(\tau_1^2 + \tau_2^2)\right] \quad ,$$

we use the FRESNEL integrals in order to compute (3.37):

$$\int\limits_{0}^{\infty} \cos\left(\frac{\pi}{2}\tau^2\right) d\tau = 1 \quad , \quad \text{and} \quad \int\limits_{0}^{\infty} \sin\left(\frac{\pi}{2}\tau^2\right) d\tau = 1 \quad . \tag{3.38}$$

The integration of the real part of the integrand in (3.37), which is represented by

$$\cos\left[\frac{\pi}{2}(\tau_1^2 + \tau_2^2)\right] = \cos(\frac{\pi}{2}\tau_1^2)\cos(\frac{\pi}{2}\tau_2^2) - \sin(\frac{\pi}{2}\tau_1^2)\sin(\frac{\pi}{2}\tau_2^2)$$

results in a value of zero. In turn, the contribution from the integrals of the imaginary part can be represented by

$$\sin\left[\frac{\pi}{2}(\tau_1^2 + \tau_2^2)\right] = \sin(\frac{\pi}{2}\tau_1^2)\cos(\frac{\pi}{2}\tau_2^2) - \sin(\frac{\pi}{2}\tau_2^2)\cos(\frac{\pi}{2}\tau_1^2) \quad .$$

The value of this expression is found to be equal to $2i$. Thus the calculation reveales that the integral (3.37) has the value

$$\int\limits_{-\infty}^{\infty}\int\limits_{-\infty}^{\infty} \exp\left[i\frac{\pi}{2}(\tau_1^2 + \tau_2^2)\right] d\tau_1 d\tau_2 = 2i \quad .$$

Substitution of the integral by $2i$ in (3.37) gives $H(f_x, f_y)$ as

$$H(f_x, f_y) = \exp(ikL)\exp\left[-i\pi\lambda L\left(f_x^2 + f_y^2\right)\right] \quad . \tag{3.39}$$

Therefore, light propagation within the limits of FRESNEL diffraction can be treated as spatial filtration, in which a particular spatial frequency is subjected only to a phase transformation.

2.2   Transparency function of a thin positive lens

Under the paraxial approximation a thin positive lens performs a transformation of the phase of an incident monochromatic wave. This follows from the fact that a plane monochromatic wave incident on the lens becomes a convergent spherical wave after refraction by the spherical surfaces of the lens. On the other hand, a spherical wave incident from a point on the first focus plane of the lens will be transformed into a plane wave after refraction.

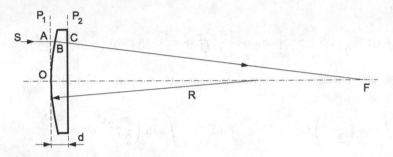

Figure 3.5.   To the phase change by a spherical thin positive lens.

Let us derive the explicit form of the transparency function of the lens, assuming a plane monochromatic wave to be normally incident on the lens (Fig.3.5). The field distribution on the front plane P_1 has the same phase, whereas light rays after passing the lens body and reaching plane P_2 are retarded. A decrease $\Delta\varphi$ of the phase appears according to the optical path length between P_1 and P_2. We denote by d the thickness of the lens at its apex, and the change in the thickness by Δd, which is dependent on the coordinates (x, y) on plane P_2 and on the focal length of the lens. n is the refractive index of the lens material. Δd close to point (x, y) is given by the relationship $(x^2 + y^2) + (R - \Delta d)^2 = R^2$, following from the spherical surface of the lens. Using the paraxial approximation we find $\Delta d \approx (x^2 + y^2)/(2R)$. Thus, the optical path length between planes P_1 and P_2 close to point (x, y) is given by

$$\Delta d + (d - \Delta d)n = dn - \Delta d(n-1) = dn - \frac{x^2 + y^2}{2R}(n-1) \quad .$$

Since the focal length is $f = R/(n-1)$, the required transparency function $t(x, y)$ takes the form

$$t(x, y) = \exp(ikdn)\exp\left(-i\pi(x^2 + y^2)/(\lambda f)\right) \quad . \tag{3.40}$$

For a given distribution $E_1(x, y)$ of the complex amplitude over the plane P_1, the function under discussion allows the calculation of the complex amplitude $E_2(x, y)$ over plane P_2 by means of a simple formula:

$$E_{P_2}(x, y) = E_{P_1}(x, y) \exp(ikdn) \exp\left(-i\pi(x^2 + y^2)/(\lambda f)\right) \quad . \quad (3.41)$$

Now we derive a relationship between $E_{P_1}(x, y)$ and the complex amplitude over the second focal plane of the lens. This kind of distribution, let it be the function $E_{F_2}(x_1, y_1)$, is calculated as the convolution of the function $E_{P_2}(x, y)$ and the response function in the form (3.30) for $L = f$ as follows:

$$E_{F_2}(x_1, y_1) = E_{P_2}(x, y) \otimes h(x, y) = \frac{\exp(ikdn)\exp(ikf)}{i\lambda f} \int\int dx\, dy\, E_{P_1}(x, y) \cdot$$

$$\cdot \exp\left(-i\pi(x^2 + y^2)/(\lambda f)\right) \exp\left(i\pi(x - x_1)^2 + i\pi(y - y_1)^2)/(\lambda f)\right) \quad .$$

A simple conversion of the phase term of the integrand allows the integral describing the FOURIER spectrum of function $E_{P_1}(x, y)$ to be re-written as

$$E_{F_2}(x_1, y_1) = \frac{\exp(ikdn)\exp(ikf)}{i\lambda f} \exp(i\pi(x_1{}^2 + y_1{}^2)/(\lambda f)) \cdot$$

$$\cdot \int\int E_{P_1}(x, y) \exp\left(-2\pi i(xx_1 + yy_1)/(\lambda f)\right) dx\, dy \quad . \quad (3.42)$$

Here the phase term $exp(i\pi(x_1{}^2 + y_1{}^2)/(\lambda f))$ contains the squared coordinates of the second focal plane.

2.3 Fourier spectrum implemented by a thin positive lens and by a double slit

Let us now deduce a formula for calculating the spatial distribution $E_{F_2}(\xi, \eta)$ on the second focal plane of a thin lens from the distribution measured over first focal plane, $E_{F_1}(x_1, y_1)$. We should keep in mind

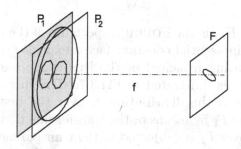

Figure 3.6. Geometrical considerations concerning the amplitude transparency function of a thin lens.

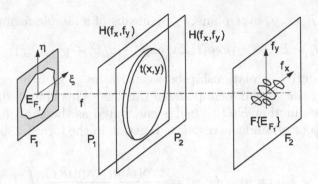

Figure 3.7. The operating principle of a thin positive spherical lens treated with FOURIER optics.

that $E_{P_1}(x, y)$ is the distribution over the plane located in front of the lens (plane P_1 in Fig.3.6). It follows from (3.39) that the operational form of (3.42) can be written as

$$E_{F_2} = \frac{\exp(ikdn)\exp(ikf)}{i\lambda f} F\{E_{P_1}\} \exp(i\pi\lambda f(f_x{}^2 + f_y{}^2)) \quad , \quad (3.43)$$

where, as before, f_x and f_y are spatial frequencies. In turn, the spatial spectrum over plane P_1, the function $F\{E_{P_1}\}$, can be found as the product

$$F\{E_{P_1}\} = F\{E_{F_1}\}H(f_x, f_y) =$$
$$= F\{E_{F_1}\} \exp(ikL) \exp\left[-i\pi\lambda L \left(f_x{}^2 + f_y{}^2\right)\right] \quad (3.44)$$

as follows from (3.35). Substitution of $F\{E_{P_1}\}$ in (3.43) by the right-hand side of (3.43) at $L = f$ gives the required formula:

$$E_{F_2} = \frac{\exp(i2kf)\exp(ikdn)}{i\lambda f} F\{E_{F_1}\} \quad , \quad (3.45)$$

which shows that E_{F_2} is the FOURIER spectrum of the function $F\{E_{F_1}\}$, multiplied by an inessential constant factor $\exp(i2kf) \exp(ikdn)/i\lambda f$.

Thus the operating principal of the lens is represented in terms of a linear filtration, as illustrated in Fig.3.7. The spectrum $F\{E\}$ of an input signal, that is the distribution E over the first focal plane F_1, passes to the plane P_1 by means of the transfer function $H(f_x, f_y)$. Then the distribution over P_1 is subjected to the transparency function of the lens, $t(x, y)$, and finally the spectrum over plane P_2 passes to plane F_2 by means of the same transfer function $H(f_x, f_y)$, which has to be calculated at $L = f$.

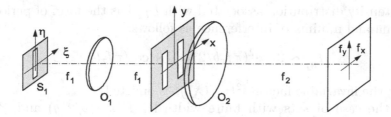

Figure 3.8. Operating principle of YOUNG's double slit experiment in terms of spatial spectra.

As an other example, let us analyze interference observed in YOUNG's double slit experiment (see Chapter 1) in terms of the operators considered above. We should remember that in this interference scheme the first vertical slit is positioned at the first focal plane of objective O_1 and the double slit at its second focal plane (Fig.3.8). The second objective O_2 is mounted close to the double slit and the interference fringes are observed on the second focal plane of this objective. A spatial spectrum of the first slit is formed on the second focal plane of O_1. At first we consider both slits, oriented along vertical direction and composing the double slit, as being infinitesimal narrow, so that one is specified by the function $\delta(x + b/2)$ and the other by $\delta(x - b/2)$. Thus, the slits are located on the horizontal axis x at points $\pm b/2$, where b is the distance between the slits. The objective O_2 generates a spatial spectrum located at the second focal plane of O_2.

According to (3.44), the spatial spectrum as a function of the frequencies f_x and f_y takes the form

$$E_{F_2} = \frac{\exp(ikdn)\exp(ikf)}{i\lambda f}\exp(i\pi\lambda f(f_x{}^2 + f_y{}^2))\cdot$$
$$\cdot (F\{\delta(x - b/2)\} + F\{\delta(x + b/2)\}) \quad . \tag{3.46}$$

The term $F\{\delta(x - b/2)\}$ represents the complex amplitude of a plane monochromatic wave in the form $\exp(-i\pi f_x b/2)\,\delta(y)$, where $\delta(y)$ results from the calculation of the spatial spectrum along the vertical direction, and the phase factor $f_x b/2$ follows from the displacement property. In a similar way, we find $\exp(i\pi f_x b/2)\,\delta(y)$ for term $F\{\delta(x - b/2)\}$. Then the complex amplitude over the second focal plane of O_2 obtained from (3.46) has to be

$$E_{F_2} = \delta(y)\frac{\exp(ikdn)\exp(ikf)}{i\lambda f}\exp(i\pi\lambda f(f_x{}^2 + f_y{}^2))\cdot$$
$$\cdot (\exp(-i\pi f_x b/2) + \exp(i\pi f_x b/2)) \quad . \tag{3.47}$$

An intensity distribution associated with E_{F_2} has the form of periodical maxima and minima of interference as follows:

$$I_{F_2} \sim 2\cos^2(\pi f_x b/2) = 1 + \cos(\pi f_x b) \quad ,$$

where the invariable factor $\delta^2(y)/(\lambda f)^2$ is omitted.

In the case of slits with finite width let $P(x - b/2, y)$ and $P(x + b/2, y)$ be the amplitude transparency functions assigned to the slits, respectively. Then the distribution E_{F_2} becomes

$$E_{F_2} = \frac{\exp(ikdn)\exp(ikf)}{i\lambda f}\exp(i\pi\lambda f(f_x^2 + f_y^2))F\{P(x,y)\}\cdot$$
$$\cdot \left(\exp\left(-i\pi f_x b/2\right) + \exp\left(i\pi f_x b/2\right)\right) \quad , \tag{3.48}$$

where $F\{P(x,y)\}$ is the spatial spectrum from a narrow slit; such a spectrum has to be positioned at the center of the second focal plane of O_2. It follows from (3.48) that in this case the intensity distribution also has an interference form: $I_{F_2} \sim |F\{P(x,y)\}|^2 [1 + \cos(\pi f_x b)]$. Here the factor $|F\{P(x,y)\}|^2$ has to have the form of a product (see (2.30)):

$$\frac{\sin^2(u)\sin^2(v)}{(uv)^2} \quad ,$$

where $u = kax/(2f), v = khx/(2f)$, a is the width of the slits and h is their height.

2.4   Image construction in general

Along with the transfer function $H(f_x, f_y)$ related to FRESNEL diffraction, a similar function $H_{lens}(f_x, f_y)$ can be assigned to a thin positive spherical lens. Such a transfer function is calculated to be the FOURIER spectrum of the amplitude transparency function in the form of (3.40) as follows:

$$H_{lens}(f_x, f_y) = F\{t(x,y)\} = \exp(ikdn)F\{\exp\left(-i\pi(x^2 + y^2)/(\lambda f)\right)\} \quad . \tag{3.49}$$

The integral on the right-hand side of (3.49) is quite similar to that in (3.36), except for the sign of the phase. Thus for function $H_{lens}(f_x, f_y)$, we hold the expression

$$H_{lens}(f_x, f_y) = \exp(ikdn)\exp\left(i\pi\lambda f(f_x^2 + f_y^2)\right) \tag{3.50}$$

to be true.

Thus using the transfer functions for the description of light propagation through the lens, as well as between input and output planes, allows the representation of all transformations of an input image in terms of modifications of its spatial spectrum. Sometimes it is more suitable to form products of the spectrum into appropriate transfer functions, instead of calculating a convolution integral.

Let us consider an input image located at plane P_0 at a distance a from the apex of a thin positive spherical lens. For a given complex amplitude E_{p_1} distributed over the input image, the spatial spectrum over plane P_1 obtained by means of (3.35) is then given as

$$F\{E_{p_1}\} = F\{E_{P_0}\}H_a(f_x, f_y) \quad , \tag{3.51}$$

where the subscript a indicates that $H(f_x, f_y)$ is calculated for $L = a$. In a similar way the spatial spectrum of an input image located at a distance b after the lens over the output plane P_3 can be written as

$$F\{E_{p_3}\} = F\{E_{P_2}\}H_b(f_x, f_y) \quad , \tag{3.52}$$

where the transfer function has to be calculated at $L = b$. In turn, the spatial spectra on planes P_1 and P_2 are related via

$$F\{E_{P_2}\} = F\{E_{p_1}\}H_{lens}(f_x, f_y) \quad . \tag{3.53}$$

On combining the terms of (3.51) – (3.53) we find between $F\{E_{P_0}\}$ and $F\{E_{p_3}\}$ the relationship

$$F\{E_{p_3}\} = F\{E_{P_0}\}H_a(f_x, f_y)H_L(f_x, f_y)H_b(f_x, f_y) \quad . \tag{3.54}$$

This equation shows that the spectrum on the output plane is formed from the spectrum on the input plane by a multiplication of $F\{E_{P_0}\}$ with the appropriate transfer functions which describe light propagation from the input to the output planes. At first we represent the spatial frequencies in the transfer function $H_a(f_x, f_y)$ in its coordinate forms $f_x = x/(\lambda a)$, $f_y = y/(\lambda a)$, where (x, y) are coordinates over plane P_1. Thus the function $H_a(f_x, f_y)$ becomes

$$H_a(x, y) = \exp(ika)\exp\left[-i\pi\left(x^2 + y^2\right)/a\right] \quad . \tag{3.55}$$

Since any point of plane P_1 has the same coordinates on P_2, the function $H_b(x, y)$ can be written as

$$H_b(x, y) = \exp(ikb)\exp\left[-i\pi\left(x^2 + y^2\right)/b\right] \quad . \tag{3.56}$$

Finally, for $H_{lens}(f_x, f_y)$, we obtain

$$H_{lens}(f_x, f_y) = \exp(ikdn)\exp\left(i\pi(x^2 + y^2)/f\right) \quad . \tag{3.57}$$

Hence, the product of the transfer functions has the form

$$H_a(x,y)H_a(x,y)H_{lens}(f_x, f_y) = \exp(ik(a+b+dn))\cdot$$

$$\cdot \exp\left[-i\pi\left(x^2+y^2\right)(1/a+1/b-1/f)\right] \quad .$$

When plane P_0 is imaged on plane P_3, these planes are conjugated and the factor $1/a + 1/b - 1/f = 0$, hence, the product under consideration takes an invariable form

$$H_a(x,y)H_a(x,y)H_{lens}(f_x, f_y) = \exp(ik(a+b+dn)) \quad .$$

Thus, it follows from (3.54) that the spatial spectra of the output images differs from the input one by this invariable phase factor. Such an ideal optical system, which inserts no limitations into spatial spectra, should thus generate an ideal output image which is a copy of the input image.

3.   Spatial filtration

Historically, first investigations related to *spatial filtration* were performed by E.ABBE (1840-1905) in 1873, when the concept of spatial spectra was put forward in order to illustrate the effect of diffraction phenomena on the resolving power of a microscope. A set of spatial filtration experiments were carried out by A.B.PORTER (1864-1909) in 1906, devoted to confirm ABBE's approach to optical image construction.

Let us discuss a simple experiment of spatial filtration. In the demonstrational experiment shown in Fig.3.9 a quadratic line grating (stripes oriented along the horizontal and vertical directions) is placed in the first focal plane of the objective O_1, and the spatial spectrum of the grid is formed on the second focal plane of the objective. The second objective

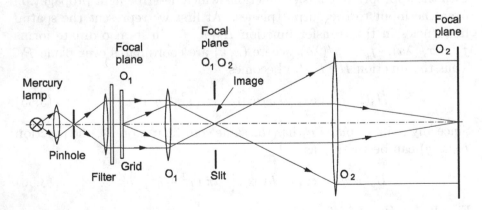

Figure 3.9.   Setup for demonstration of spatial filtration.

O_2, $f_2 = 130$ cm, is placed in such a way that its first focal plane is superimposed to the second focal plane of O_1. O_2 forms an output image on its second focal plane. The image of the grating and the spectrum associated with it are shown in Figs.3.10,a,b.

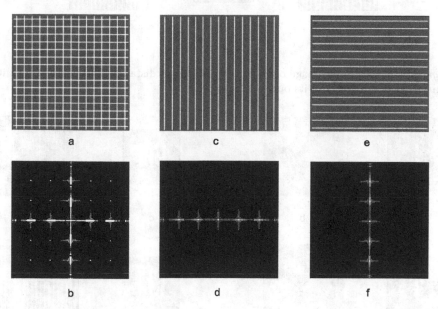

Figure 3.10. Images of line grids and its spectra. The original image (a) and spectrum (b); the image found after filtration by a slit positioned horizontally (c) and vertically (e) and spatial spectra associated with these images (d) and (f), respectively.

Now we can perform the spatial filtration of the spectrum, thus modifying the final image. For example, let the spatial spectrum be modified by a narrow vertical slit in such a way that only the central diffraction maxima along the horizontal direction are open. Then the output image will take the form of bright equidistant lines along the vertical direction. This image and the spectrum associated with it are shown in Fig.3.10,c,d. In turn, with the slit oriented vertically, the output image is a set of horizontal lines. The output image and the spectrum are shown in Fig.3.10,e,f.

As another example, let us use a grid composed of two parts with different distances between the grid lines as an object, as shown in Fig. 3.11,a. Fig. 3.11,b shows the spatial spectrum and Fig.3.11,c the output image. Let us now perform spatial filtering using the masks shown in Fig.3.12,a,c,e,g. The corresponding output images are shown in Fig.3.12,b,d,f,h. Such spatial filtering techniques can be used in opti-

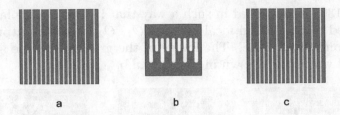

Figure 3.11.   An input image of two grids of different distances between lines (a); its spatial spectrum (b); and its output image (c).

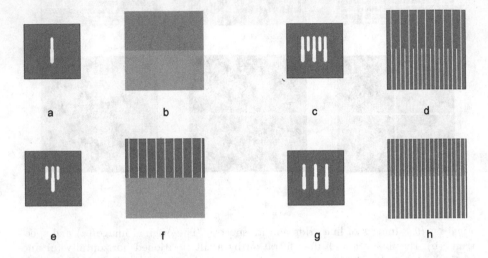

Figure 3.12.   Filter masks applied to the spectrum of Fig. 3.10A (a,c,e,g) and observed images (b,d,f,h)

cal image processing to remove undesired periodic structures from noisy images.

3.1   The resolving power of a microscope

The fact that diffraction phenomena affect image formation in microscopes was put forward in the ABBE *theory*, which treated the resolving power of a microscope in terms of diffraction. Since microscopes form a large scale image of tiny objects, high quality is needed in the output image in order to show fine details of an original object. However, the finer the detail of the image, the greater the diffraction angle of rays diffracted on the object. It is clear that rays, diffracted so much that they cannot enter the entrance pupil of the microscope objective, will not take part in the output image formation. This fact leads to a loss

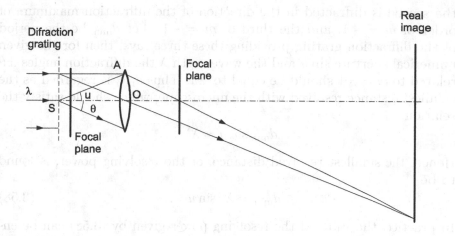

Figure 3.13. Illustration of spatial filtration by a microscope objective with its numerical aperture u. The angle θ is the diffraction angle for the first diffraction maximum of the smallest detail of the object image.

of fine details of the output image and to a limitation of the resolving power of the microscope.

Let an amplitude diffraction grating of period d be the object and let a plane monochromatic wave with wavelength λ incident on the grating. The object is placed close to the first focal plane of the objective. The microscope objective constructs a real enlarged image of the grating (Fig.3.13). The spatial spectrum S formed on the second focal plane takes the from

$$S = P \cdot F\{O\} \quad ,$$

where $F\{O\}$ is the FOURIER spectrum of the object, and P is the entrance pupil function of the objective. The real image $O^{/}$ is given by the convolution

$$O^{/} = F\{P\} \otimes O \quad .$$

Due to the finite dimension of the entrance pupil, the function $F\{P\}$ should differ from a two-dimensional δ function, and for this reason $O^{/}$ must loose fine details present in the object O. In the simple case schematically shown in Fig.3.13, the entrance pupil is assumed to be the mounting of the objective. We denote the entrance aperture of the objective, the angle OSA, by u. The quantity $\sin u$, the so-called *numerical aperture* of the microscope objective, provides the resolving power of the microscope. In order to make an estimation of the resolving power we assume that the simplest periodical structure in the output image will appear under propagation of only three rays: one is non-diffracted, the second is diffracted in the direction of the diffraction maximum of order of $m = +1$, and the third of $m = -1$. Let $d_{\min}$ be the period of the diffraction grating providing these three rays, then for the given numerical aperture $\sin u$ and the wavelength λ the diffraction angles $\pm\theta$ related to $m = \pm 1$ should be equal to $\pm u$. Thus $d_{\min}$ is regarded as the smallest distance resolved with the microscope, which should satisfy the relation

$$d_{\min} \sin u = \lambda \quad .$$

Hence, the smallest resolved distance, or the resolving power, is found to be

$$d_{\min} = \lambda / \sin u \quad . \tag{3.58}$$

In practice, the value of the resolving power given by (3.58) can be enhanced by using an immersion liquid of the refractive index n. The space between the object and the objective is filled with the immersion liquid that provides an increase of the numerical aperture up to the magnitude $n \sin u$, which provides the resolving power of $d_{\min} = \lambda / (n \sin u)$.

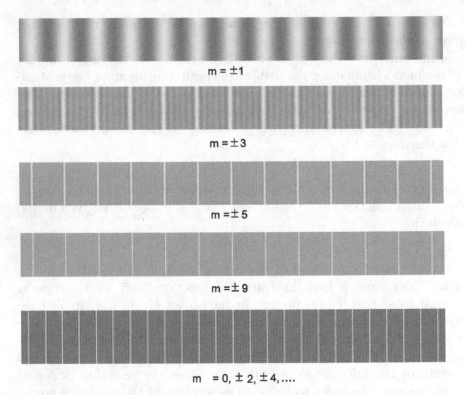

$m = \pm 1$

$m = \pm 3$

$m = \pm 5$

$m = \pm 9$

$m = 0, \pm 2, \pm 4, \ldots.$

Figure 3.14. A set of images of the amplitude diffraction grating found to be at different amount of opened diffraction maxima specified by index m.

The operating principle of the microscope can be demonstrated by using the experimental setup discussed above, Fig.3.9. An amplitude diffraction grating with a period $d = 2$ mm is the object for observation in such a microscope model. The grating is positioned horizontally and produces a vertical diffraction structure in the second focal plane of objective O_1. In order to shut down maxima of the spatial spectrum, a slit with variable width is used. Firstly the aperture of the slit is large enough to transmit several maxima, and we get a sharp picture of the grating. Then we close the slit and allow the transmission of only three orders of the spatial spectrum, $m = 0, -1, +1$. This gives a sinusoidal intensity distribution of the now basely resolved grating image. When we allow only the central diffraction maxima to pass, any structure in the image vanishes. Images for a different amount of the diffracted rays are shown in Fig.3.14.

One can further modify the output image by means of a mask, which for example shuts down the odd maxima of the FOURIER spectrum. Using this mask provides the interference of all the rays incident from even maxima of the spectrum, which results to a doubled structure of the output image (Fig.3.14).

3.2   The phase contrast Zernike microscope

An another example of spatial filtration is based on an idea suggested in 1935 by F.ZERNIKE (1888-1966) [4]. Great difficulties arise in the observation of fine structures of transparent biological objects by means of a microscope, because images of such transparent objects have low intensity contrast. According to the idea put forward by ZERNIKE such objects form a phase modulation of the transparent light, whereas the amplitude remains nearly unaffected. Objects of this sort produce a phase disturbance in the incident light along with nearly invariable amplitude. Thus the transparency function can be represented in the form:

$$t(x,y) = t_0 \exp[i\varphi(x,y)] \quad ,$$

where t_0 is the amplitude transparency coefficient of the object image. We call attention to the fact that here the function $t(x,y)$ describes the transparency properties of the image constructed by the phase contrast microscope. For ordinary biological preparations we now assume phase variations $\varphi(x,y)$ over the image of the object to be much less than unity:

$$\varphi(x,y) \ll 1 \quad .$$

If this is the case, the transparency function becomes

$$t(x,y) \approx t_0[1 + i\varphi(x,y)] \quad , \tag{3.59}$$

which allows us to neglecting terms of the order φ^2 and higher. In other words the following has to be true:

$$\varphi^2 \ll |\varphi| \quad .$$

Under the assumption above the intensity of light passing through such a transparent object has an invariable value as follows:

$$I_t \sim I_0 t t^* = I_0 t_0^2 (1 - \varphi^2) \approx I_0 t_0^2 \quad , \tag{3.60}$$

where I_0 is the intensity of the incident light.

The method proposed by ZERNIKE, which is called the *phase-contrast technique*, uses a small thin transparent film placed at the center of the second focal plane of the microscope objective. The film causes an extra phase shift for low spatial harmonics of the spatial spectrum of the observed object. For a given wavelength the phase shift introduced by inserting such a film should be either $\pi/2$ or $3\pi/2$ in order to provide a high contrast in the observed image. Using this film, the zero spatial frequency, which contains most part of the light intensity, will gain an additional phase difference of $\pi/2$ (or $3\pi/2$) with respect to rays of higher frequencies, which are responsible for the formation of the details of the image. For example, let this phase shift be equal to $\pi/2$, then the transparency function will take the form

$$t \approx t_0 \left[\exp(i\frac{\pi}{2}) + \exp(i\frac{\pi}{2})\varphi(x, y) \right] = i t_0 [1 + \varphi(x, y)] \quad .$$

It is clear, that now, in contrast to (3.60), the intensity of the image should include a linear term with $\varphi(x, y)$ as follows:

$$I \sim I_0 t t^* = I_0 t_0^2 [1 + 2\varphi(x, y)] \quad . \tag{3.61}$$

Hence, the contrast of the image should be increased. Since the phase term $2\varphi(x, y)$ is added to unity in (3.61) it is usually called a positive contrast. In the case, where the film inserts a phase shift of $3\pi/2$ the amplitude distribution becomes

$$t \approx t_0 \left[\exp(i\frac{3\pi}{2}) + \exp(i\frac{\pi}{2})\varphi(x, y) \right] = -i t_0 [1 - \varphi(x, y)] \quad .$$

The intensity distribution takes the form

$$I \sim I_0 t t^* = I_0 t_0^2 [1 - 2\varphi(x, y)] \quad .$$

This is the case of a so-called negative contrast, where the bright regions of the image are located at the positions related to dark regions of the image found with positive contrast and vise versa.

Figure 3.15. The original image of a phase grating (a) and the image (b) obtained with closed central ray in the spatial spectrum of the grating.

Using a film which absorbs light of a desired wavelength allows the contrast to be higher than in the previous cases. In this case the transparency function will take the form

$$t \approx t_0 \left[\kappa \exp(i\frac{\pi}{2}) + \exp(i\frac{\pi}{2})\varphi(x,y) \right] = it_0[\kappa + \varphi(x,y)] \quad ,$$

where $\kappa < 1$ stands for light absorption by the film. In comparison to the previous cases the image intensity appears to be modulated to a higher degree and becomes

$$I \sim I_0 tt^* = I_0 t_0^2 \kappa[\kappa - 2\varphi(x,y)] \quad .$$

Since $\kappa < 1$, the intensity variations caused by the phase modulation via $\varphi(x,y)$ will take place against the background level of κ instead of unity.

A phase diffraction grating causes a phase modulation but (nearly) no amplitude modulation of an incident light beam. In order to demonstrate the phase-contrast method, a phase grating is placed in the first focal plane of the objective of the setup shown in Fig.3.9. The image is observed on the second focal plane of objective O_2. It is seen from Fig.3.15,a that this image has a very low contrast. By shutting down the non-diffracted rays of the spatial spectrum by a small opaque disk the contrast of the input image is increased as shown in Fig.3.15,b. The action of the disk on the input image is that the low spatial frequencies are suppressed. In this case the intensity of the image becomes

$$I \sim I_0 tt^* = I_0 \varphi^2(x,y) \quad , \qquad (3.62)$$

where, as before, $\varphi(x,y)$ describes periodical variations of the object.

SUMMARY

With the help of the formalism of Fourier integrals image construction can be seen as a linear filtration process. Beginning with given object properties, application of linear operators, describing the action of the inserted optical elements, allows to find the properties of the image.

Another widely used application is spatial filtering. Based on ABBE's theory of image formation in microscopes we have seen that in the second focal plane of an objective a spatial spectrum exists which is responsible for the image formation. Thus, influencing the spatial spectrum, e.g. by masking special parts, leads to a change in the image. This fact can be used to enhance the image quality. Influencing the phase of central parts of the spatial spectrum can lead to substantially increased intensity contrast in the image, as used in the phase contrast microscope.

PROBLEMS

3.1. Let a plane monochromatic wave fall on an aperture which is placed on the first focal plane of a positive thin spherical lens. Represent the angles of diffraction in terms of spatial frequencies over the second focal plane of the lens.

3.2 Let a plane monochromatic wave fall on a diffraction grating, which has an amplitude transparency function $t(x, y) = 1 + a\cos(\Omega_0 x)$, with $a < 1$. Calculate the spatial spectrum after the grating.

3.3 A plane monochromatic wave of wavelength λ passes through a diffraction grating. The grating has a transparency function $t(x) = 1 + a\cos(\Omega_0 x)$, with $a < 1$. Find the intensity distribution on a plane parallel to the grating at a distance Δz.

3.4 A plane monochromatic wave of wavelength λ passes through two identical gratings which are separated by the distance Δz. The transparency function of each grating is $t(x) = 1 + 0.25\cos(2\pi x/d)$. Find the distance which produces a maximal value of the total intensity of the first diffraction maxima of the outgoing light. Assume a wavelength $\lambda = 600$ nm, and a period of the gratings of 0.1 mm.

SOLUTIONS

3.1. Since the lens implements a spatial spectrum under the conditions of the paraxial approximation, for each point (x_1, y_1) of the second focal plane of the lens the following relationships has to be true:

$$x_1/f = \sin \theta_x \quad , \qquad y_1/f = \sin \theta_y \quad ,$$

where θ_x and θ_y are the angles of diffraction, and f is the focal length of the lens. On the other hand the propagation of one spatial harmonic associated with the spatial frequencies f_x, f_y is given by the complex amplitude:

$$\exp\left(-ik(xx_1 + yy_1)/f\right) = \exp\left(-i\pi(\lambda f)(f_x y + f_y y)\right) \quad ,$$

with $f_x = x_1/(\lambda f)$, and $f_y = y_1/(\lambda f)$. Hence the angles of diffraction have to be represented by f_x and f_y in the from

$$\sin \theta_x = \lambda f_x \quad , \qquad \sin \theta_y = \lambda f_y \quad .$$

3.2 We represent the function $t(x, y)$ in terms of complex amplitudes as follows:

$$t(x, y) = 1 + \frac{a}{2} \exp\left(i2\pi f_{x0} x\right) + \frac{a}{2} \exp\left(-i2\pi f_{x0} x\right) \quad ,$$

which gives the spatial spectrum in the form of three harmonics:

$$F\{t(x, y)\} = F\{1\} + \frac{a}{2} F\{\exp\left(i2\pi f_{x0} x\right)\} + \frac{a}{2} F\{\exp\left(-i2\pi f_{x0} x\right)\} \quad .$$

The first term represents a non-diffracted wave in the form of a δ function: $F\{1\} = \delta(f_x)$, where $\delta(f_x)$ indicates a plane wave propagating along the direction $f_x = 0$, and the angle of diffraction is equal to zero. The term $(a/2)F\{\exp\left(i2\pi f_{x0} x\right)\}$ specifies a plane wave of amplitude $a/2$ propagating along the direction $f_x = f_{x0}$ at the angle of diffraction $\sin \theta = -\lambda f_{x0}$. The third item represents the same wave, but diffracted at the angle $\sin \theta = \lambda f_{x0}$.

3.3. Since the periodic structure of the grating is along the x-direction, let us calculate the desired complex distribution along the x-direction only by means of the convolution integral

$$\int_{-\infty}^{+\infty} t(x)h(x - x_1)dx = \frac{\exp(ik\Delta z)}{i\lambda\Delta z} \int_{-\infty}^{+\infty} [1 + a\cos(\Omega_0 x)] \cdot$$

$$\cdot [1 + a\cos(\Omega_0 x)] \exp\left(ik(x - x_1)^2/(2\Delta z)\right) dx \quad ,$$

where $h(x)$ is the impulse response function corresponding to FRESNEL diffraction, and x_1 is the x-coordinate on the plane at a distance Δz from the grating. We call attention to the fact that the integration of the term

$$\int_{-\infty}^{+\infty} \exp\left(ik(x - x_1)^2/(2\Delta z)\right) dx$$

can be reduced to the form of a FRESNEL integral, so we replace it by A. Since $\cos(\Omega_0 x) = 0.5 \exp(i\Omega_0 x) + 0.5 \exp(-i\Omega_0 x)$, we can reduce the integral containing the cosine term to

$$\frac{a}{2} \int_{-\infty}^{+\infty} \exp\left(ik(x - x_1)^2/(2\Delta z) + i\Omega_0 x\right) dx+$$

$$+\frac{a}{2} \int_{-\infty}^{+\infty} \exp\left(ik(x - x_1)^2/(2\Delta z) - i\Omega_0 x\right) dx \quad .$$

Let us transform the phase of the first integral:

$$\frac{k}{2\Delta z}(x - x_1)^2 + \Omega_0 x = \frac{k}{2\Delta z}\left(x - x_1 + \frac{2\Omega_0 \Delta z}{k}\right)^2 +$$

$$+\Omega_0 x_1 - \frac{\Omega_0^2 \Delta z}{2k} \quad .$$

It is clear that after integration of variable x the first term of the phase results in the similar form of the FRESNEL integral, that is A. So the integration of the first two terms gives the result

$$A + A\frac{a}{2} \exp(i\Omega_0 x_1) \exp(-i\Omega_0^2 \Delta z/(2k)) \quad .$$

The integration of the third term containing the magnitude $-i\Omega_0 x$ in its phase results in a similar expression: $A \exp(-i\Omega_0 x_1) \exp(-i\Omega_0^2 \Delta z/(2k))$. Hence, the required distribution of the complex amplitude is given by

$$A\left[(1 + a\cos(\Omega_0 x_1) \exp(-i\Omega_0^2 \Delta z/(2k)))\right] \quad .$$

Now the desired intensity distribution $I(x)$ on the plane at a distance Δz from the phase amplitude grating can be calculated as being proportional to the squared amplitude:

$$I(x) \sim 1 + 2a\cos(\Omega_0 x_1)\cos(\Omega_0^2 \Delta z/(2k)) + a^2 \cos^2(\Omega_0 x_1) \quad .$$

Since $a < 1$, the third item can be neglected, and we get the desired the intensity as

$$I(x) \sim 1 + 2a\cos(\Omega_0 x_1)\cos(\Omega_0^2 \Delta z/(2k)) \quad .$$

It follows from the explicit form of $I(x)$ that the image takes the from of a periodical structure with the same spatial period as in the function $t(x)$. For the given value Ω_0 distances Δz exist, which provide a high contrast of the image; this is true for $\cos(\Omega_0^2 \Delta z/(2k)) = \pm 1$. But other distances exist which result in the disappearance of any periodical structure; for $\cos(\Omega_0^2 \Delta z/(2k)) = 0$.

3.4 It follows from the solution of the previous problem that the distribution of the complex amplitude in the vicinity of the second grating is given by

$$E(x_1) \sim 1 + 0.25 \cos(2\pi x_1/d) \exp(-i(2\pi/d)^2 \Delta z/(2k)) \quad,$$

where x_1 is the coordinate along the periodical structure of the grating. The distribution of the complex amplitude after the second grating gets the form

$$E(x_1)t(x_1) \sim [1 + 0.25 \cos(2\pi x_1/d)] \cdot$$
$$\cdot [1 + 0.25 \cos(2\pi x_1/d) \exp(-i(2\pi/d)^2 \Delta z/(2k))] =$$
$$= 1 + \frac{1}{4} \cos(2\pi x_1/d)(1 + \exp(-i(2\pi/d)^2 \Delta z/(2k))) +$$
$$+ \frac{1}{16} \cos^2(2\pi x_1/d) \exp(-i(2\pi/d)^2 \Delta z/(2k)) \quad.$$

It is the second term of the distribution which will produce two first diffraction maxima. They both have the amplitude

$$\frac{1}{4}(1 + \exp(-i(2\pi/d)^2 \Delta z/(2k))) \quad.$$

Hence, the relative intensity associated with rays propagating after the second grating along directions $m = \pm 1$ is given by

$$\frac{1}{16}(1 + \exp(-i(2\pi/d)^2 \Delta z/(2k)))(1 + \exp(i(2\pi/d)^2 \Delta z/(2k))) =$$
$$= \frac{1}{4} \cos^2((2\pi/d)^2 \Delta z/(4k))) \quad.$$

It is clear that a maximal value of the intensity appears at $(2\pi/d)^2 \Delta z/(2k)) = \pm \pi$, which gives for Δz

$$\Delta z = \frac{2d^2}{\lambda} \quad.$$

Substitution of the numerical values results in $\Delta z \approx 3.3$ cm.

Chapter 4

HISTORY OF QUANTA

Retrospectively, we owe the origin of quantum physics primarily to the investigation of the spectral distribution of the radiation emitted by heated bodies. This distribution could not be described with formulas derived from classical physical treatments. But in 1900 MAX PLANCK (1858-1947) was able to derive the correct law for the *black body radiation* by the assumption that the energy within the radiation is "quantized". This assumption can be treated as the beginning of quantum physics. As the radiation emitted by a "black" body takes a special place in understanding and introducing the quantum ideas in optics, we shall analyze it in more detail after some introducing remarks.

The light emitted by any source carries energy. When the light and its energy is created by heating of a material body, the radiation is called *thermal radiation*. Thermal radiation takes a special place among other types of radiation as it is the only one that can be in equilibrium with its environment. Therefore the problems of thermal radiation are closely connected to the thermodynamics of the emitting bodies. It was PLANCK's theory which introduced the quantum statistical approach and the representation of basic thermodynamic quantities, such as entropy and energy. In contrast to the classical thermodynamical treatment, which is connected with a continuous variation of these quantities, PLANCK's theory led to the idea of *energy quanta*.

1. Black body radiation

Radiation incident on the surface of a body is partially reflected and partially scattered by the surface and partially transmitted through it. We restrict ourselves to the case of an entirely opaque body, assuming that transmission of radiation through the body is impossible. If this is the case, one can characterize the surface of such a body with only the *absorbability* $\alpha(\nu)$. The absorbability depends on the frequency of the incident radiation and also depends on the temperature of the body. A quantitative measure of the absorbability is the fraction of absorbed energy compared to the total incident radiation energy (Fig.4.1). On the other hand, a body also emits radiation energy. We introduce the *emissivity* $e(\nu)$ as the amount of light with a certain frequency ν which is emitted from a unit area of the body in all directions.

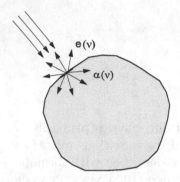

Figure   4.1 Illustrating emissivity and absorbability of a body

Experience shows that the higher the absorbability $\alpha(\nu)$ of a body within a certain frequency range, the higher the value of the emissivity $e(\nu)$ will also be. The fraction $e(\nu)/\alpha(\nu)$ is not dependent on the substance of the body but is an universal function of frequency and temperature:

$$\frac{e(\nu)}{\alpha(\nu)} = f(\nu, T) \quad . \tag{4.1}$$

The relationship (4.1) was established by G.KIRCHHOFF (1824-1887) in 1860 and is called the KIRCHHOFF *law* since that time. The KIRCHHOFF law constitutes that the fraction $e(\nu)/\alpha(\nu)$ is constant for different materials, whereas the quantities $e(\nu)$ and $\alpha(\nu)$ may vary over a wide range.

Traditionally, the basic model used for establishing the laws of thermal radiation is the so-called *black body*, which can be thought as a material substance enclosed in an adiabatic envelope at temperature T. As there is no energy transmitted through the adiabatic envelope, thermodynamic equilibrium is established within the envelope. For this reason the adiabatic envelope may be also treated as a cavity filled with electromagnetic radiation. The inner walls of the cavity are able to absorb and emit light energy. Absorption and emission processes provide an energy balance between the radiation and the substance. Because no radiation is accumulated inside the envelope, the absorbability of the substance has to be equal to one: $\alpha = 1$. It follows from (4.1) that the emissivity of the black body is therefore the universal function $f(\nu, T)$:

$$e(\nu) = f(\nu, T) \quad . \tag{4.2}$$

Any substance which possesses the similar property may be called a black body, and radiation emitted by such a substance is black body radiation, or *equilibrium radiation*.

It was shown by KIRCHHOFF that the function $f(\nu, T)$ has to be proportional to the spectral density of the black body radiation, which we denote by $\rho(\nu)$:

$$\rho(\nu) \sim f(\nu, T) \quad . \tag{4.3}$$

The function $\rho(\nu)$ represents the amount of radiation energy per unit volume and per unit frequency range. Thus, the study of black body radiation is connected to the establishment of the (up to now unknown) function $f(\nu, T)$. Experimentally, the emissivity $e(\nu)$ of a black body can be found by measurements of the radiation passing through a small aperture in the wall of an adiabatic envelope considered to be a black body (Fig.4.2). The amount of energy which passes out of the cavity has to completely be compensated by heating the external walls of the envelope to provide equilibrium conditions between the radiation and the walls inside the envelope. In such a way the dependency of the emissivity $e(\nu)$ on the frequency ν was studied.

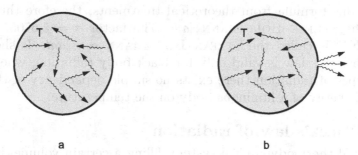

a   b

Figure 4.2. A black body (a) and its model proper for measurements (b)

Now we discuss fundamental phenomenological results that were obtained before PLANCK found the right law of black body radiation. In 1893, based on thermodynamics concepts, M.K.WIEN (1866-1936) established that the spectral density of the equilibrium radiation must be dependent on the 3rd power of the frequency and on a function of the ratio ν/T,

$$\rho(\nu) = \nu^3 f(\nu/T) \quad ,$$

and later he developed a formula for the spectral density, which is called
WIEN's *law*:

$$\rho(\nu) = C\nu^3 \exp(-\beta\nu/k_B T) \quad , \tag{4.4}$$

where $k_B = 1.3806568 \times 10^{-23}\,\mathrm{J\,K^{-1}}$ is the BOLTZMANN *constant*, and β
and C are constants. This law is in agreement with experimental data
only in the high-frequency range (ultraviolet radiation).

A classical treatment of the interaction between the electrons of a
substance and the radiation leads to the formula

$$\rho(\nu) = B\nu^2 k_B T \quad , \tag{4.5}$$

where B is a constant. This formula is known as RAYLEIGH-JEANES'
law (J.W.RAYLEIGH, 1842-1919 and J.H.JEANES, 1877-1946). This
law fits to the experimental observation only within the low–frequency
range (infrared radiation), but deviates from experimental results more
and more with increasing frequency.

Based on his analysis of experimental observations, in 1879 J.STEFAN
(1835-1893) established that the total radiated energy U of the surface of
a black body should be proportional to the fourth power of temperature:

$$U = \sigma T^4 \quad . \tag{4.6}$$

Using thermodynamic considerations, L.BOLTZMANN (1844-1906) later
derived this formula from theoretical treatments, therefore this law is
called the STEFAN–BOLTZMANN *law*. The factor $\sigma = 5.67051 \times 10^{-8}$
$\mathrm{W\,m^{-2}\,K^{-4}}$ is called the STEFAN–BOLTZMANN *constant*. It should be
noted that this law is valid only for black body radiation, whereas for
other types of radiation there exists no simple dependency of the total
radiated energy of a luminous body on the temperature.

2.   Planck's law of radiation

A closed thermodynamical system, filling a certain volume, is represented by its total energy and by the total amount of particles inside.
According to PLANCK's theory a substance within the adiabatic envelope may be regarded as consisting of oscillators. It is assumed that all
oscillators act independently from each other. For this reason we discuss
the radiating properties of one group of oscillators, all having the same
resonance frequency ν.

One oscillator treated as an elementary unit of a closed thermodynamic system may be described by its average energy $\langle w \rangle$, its entropy
s, and its temperature T (which has the same value as the temperature
of the adiabatic envelope). Using the laws of classical electrodynamics

PLANCK derived the following relationship between the spectral density of the equilibrium radiation $\rho(\nu)$ and the average energy $\langle w \rangle$:

$$\rho(\nu) = \frac{8\pi}{c^3} \nu^2 \langle w \rangle \quad . \tag{4.7}$$

In the context of thermodynamics, a relationship between the average energy of one oscillator and its entropy was known in the form

$$\frac{1}{T} = \frac{\partial s}{\partial \langle w \rangle} \quad . \tag{4.8}$$

This expression, being valid under thermodynamic equilibrium conditions, permits the calculation of $\langle w \rangle$, provided that the entropy s as a function of T is known. PLANCK found that s calculated by means of a statistical method which had been proposed by BOLTZMANN, gives rise to the magnitude of $\langle w \rangle$, which in turn led to the correct function $\rho(\nu)$.

In order to understand the idea suggested by PLANCK, let us consider that the total energy W of the radiation with the required frequency ν is distributed over a finite number N of oscillators. As before, we also assume that all these oscillators have the same frequency ν. For a given moment, one particular distribution of this energy W over N oscillators is called the *energetic state* of the oscillators. At the following moment, one can find another energetic state which may not be similar to that at the previous moment. A set of energy states, where one energy distribution can be exchanged with another one, is a simple statistical model of a substance in thermal balance with the radiation field.

There exist two different ways to calculate $\langle w \rangle$. A first way is to assume that the amount of total possible different energetic states is principally infinite. With a finite number N of oscillators, the idea of an infinite amount of different energetic states is equivalent to apply a continuous distribution of the energy W over the N oscillators. A second way is to treat that this amount of energetic states is rather huge, but of finite magnitude. Here we assume that the energy W can be represented in terms of small identical portions, called *energy quanta*. It is the discrete division of W into small portions that provides a finite amount of possible different energetic states. Such a finite amount of energetic states can be calculated by known mathematical methods. If such calculations led to a disagreement with experimental data, the partitioning of energy should be made smaller and would finally tend to zero to provide continuous energy states, corresponding to classical physics.

Let the system be composed of oscillators and the radiation be in a non–equilibrium state at an initial moment. According to the second law of thermodynamics, this closed thermodynamic system will tend towards a thermodynamic equilibrium during the following moments. The assumption of a finite amount of energetic states implies that the system has to pass through *all possible different* energetic states until thermodynamic equilibrium is reached. In other words, while reaching thermodynamic equilibrium, at any moment the system may be found in a state which is either one of the prior states, or is a new state. When thermodynamic equilibrium has been reached, then new energetic states will never be found. Hence, under the condition of the thermodynamic equilibrium, all possible different energetic states have been passed by the closed system.

To illustrate this important statement, let us assume that we have a chance to look at the states of the system at certain moments in time, marking these states by Z_p in series of moments t_k. Then, when starting from t_0, our sequence of states could look as follows:

$$Z_1(t_0), Z_2(t_1), Z_3(t_3), Z_2(t_4), \; ... \; Z_m(t_k), Z_{m-7}(t_{k+1}), Z_3(t_{k+2}), \; ...$$

where the order of states is of little significance. A certain state can be repeated many times. The more important fact is that all previous states will be repeated after moment t_k, provided that $Z_m(t_k)$ is the last energetic state unknown to us among the group of states

$$Z_1, Z_2, Z_3, ... Z_m \quad .$$

This group of states should be regarded as representing all possible different states. It also means that the system has already achieved its thermodynamic equilibrium at the moment t_k. The entropy of any closed thermodynamic system shows a similar evolution. In other words, the fact that the amount of known different energetic states gets larger while the system tends towards statistical equilibrium permits the assumption that the amount of possible different states is directly proportional to a function of the entropy of the system. The statement above is quite valid for the group of oscillators under consideration. Therefore, if we calculate an expression for the amount of possible different energetic states, additionally we will know the entropy under thermodynamic equilibrium. Thus, our assumption of discrete small portions of the energy gives a simple alternative way for calculating the entropy of our system.

Let S be the entropy of the ensemble of N oscillators, and let P be the total amount of different energetic states of these oscillators. According to PLANCK's theory, the relationship between S and P has to be

$$S = k_B \ln P + const. \tag{4.9}$$

For the given entropy S, both values s and $\langle w \rangle$, associated with one oscillator, may be found as follows:

$$s = S/N \quad , \qquad \langle w \rangle = W/N \quad .$$

Now we permit the energy W to be represented by the number M of small identical portions, or energy quanta ε:

$$W = M\varepsilon \quad .$$

The problem to find P is therefore the task: What is the amount of different ways to distribute M units over N cells?

First, we number the oscillators and quanta:

$$o_1, o_2, \ldots, o_N \quad , \qquad q_1, q_2, \ldots, q_M \quad ,$$

and select the first oscillator o_1. At the beginning we assume that all other oscillators and all quanta are identical objects, without distinguishing between oscillators and quanta. So, the total amount of objects is $M + N - 1$. Now we arrange all these objects in random order to the right of o_1, for instance, in the following sequence:

$$o_1, q_8, q_9, q_{27}, o_6, o_{14}, q_{11}, q_{28}, \cdots \quad . \tag{4.10}$$

The combination (4.10) has the following meaning: We place all the quanta corresponding to the first oscillator to the right of it. In our case, the 8th, 9th, and 27th quanta will be located within the 1st oscillator; the 6th oscillator will be empty; in the 14th oscillator we place the 11th and 28th quanta, and so on. It is clear why the first member of the series is an oscillator. If the first term in (4.10) would be a quantum, then, according the rule above, this quantum would not belong to an oscillator. The first few oscillators and quanta of two sequences are shown in Fig.4.3. The right-hand sides of the figures illustrate how the oscillators must be filled up by quanta.

The total amount of possible combinations of $M + N - 1$ objects is

$$(M + N - 1)! \quad .$$

We have to take into account that every energetic state has already been calculated several times. For example, two combinations o_1, q_2, o_2, q_1 and o_2, q_1, o_1, q_2 are assumed to be identical with respect to their energy, because these states differ from each other only in transposition of their terms. In order to calculate all possible combinations which result in energetically identical states, we find the total amount of permutations for the oscillators to be $(N - 1)!$. Similarly, due to identity of quanta,

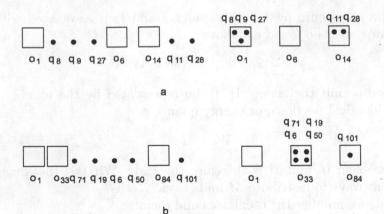

Figure 4.3. Two possible cases of quanta filling the oscillators, (a) and (b). In both cases an oscillator should be placed on the left-hand side of the sequence "oscillators - quanta".

the total amount of permutations connected with quanta has to be $M!$. Thus, we have to regard $(N-1)!M!$ states to be energetically identical among the total of $(M+N-1)!$ states. Therefore, the total amount of different states P has to be equal to

$$P = \frac{(M+N-1)!}{M!(N-1)!} \ . \tag{4.11}$$

When substituting P in (4.9) with the right side of (4.11), for the entropy of N oscillators we obtain

$$S = k_B \ln \frac{(M+N-1)!}{M!(N-1)!} + const.$$

The last expression may be simplified by approximating with the STIRLING formula

$$\ln n! \approx n \ln n - n \ ,$$

which is valid for $n \gg 1$, and using $(M+N-1) \approx (M+N)$, $(N-1) \approx N$ we get

$$S \approx k_B[(M+N)\ln(M+N) - N\ln N - M\ln M] + const. =$$

$$= k_B N \left[\left(1+\frac{M}{N}\right) \left(\ln N + \ln\left(1+\frac{M}{N}\right)\right) - \ln N - \frac{M}{N}\ln M\right] + const. =$$

$$= k_B N \left[\left(1+\frac{M}{N}\right) \ln\left(1+\frac{M}{N}\right) - \frac{M}{N}\ln\left(\frac{M}{N}\right)\right] + const.$$

Since $M = W/\varepsilon$ and $N = W/\langle w \rangle$,we get $M/N = \langle w \rangle /\varepsilon$; thus for S we find

$$S \approx k_B N \left[\left(1 + \frac{\langle w \rangle}{\varepsilon} \right) \ln \left(1 + \frac{\langle w \rangle}{\varepsilon} \right) - \frac{\langle w \rangle}{\varepsilon} \ln \left(\frac{\langle w \rangle}{\varepsilon} \right) \right] + const.$$

The entropy of one oscillator is $s = S/N$; therefore we get

$$s \approx k_B \left[\left(1 + \frac{\langle w \rangle}{\varepsilon} \right) \ln \left(1 + \frac{\langle w \rangle}{\varepsilon} \right) - \frac{\langle w \rangle}{\varepsilon} \ln \left(\frac{\langle w \rangle}{\varepsilon} \right) \right] + const.$$

By using the relationship (4.8) between $\langle w \rangle$, T and s, we find the expression

$$\langle w \rangle = \frac{\varepsilon}{\exp(\varepsilon/k_B T) - 1} \quad . \tag{4.12}$$

Based on formula (4.7) we may now write an explicit form of ρ_ν, called now PLANCK's *distribution*:

$$\rho_\nu = \frac{8\pi\nu^2 \varepsilon}{c^3} \frac{1}{\exp(\varepsilon/k_B T) - 1} \quad . \tag{4.13}$$

We finally have to consider that the distribution (4.13) can be transformed to a shape similar to the WIEN formula (4.4) under the assumption $\varepsilon \gg k_B T$ (which causes $\exp(\varepsilon/k_B T) \gg 1$):

$$\rho_\nu \approx \frac{8\pi\nu^2 \varepsilon}{c^3} \exp(-\varepsilon/k_B T) \quad .$$

We can enforce the equality with Eq.(4.4) if we set $\varepsilon = \beta\nu$ and $\frac{8\pi\beta}{c^3} = C$, but we will further use h instead of β:

$$\varepsilon = h\nu = \hbar\omega \quad , \tag{4.14}$$

where $h = 2\pi\hbar = 6.6260755 \times 10^{-34}$ J s ($\hbar = h/2\pi \approx 1.05459 \cdot 10^{-34}$ J s) is a fundamental constant which is called PLANCK's *constant*.

In this way the law for the spectral density of the equilibrium radiation was found by PLANCK. The function $\rho(\nu)$ is finally given by the following expression:

$$\rho(\nu) = \frac{8\pi h\nu^3}{c^3} \frac{1}{\exp(h\nu/k_B T) - 1} \quad . \tag{4.15}$$

Three distributions of $\rho(\nu)$ calculated at $T = 1800$ K, 2400 K, and 3000 K are shown in Fig.4.4.

Each of these distributions has a maximum at the frequency $\nu_{max} = Tc/b$, which is equivalent to the expression

$$\lambda_{max} T = b \quad , \tag{4.16}$$

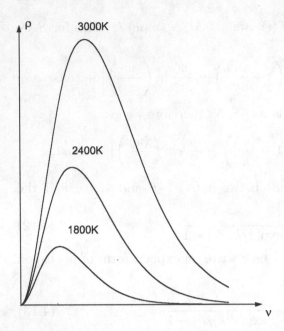

Figure 4.4 Three distributions of light energy density versus frequency for three different values of temperature.

where $b = 1.265\hbar c/k = 0.0029\ m.\mathrm{K}$. The relation (4.16) is known as WIEN's *law of maximum shift.*

In the low-frequency range, PLANCK's distribution (4.15) can be transformed into the formula of RAYLEIGH and JEANES (4.5), which was derived assuming the classical representation of the interaction between the ensemble of oscillators and radiation. When $h\nu \ll k_BT$, the exponential term in (4.15) can be approximated by $\exp(h\nu/k_BT) - 1 \approx h\nu/k_BT$. Further, from (4.12) we get $\langle w \rangle \approx k_BT$, and using these approximations, formula (4.15) becomes equal to (4.5).

According to BOLTZMANN's law of the equipartition of energy between the degrees of freedom, the energy $k_BT/2$ is given to every degree of freedom of an atom or molecule under thermal equilibrium. The common case of equilibrium radiation is the case of unpolarized light. One can assume such light as being composed of two beams with two orthogonal linear polarization directions (compare Part 1, Chapter 3). This means that an oscillator under the action of light has two degrees of freedom, each taking $k_BT/2$; therefore the average energy of the oscillator has to be equal to $\langle w \rangle = k_BT$.

The famous Austrian physicist L.BOLTZMANN was the first who founded and developed statistical methods in thermodynamics [2]. Formula (4.6) for the entropy was derived by BOLTZMANN in his work on the thermodynamics of gases and statistical mechanics [3]. The concept of discrete variables was also successfully used by BOLTZMANN, for the first time in his investigations of statistical mechanics. Nevertheless, even if discrete variables were applied to calculations in intermediate stages, at the end of the calculations such discrete variables usually tended to zero. This step allowed these variables to be continuous.

3. Formulae for equilibrium radiation

A set of important equations can be derived from PLANCK's distribution (4.15). We find the total energy density $\rho(T)$ by integration of PLANCK's distribution over all frequencies ν:

$$\rho(T) = \frac{8\pi h}{c^3} \int_0^\infty \frac{\nu^3 d\nu}{\exp(h\nu/k_B T) - 1} \quad .$$

A new variable $x = h\nu/k_B T$ permits to transform this integral into

$$\rho(T) = \frac{8\pi (k_B T)^4}{c^3 h^3} \int_0^\infty \frac{x^3 dx}{\exp(x) - 1} \quad .$$

This integral is equal to $\pi^4/15$. It can be seen that the total energy density is proportional to the 4th power of the temperature:

$$\rho(T) = \frac{8\pi^5 k_B^4}{15 c^3 h^3} T^4 = \sigma' T^4 \quad , \tag{4.17}$$

similar to the STEFAN–BOLTZMANN law (4.6).

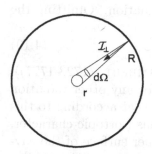

Figure 4.5 For calculations of $I_\perp$ from a unit element of the radiating wall of a black body.

Now the properties of the radiating surface of a black body can be described by formula (4.17). We shall find the energy $I_\perp$ emitted by a unit element of the radiating surface per unit time into a unit element of solid angle. Let us consider a spherical radiating cavity of radius R, being in thermodynamic equilibrium with the radiation at temperature T (Fig.4.5). We calculate the radiation passing through a small sphere of radius r ($r \ll R$) at the center of the cavity. From every point of the inner surface of the cavity, the small sphere is observed at the solid angle $\Omega = \pi r^2/R^2$. From a unit area of the cavity, radiation passes through this small sphere with the cross section πr^2 with the velocity of light c. This radiation carries the energy $\mathcal{I}_\perp \Omega$ per unit time. Thus, the energy density of the radiation is

$$\frac{\mathcal{I}_\perp \Omega}{\pi r^2 c} = \frac{\mathcal{I}_\perp}{R^2 c} \quad,$$

whereas the total energy density caused by emission of the full inner surface of the cavity is $4\pi R^2$ times greater :

$$\rho = \frac{4\pi R^2 \mathcal{I}_\perp}{R^2 c} = \frac{4\pi}{c} \mathcal{I}_\perp \quad. \qquad (4.18)$$

The energy emitted by a surface element depends on the angle between the normal vector of the surface element and the direction of radiation. Two elements of equal area S of the cavity at a distance L from each other are shown in Fig.4.6. The surfaces of the elements include the angle θ. These elements exchange energy under the thermodynamic equilibrium condition. The energy flux of the first element should be equal to that of the second one. The first element is viewed from the second element at the solid angle S/L^2, and the second element from the first at the solid angle $S\cos\theta/L^2$. It follows from the balance of fluxes that

$$\mathcal{I}_\theta \frac{S}{L^2} = \mathcal{I}_\perp \frac{S\cos\theta}{L^2} \quad,$$

where the index θ in $\mathcal{I}_\theta$ specifies the angle of radiation. Omitting the factor S/L^2 we get

$$\mathcal{I}_\theta = \mathcal{I}_\perp \cos\theta \quad. \qquad (4.19)$$

Formula (4.19) is called LAMBERT's *law* (J.H.LAMBERT, (1728-1777)), which is exactly valid for black bodies, whereas for any other radiation sources it is rather approximately valid. We note that according to this law the equilibrium radiation of the black body has isotropic character, which means, that in the free space inside the inner surface of the cavity the flux of radiation has the same magnitude in every direction of propagation. According to (4.18) the intensity $\mathcal{I}_\perp$ depends only on the magnitude of ρ, independent of where the luminous surface element is located. For this reason the second element in Fig.4.6 emits the same intensity $\mathcal{I}_\perp$ in the direction of its normal vector, 2-3. Hence, the dependency in (4.19) holds for the change in $\mathcal{I}_\theta$ on angle θ measured from the normal of any radiating element.

In the opposite case of anisotropic radiation, where the concept of light beams is usually used, the energy of a light beam concentrates around the direction of propagation. For a given value $\mathcal{I}_\perp$, its dependency $\mathcal{I}_\theta$

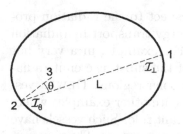

Figure 4.6 Illustrating the
LAMBERT law.

will therefore not follow Eq. (4.19), due to the limited angular width of
the light beam.

The energy U emitted by a unit area per unit time over all solid angles
is

$$U = \int \mathcal{I}_\theta d\Omega = 2\pi \int \mathcal{I}_\perp \cos\theta \sin\theta d\theta \quad ,$$

where $2\pi \sin\theta d\theta = d\Omega$ is the element of the solid angle. Integration over
the whole semi–sphere, where θ changes from 0 to $\pi/2$, gives:

$$U = 2\pi \int_0^{\pi/2} \mathcal{I}_\perp \cos\theta \sin\theta d\theta = \pi \mathcal{I}_\perp \quad .$$

Substitution of $\mathcal{I}_\perp$ from (4.18) gives

$$U = \frac{c}{4}\rho \quad . \tag{4.20}$$

Thus the formula (4.20) expresses the full flux U from the unit area of
a radiating surface with energy density ρ over the whole spectral range.
According to (4.17) ρ is proportional to T^4, thus U will take the form
of the STEFAN–BOLTZMANN law (4.6),

$$U = \frac{2\pi^5 k_B^4}{15c^2 h^3} T^4 \quad , \tag{4.21}$$

where the constant σ from (4.6), expressed in terms of c, k_B, and h, is
given by

$$\sigma = \frac{2\pi^5 k_B^4}{15c^2 h^3} \quad . \tag{4.22}$$

The fact that the energy flux from a unit area of a black body is proportional to the 4th power of the temperature plays an important role
concerning the energy balance of heated bodies at high temperatures.
Under conditions of low temperatures, convection and thermal conductivity cause the main part of energy loss, whereas the losses due to
radiation are rather negligible. Nevertheless, when the temperature increases, convection and thermal conductivity, both being proportional

to T, become more or less negligible with respect to the radiation process, because this process depends on T^4. Energy transport by radiation is a basic process for very hot substances. For example, in a very hot plasma, electrons located in hotter regions of the substance emit radiation, which then is absorbed by electrons in cooler regions. DEWAR vessels (J.DEWAR (1842-1923)) or thermos flasks are other examples where energy exchange via radiation plays an important role. Such vessels have a double wall; the space between the walls is evacuated. Thus, energy transport due to convection and thermal conductivity plays a minor role, and the radiation process dominates. For this reason, the walls of such cells are often provided with reflecting covers.

Every element of the cavity experiences the radiation pressure of the electromagnetic waves. Let us consider a unit element of the wall of the cavity. The power element dN carried by the radiation under angle θ into an element of solid angle $d\Omega$ is equal to $dN = \mathcal{I}_\theta d\Omega$. This power leads to a change of the normal projection of the pressure of the radiation, according to the relation

$$cdp = \cos\theta dN = \mathcal{I}_\theta \cos\theta d\Omega \quad .$$

Using LAMBERT's law, $d\Omega = 2\pi \sin\theta d\theta$, one can write for the contribution to the normal projection of the pressure: $cdp = 2\pi\mathcal{I}_\perp \cos^2\theta \sin\theta d\theta$. The total change in the pressure is then given by the integral

$$p = \frac{2\pi}{c}\mathcal{I}_\perp \int_0^{\pi/2} \cos^2\theta \sin\theta d\theta = \frac{2\pi}{3c}\mathcal{I}_\perp \quad .$$

Since, according to (4.18), $\mathcal{I}_\perp = \rho c/(4\pi)$, this portion of the change of the pressure is equal to $p = \rho/6$. Using the equilibrium condition for emitted and absorbed energy by the area element, the portion of pressure change due to emission is equal to the change due to absorption. Therefore, the total pressure of the black body radiation is given by

$$p = \frac{\rho}{3} \quad . \tag{4.23}$$

In the case of classical particles like atoms of an ideal gas of energy density u, thermodynamics gives the well known formula for pressure $p = 2u/3$, differing from eq.4.23 by a factor 2. This difference follows from the classical treatment of the gas particles in contrast to the relativistic treatment of the photons of the radiation.

4. Einstein's hypothesis of light quanta

We have seen the principal idea of PLANCK's theory is that the amount of energy states in the system modeling the substance and the radiation should be of finite magnitude. This assumption permits the solution of the problem of entropy statistically by involving the concept of quanta. However, there is a contradiction which was noted by ALBERT EINSTEIN (1879-1955). Really, the two parts of the closed system - the radiation on one side and the material substance on the other side - are treated in principally different ways. The relationship (4.7) was derived from classical concepts, assuming that the classical consideration of entropy of radiation is correct. In contrast, the other part of the entropy of the same system, connected with the oscillators (the expression (4.15)), was treated in terms of the entropy with the quantum approach.

Taking into account this fact EINSTEIN suggested in 1905 a new definition of entropy for the equilibrium radiation [5]. The main result of this work was the introduction of *light quanta*. According to EINSTEIN, light consists of energy quanta $\varepsilon = h\nu$ propagating with light velocity which interact with the electrons of a material substance as elementary particles. Based on his hypothesis, EINSTEIN explained a number of fundamental experimental phenomena, for instance, the *photoeffect*.

5. Photoeffect

This effect, originally known as the HALLWACHS effect, was investigated in 1888 by W.L.F.HALLWACHS (1859-1932) due to a suggestion of H.HERTZ (1857-1894), who noticed that the appearance of electric

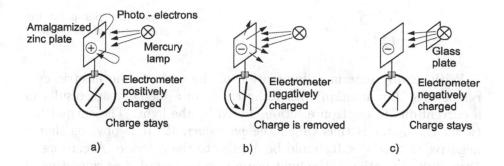

Figure 4.7. HALLWACHS' effect. If the metal plate carries a positive charge, the photoelectrons are attracted by the plate, and the total charge does not change (a). If the plate carries negative charge, the total charge vanishes (b). If the UV-part of the radiation is blocked by a glass plate, no photoelectrons arise (c). For measuring the charge we use an electrometer.

sparks is encouraged by the action of ultraviolet light. A metal plate (usually amalgamized zinc) was mounted on an electrometer and irradiated with the light of a mercury lamp (everything in air). If the electrometer carries positive charge, then it has a shortage of electrons, and irradiation does not change the charge. Otherwise, if the electrometer carries negative charge, irradiation causes the charge to vanish. In this case, electrons which are set free from the metal surface are pushed away from the plate, and the plate looses its charge. If a glass plate is inserted between lamp and metal plate, the effect vanishes, since the glass plate blocks the ultraviolet part of the light emitted by the mercury lamp (Fig. 4.7c). This effect was later called *photoeffect*.

A setup for the observation the photoeffect, or the photoelectron emission caused by irradiation of a metal surface by light, is shown in Fig.4.8. Two metallic plates, specified as the anode and cathode, are placed inside a quartz vacuum cell. An electric voltage is applied between the anode and cathode by means of an external electric circuit.

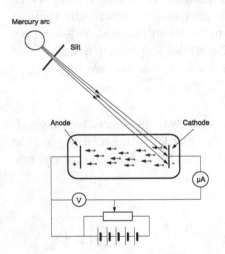

Figure 4.8 Setup for the observation of the photoeffect.

Without any external illumination of the cathode, no electric current exists. Illumination with radiation from a mercury arc results in a current due to electron ejection caused by the light. The current between the plates is thus called the *photocurrent*. It is obvious that a negative voltage $V < 0$ should be applied to the cathode. According to EINSTEIN's hypothesis, the light beam may be regarded as consisting of light quanta. Each of the quanta may by absorbed by a single bound electron of the metal, which results in the electron leaving the metal. Such an electron is called a *photoelectron* (see Fig.4.9). The light energy falling on one unit area normal to the metallic surface per unit time is proportional to the light intensity. For a given space density n of light

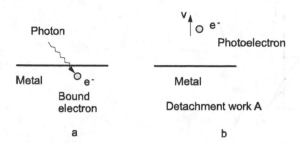

Figure 4.9 Initiation of one photoelectron. Light quant energy $h\nu$ is absorbed by a bound electron (a). This electron is set free with the kinetic energy $E_{kin} = h\nu - A$ (b).

quanta with the frequency ν, the light intensity is measured in terms of light quanta by $cnh\nu$. Because of this fact, the number of photoelectrons, leaving the surface per unit time (and therefore the photocurrent i) should be directly proportional to the incident light intensity. Indeed, experimental observations show that for a given voltage (which has to be chosen high enough) the maximum value of i is directly proportional to light intensity I (Fig.4.10,b). Further, for a given intensity I, it is in good agreement with the considerations given above that the photocurrent i increases when the negative voltage V is increased from zero until a maximum value of i is reached, as shown in Fig.4.10,a.

The fact that the amount of photoelectrons is directly proportional to the light intensity can be followed also from the classical treatment of the photoeffect. However, a phenomenon exists which is in contradiction to the classical model. For a given intensity I and frequency ν the photocurrent is not zero for $V = 0$, and to force $i = 0$, application of a certain positive voltage V_r to the cathode is necessary. This limiting voltage is called the *retarding voltage*.

According to EINSTEIN's hypothesis, the energy of one quantum $h\nu$ is absorbed by a bound electron, which gets the total energy amount $w = h\nu$. An increase of this energy by increasing the frequency ν of light results in a higher retarding voltage V_r. To leave the metallic surface, the electron should surmount the potential barrier existing on the boundary "metallic surface – air". To surmount this barrier the electron needs an energy A. This means that the quantum energy of light causing the photoeffect should be equal to or greater than A:

$$h\nu \geq A \quad .$$

The excess energy over A, Δw, is given to the electron as kinetic energy

$$\Delta w = mv^2/2 \quad .$$

Hence, the energy balance has to be expressed in the form

$$h\nu = \Delta w + A = \frac{mv^2}{2} + A \quad . \tag{4.24}$$

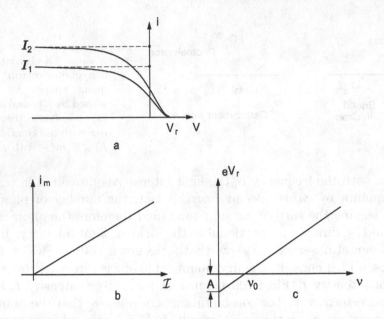

Figure 4.10. Photoeffect. a) for a given frequency ν of the light with intensity I_1, the photocurrent has a certain constant value at reasonable negative voltage V. At zero voltage, a certain current is still flowing. Forcing this current to zero, a positive voltage V_r is necessary. For higher intensity I_2 the current is higher, but V_r remains the same. b) The maximum current i_m is proportional to the light intensity. c) Kinetic energy of the photoelectrons in dependence on the light frequency. Photoelectrons are set free for $\nu > \nu_0$. Prolongation of the line to $\nu = 0$ gives us the work A.

Any electron can therefore leave the metal if $h\nu > A$. Otherwise, the electron will remain inside the metal, if its total energy is smaller than A: $h\nu < A$. Under the threshold condition $h\nu = A$, an electron can leave the metal with nearly zero velocity. Such a condition provides the minimal frequency ν_{min} of light that causes the photoeffect, according to the simple relationship

$$\nu_{min} = A/h \quad .$$

For a given material of the metallic cathode, the only parameter which decides if the photoeffect for a given light frequency occurs is the work A, which decides if ν is smaller or larger than ν_{min}. If $\nu < \nu_{min}$, no photoeffect is observed, independent of the intensity I of the light. In other words, for a given substance the occurrence of photoelectrons depends only on the energy of the quanta, or on the light frequency ν. For most metallic surfaces, typical values of ν_{min} correspond to the ultraviolet region of light frequencies. For this reason a quartz cell is usually used to observe the photoeffect, since quartz provides good transmission of ultraviolet radiation.

In turn, for a given material of the photocathode, the increase in the light frequency gives rise to a proportional increase in the kinetic energy Δw_{max} of the electrons which leave the metal surface. A measure of this energy is the retarding voltage V_r which is needed to cut off the photocurrent. If i just reaches zero, the following equation is valid: $w_{max} = eV_r$. Hence, the balance equation (4.24), corresponding to this case, has the form

$$eV_r = h(\nu - \nu_{min}) \quad . \tag{4.25}$$

It implies that the maximum energy of the photoelectrons obeys a linear dependency on the light frequency (Fig.4.10,c).

6.   Spontaneous and induced radiation

In 1916, EINSTEIN suggested a new approach to the problem of radiation emitted by quantum particles [4]. At the time of this work just some years ago N.BOHR (1882-1962) was able to explain the emission wavelengths of the hydrogen atom using a kind of quantum representation. At the same time, statistical treatments, applicable for example to BRAUN's motion of particles, were generalized by EINSTEIN to describe the radiating processes performed by PLANCK's oscillators.

Let us consider a monochromatic oscillator in a radiation field, and let w be the instantaneous magnitude of the energy of the oscillator. We should find its energy after a time span τ, which is short with respect to the period of the oscillator. The relative increment or decrement of the energy Δw is then sufficiently small with respect to w. According to EINSTEIN there are two possible kinds of changes of the energy. First, the change

$$\Delta_1 w = -Aw\tau \quad , \tag{4.26}$$

due to so called *spontaneous emission,* where A is a positive factor. Secondly, a change $\Delta_2 w$ associated with the work applied to the resonator by an external electromagnetic field of radiation. This change happens due to so called *induced processes,* since they are caused by the electric field of the radiation. We speak of (induced) *absorption* when the field transfers energy to the oscillator. In this case the energy density of the electromagnetic field decreases. *Induced emission* increases the energy density of the electromagnetic field and decreases the energy of the oscillator. All induced processes become more probable with increasing energy density of radiation but are chaotic with respect to sign (absorption or emission) and magnitude. Speculations based on electrodynamics and statistics give the following form for the mean magnitude of $\Delta_2 w$:

$$\langle \Delta_2 w \rangle = B\rho\tau \quad , \tag{4.27}$$

where B is a constant factor. A relationship between the factors A and B can be calculated in the following way: In statistics, treating a huge amount of physically identical oscillators, these factors result from the averaging procedure applied to the energies w of these oscillators. Under conditions of a thermal equilibrium, both, spontaneous and induced processes, occur with equal probability, that means, in such a way that the average energy $\langle w \rangle$ remains invariable:

$$\langle w + \Delta_1 w + \Delta_{21} w \rangle = \langle w \rangle \quad .$$

We substitute $\Delta_1 w$ from (4.26) and $\langle \Delta_2 w \rangle$ from (4.27) into the last expression and perform the averaging procedure. We find that the following expression is valid:

$$\langle w \rangle = \frac{B}{A}\rho \quad . \tag{4.28}$$

We consider now a gas of identical atoms and radiation under the condition of the thermal equilibrium. Let every atom only be in discrete states Z_1, Z_2, Z_3,... with energies w_1, w_2, w_3, ... Then, according to the BOLTZMANN principle, the state Z_n occurs with the probability P_n, which follows from the BOLTZMANN *distribution* in the form

$$P_n = p_n \exp(-w_n/k_B T) \quad , \tag{4.29}$$

where p_n is a constant, which is called the statistical weight of the nth state, and which does not depend on the temperature T.

We assume that every atom can go from state Z_n to state Z_m by absorbing light of frequency $\nu = \nu_{nm}$, and from state Z_m to state Z_n by radiating the same frequency. In general case, such transfers can take place for any combination of indices n, m. Under the condition of thermal equilibrium, the statistical equilibrium takes place with respect to each elementary process of radiation and absorption. For this reason we restrict our discussion to only one process described by one pair of indices m, n.

Under thermal equilibrium, the average amount of atoms transferred per unit time from Z_n to Z_m due to absorption should be equal to the average amount of atoms transferred by emission from Z_m to Z_n.

According to EINSTEIN, the spontaneous transition from Z_m to Z_n is connected with the emission of one energy quantum $w_m - w_n = h\nu_{mn}$.

Such a transfer happens randomly without any outside causes. The amount of transfers per unit time is given by

$$A_m^n N_m \quad , \tag{4.30}$$

where the constant A_m^n is associated with the states Z_m and Z_n, and N_m is the number of atoms in state Z_m.

Induced transitions from Z_n to Z_m are each accompanied by absorption of the energy $h\nu_{mn} = w_m - w_n$, and the number of absorption processes per unit time is given by

$$B_n^m N_n \rho \quad , \tag{4.31}$$

where the constant B_n^m has the same meaning as B in (4.27). Induced transfers from Z_m to Z_n are connected with the emission of one energy quantum $h\nu_{mn}$. The number of induced emissions per unit time is described, in a similar way, by

$$B_m^n N_m \rho \quad ,$$

where the factor B_m^n is similar to B_n^m. The requirement that statistical equilibrium has to be fulfilled for any pair of states n, m allows us to write

$$A_m^n N_m + B_m^n N_m \rho = B_n^m N_n \rho \quad . \tag{4.32}$$

Formula (4.29) gives

$$\frac{N_n}{N_m} = \frac{p_n}{p_m} \exp[(w_m - w_n)/k_B T] \quad ,$$

and with (4.32) we get the expression

$$A_m^n p_m = \rho \left(B_n^m p_n \exp[(w_m - w_n)/k_B T] - B_m^n p_m \right) \quad , \tag{4.33}$$

where ρ is the energy density of the radiation with the required frequency ν:

$$\nu = \nu_{mn} = \nu_{nm} = (w_m - w_n)/h \quad ,$$

which is associated with the transitions $Z_n \to Z_m$ and $Z_m \to Z_n$. Equation (4.33) gives the dependency $\rho = \rho(T)$ for the given constants $A_m^n p_m$, $B_n^m p_n$, and $B_m^n p_m$ from (4.30, 4.31) and given energy of the atomic states w_m and w_n.

We assume that ρ tends to be infinite for infinite temperature T ($\rho \longrightarrow \infty$ for $T \longrightarrow \infty$). This will be true if

$$B_n^m p_n = B_m^n p_m \quad . \tag{4.34}$$

Then (4.33) takes the form

$$A_m^n = \rho B_m^n \left(\exp[(w_m - w_n)/k_B T] - 1 \right) \quad .$$

Therefore, for the energy density ρ corresponding to the frequency $\nu_{mn} = (w_m - w_n)/h$, we find the expression

$$\rho = \frac{A_m^n / B_m^n}{\exp(h\nu_{mn}/k_B T) - 1} \, , \qquad (4.35)$$

which is equivalent to PLANCK's formula (4.15) with

$$A_m^n / B_m^n = 8\pi h \nu_{mn}^3 / c^3 \quad . \qquad (4.36)$$

6.1   Population

Now we consider another way for the description of spontaneous and induced radiation. Let a cavity be surrounded by mirrors and an atom be placed inside the cavity. This atom is able to emit and to absorb electromagnetic radiation. We also consider only two states Z_1 and Z_2 of the atom. Z_1 is the lower level with energy w_1, and Z_2 is the excited level with energy w_2. Then we fill the cavity with radiation of the frequency $\nu = (w_2 - w_1)/h$. The number of quanta of radiation which fills the cavity depends on the state of the atom. We denote the number of quanta when the atom is in its excited level Z_2 by n. If the atom is in state Z_1, this amount will be enhanced by one due to one emitted quantum, hence the new number of quanta will become $n + 1$. During a long period of time t, the atom has undergone transitions from state Z_1 to Z_2 and back many times. Let N_{1-2} be the number of transitions $Z_1 \rightarrow Z_2$ during a time t_1, which is shorter than the full time of observation t. A single transition $Z_1 \rightarrow Z_2$ can be described by the number of transfers $\Delta p_{1-2}^{(1)}$ per unit time, referring to one quantum. When the atom is in state Z_1, the number of quanta is equal to $n + 1$, and the total amount N_{1-2} will be proportional to $n + 1$ and t:

$$N_{1-2} = \Delta p_{1-2}^{(1)}(n + 1)t \quad . \qquad (4.37)$$

If the period of observation t is sufficiently long, the number of transitions $Z_1 \rightarrow Z_2$ will be equal to the number of back transitions $Z_2 \rightarrow Z_1$ (or it can differ by one, which is not important for a long time of observation). Therefore, we can represent the number N_{2-1} in terms of the constants $\Delta p_{1-2}^{(1)}$ and the number $(n + 1)$, connected to the transitions N_{1-2}, as

$$N_{2-1} = N_{1-2} = \Delta p_{1-2}^{(1)}(n + 1)t = \Delta p_{1-2}^{(1)}nt + \Delta p_{1-2}^{(1)}t \quad . \qquad (4.38)$$

It can be seen that the number of the transitions N_{2-1} accompanied by radiation consists of two terms. The first term, which is due to induced transitions, depends on the number of quanta n, whereas the second one, describing spontaneous transitions, does not. If one denotes the number of spontaneous by $N_{2-1}^{(sp)} = \Delta p_{1-2}^{(1)} t$ and the number of induced transitions by $N_{2-1}^{(in)} = \Delta p_{1-2}^{(1)} nt$, their ratio is equal to

$$N_{2-1}^{(in)} / N_{2-1}^{(sp)} = n \quad . \tag{4.39}$$

Now, with a lot of identical atoms placed into the cavity, the expression (4.38) holds for each atom, provided that the atoms do not interact. This means that, for all atoms, the ratio of the numbers of appropriate transitions should take a form like that in (4.39). Since the ratio (4.39) does not contain a time of observation t, this ratio therefore may be regarded to be the ratio of mean magnitudes, or probabilities $P_{2-1}^{(in)}$ and $P_{2-1}^{(sp)}$, of the transitions:

$$P_{2-1}^{(in)} / P_{2-1}^{(sp)} = n \quad . \tag{4.40}$$

Since the number of quanta n is proportional to the energy density ρ_ν of the radiation, connected to frequency $\nu = (w_2 - w_1)/h$, formula (4.40) may be transformed to a form like (4.36):

$$\frac{P_{2-1}^{(in)}}{P_{2-1}^{(sp)}} = \frac{c^3}{8\pi h\nu^3} \rho_\nu \quad . \tag{4.41}$$

Formula (4.41) is a general expression and will be valid even if the radiation can not be treated as an equilibrium black body radiation.

The analysis of induced and spontaneous radiation also permits establishment of links between the probabilities of absorption and radiation. Let N_1 and N_2 be the numbers of atoms in states Z_1 and Z_2, respectively. These numbers, referred to the total number of atoms N ($N = N_1 + N_2$), are usually called the *populations* N_1/N and N_2/N of the atomic energy levels. The number of transitions N_{1-2} caused by induced absorption during the time of observation t will then be proportional to N_1, and to the amount of quanta n

$$N_{1-2}^{(in)} = N_1 \Delta p_{1-2}^{(1)} nt \quad . \tag{4.42}$$

The transitions from the excited state Z_2 consist of spontaneous transitions (let their amount be $N_{2-1}^{(sp)}$) and of induced transitions (amount

$N_{2-1}^{(in)}$). By analogy we can write the following expressions for $N_{2-1}^{(sp)}$ and $N_{2-1}^{(in)}$:

$$N_{2-1}^{(sp)} = N_2 \Delta p_{2-1}^{(1)sp} t \quad , \tag{4.43}$$

$$N_{2-1}^{(in)} = N_2 \Delta p_{2-1}^{(1)in} nt \quad . \tag{4.44}$$

Under the conditions of the thermal equilibrium the number of absorptions and emissions must equal:

$$N_{1-2} = N_{2-1}^{(sp)} + N_{2-1}^{(in)} \quad ,$$

which gives

$$N_1 \Delta p_{1-2}^{(1)} n = N_2 \Delta p_{2-1}^{(1)in} n + N_2 \Delta p_{2-1}^{(1)sp} \quad . \tag{4.45}$$

With increasing temperature ($T \to \infty$), the difference in populations N_1 and N_2 disappears and the number of quanta increases to a huge magnitude, which results in $N_1 \approx N_2$ and $N_2 \Delta p_{2-1}^{(1)in} n \gg N_2 \Delta p_{2-1}^{(1)sp}$. Thus equation (4.44) gives

$$\Delta p_{1-2}^{(1)} = \Delta p_{2-1}^{(1)in} \quad . \tag{4.46}$$

Since both magnitudes do not depend on temperature, this equality holds in any case. Now, the probabilies of emission calculated per unit quantum, $n = 1$ in (4.44), are equal as follows from (4.43) and (4.44):

$$\Delta p_{2-1}^{(1)in} = \Delta p_{2-1}^{(1)sp} \quad ,$$

what together with (4.46) results to the formula

$$\Delta p_{2-1}^{(1)in} = \Delta p_{2-1}^{(1)sp} = \Delta p_{1-2}^{(1)} \quad . \tag{4.47}$$

This formula shows that probabilities of induced emission $\Delta p_{2-1}^{(1)in}$ and spontaneous emission $\Delta p_{2-1}^{(1)sp}$ are both equal to that of the induced absorption $\Delta p_{1-2}^{(1)}$, provided that the all magnitudes refer to one quantum.

With a lot of quanta, the ratio of the probability for induced radiation $P_{2-1}^{(in)}$ to that of induced absorption $P_{1-2}^{(in)}$ should be equal to the ratio of the numbers of transitions $N_{2-1}^{(in)}$ and $N_{1-2}^{(in)}$, which gives the expression

$$P_{2-1}^{(in)}/P_{1-2}^{(in)} = N_{2-1}^{(in)}/N_{1-2}^{(in)} \quad .$$

Substitution of the magnitudes on the right hand side by the right hand sides of (4.42) and (4.44) gives

$$\frac{P_{2-1}^{(in)}}{P_{1-2}^{(in)}} = \frac{N_2}{N_1} = \frac{N_2/N}{N_1/N} \quad . \tag{4.48}$$

We see the ratio of probability for induced emission to that of induced absorption is equal to the ratio of the population of the excited level to that of the lower level.

Usually, the population decreases with increasing energy of the states, hence $N_2 < N_1$. Therefore, the induced emission is usually weaker than the light absorption. Moreover, for a great amount of thermal sources working in the optical region of radiation, where the ratio N_2/N_1 has the order of magnitude $\exp(-h\nu/k_B T)$, the probability of spontaneous emission is negligible with respect to the probability of absorption, as well as with the probability of induced emission. Nevertheless, if conditions for the inequality $N_2 > N_1$ are fulfilled, then atoms, being in resonance with a required frequency, will amplify the radiation interacting with these atoms. A detailed analysis shows that radiation emitted due to induced transfers has the same polarization state and it is in phase with the radiation which causes these transfers. Thus the radiation initiated by induced transitions is coherent with respect to the radiation which causes induced emission. The operating principle of lasers is induced emission.

SUMMARY

Quantum ideas were mainly developed due to the statistical approach, which was applied to problems of the interaction of light with matter. The explanation of the photoeffect by the fact that an energy portion $h\nu$ is transferred to one electron resulted in a general acceptance of the picture of energy quanta as a real model of light energy. As it results from PLANCK's radiation law (4.15), in the limiting case $h\nu \gg k_B T$ thermal radiation follows a quantum picture to a higher degree rather than a classical picture. A treatment of the quantum features of the interaction of light with matter is necessary for a detailed understanding of the operation principles of most photodetectors.

PROBLEMS

4.1 The universe, having an age $t_1 \approx 10^{10}$ years, is filled with relict black body radiation at $T_1 \approx 3$ K. Beginning with the age, when the temperature of the relict radiation was $T_0 \approx 3000$ K and neutral atoms were formed, the radiation had a weak interaction with the atoms, expanding together. Estimate the age t of the universe at which the neutral atoms were formed. Use the fact that the speed of linear expansion of the universe may be regarded constant.

4.2 . A filament is heated by an electric current I to a temperature of $T = 1500$ K and emits light at $\lambda = 500$ nm, assuming that the filament radiation is black body radiation. The current I is then increased by 1%. Estimate the change in light flux.

4.3 Photons of the Sun spectrum, which have energies $W_0 \leq 1.3 \times 10^{-19}$ J ($\lambda_0 \geq 1.5$ μm), are reflected and absorbed by the layers of an optical heat-reflecting filter. This filter is practically opaque for photons with energies less than W_0. Estimate the fraction of reflected (and absorbed) light. The Sun can be regarded to emit black body radiation at a temperature $T = 5300$ K.

4.4 Now photons of the Sun spectrum, which have energies $W_0 \geq 6 \times 10^{-19}$ J ($\lambda_0 \leq 332$ nm), are absorbed by the layers of an optical filter, blocking the ultraviolet spectrum. This filter is practically transparent for photons with energies less than W_0. Estimate the fraction of the absorbed light.

4.5 An excited atom of excitation energy $W = 1.6 \cdot 10^{-19}$ J is surrounded by equilibrium radiation at temperature $T = 3000$ K. Estimate the ratio of the induced probability to the spontaneous probability of undergoing a transition to its ground state.

4.6 Find the region of frequencies where at $T = 293$ K the probability of spontaneous transitions is more than 100 times larger than the probability of induced transfers.

SOLUTIONS

4.1. One can assume that at the time when neutral atoms were formed the relict radiation was at thermodynamic equilibrium due to its interaction with the atoms, so that the laws of black body radiation held during every small period of the universe expanding. Nevertheless, every dimension in the universe was undergoing an infinitesimal change with time, which is also valid for the wavelengths of the relict radiation. Taking into account the assumptions mentioned above, we can consider that the wavelength $\lambda_{\max}$ corresponding to the maximum of PLANK's curve was also changing with time. As every linear measure $\lambda_{\max}$ was changing with a constant rate:

$$\frac{\Delta \lambda_{\max}}{\Delta t} = const. \quad ,$$

which is equivalent to

$$\frac{\lambda_{\max}(t)}{\lambda_{\max}(t_1)} = \frac{t}{t_1} \quad .$$

Since the laws of black body radiation still were valid during the expansion, one can substitute $\Delta \lambda_{\max}$ by the change in temperature, according to the law $\lambda_{\max} T = const$. Therefore, in the latter formula we can write

$$\frac{T_1}{T_0} = \frac{t}{t_1} \quad .$$

This gives the estimation for t: $t \sim (T_1/T_0)t_1 \sim 10^7$ years.

4.2 Let us find the factor $h\nu/k_B T$ at $\lambda = 500$ nm and $T = 1500$ K: $h\nu/k_B T \approx 20$. Since $h\nu/k_B T \gg 1$, the mean number of quanta emitted by the filament (which is regarded to be a black body) can be evaluated by means of the formula $\bar{n}_\lambda \approx \exp(-h\nu/k_B T)$, whereas the energy flux is proportional to the magnitude $\bar{n}_\lambda h\nu = h\nu \exp(-h\nu/k_B T)$. If T_1 $(T_1 > T)$ is the new temperature caused by the change in the current, the ratio of magnitudes of the energy flux is given by $\Phi_1/\Phi = \exp(-h\nu/k_B T_1)/\exp(-h\nu/k_B T)$. This expression can be transformed into

$$\Phi_1/\Phi = \exp[(h\nu/k_B T)(\Delta T/T)] \quad .$$

After raising the applied current, the electrical power introduced to the filament changes according to the law $I^2 R$, where R is the resistance of the filament (assumed to be constant). Since this power is emitted by the black body, we can write: $I^2 \sim T^4$. Therefore, the link between the change in the current and the temperature takes the form $(T_1/T)^2 = I_1/I$, or $[(T+\Delta T)/T]^2 = (I+\Delta I)/I$. The right hand of the last equality is equal to $1 + 0.01$, which allows us to regard the increment ΔT as small. Hence, the left hand side may be presented by $[(T + \Delta T)/T]^2 \approx 1 + 2\Delta T/T$. Since $1 + 2\Delta T/T = 1 + 0.01$, we get $\Delta T/T = 0.005$. Thus, substitution of the ratio $\Delta T/T$ and of the numerical values of h, ν, k, and T_0 in the expression for Φ_1/Φ gives $\Phi_1/\Phi \approx \exp(20 \times 0.005) \approx 1.1$. Hence, the flux at $\lambda = 500$ nm rises approximately by 10%.

4.3 Let us first find the frequency ν_0 associated with the energy $W_0 = 2 \cdot 10^{-19}$ J and compare that with the frequency of the maximum of the Sun light spectrum ν_1. Using $W_0 = h\nu_0$, we get $\nu_0 \approx 3 \times 10^{14}$ Hz. Then, using $b = \lambda_1 T$ with $b = 0.0029$ m·K we find $\lambda_1 = 550$ nm, and

$\nu_1 = c/\lambda_1 = 5.5 \times 10^{14}$ Hz (at $T = 5300$ K for photons emitted at the maximum of PLANCK's function). Since $\nu_0 < \nu_1$, we believe that the band of transparency of the filter lies in the low frequency region of the Sun spectrum, where the RAYLEIGH-JEANS approximation should be valid. This implies the energy W_a of reflected (and absorbed) light has to be proportional to the integral

$$W_a \sim \frac{8\pi}{c^3}k_BT \int_0^{\nu_0} \nu^2 d\nu = \frac{8\pi\nu_0^3}{3c^3}k_BT \ .$$

According to (4.17) the total energy W falling on the filter, has to be proportional to

$$W \sim \frac{8\pi^5(k_BT)^4}{15c^3h^3} \ .$$

Hence the fraction of energy of the reflected (and absorbed) light is given by

$$\frac{W_a}{W} = \frac{5}{\pi^4}\left(\frac{W_0}{k_BT}\right)^3 \ .$$

Substitution of the numerical values gives the following estimation: $W_a/W \approx 0.30$.

From Fig. 4.11 we can see that the RAYLEIGH-JEANS approximation is overestimating the energy contained in the low-frequency range under consideration. Numerical integration of PLANCK's function gives $W_a/W \approx 0.15$.

4.4 The solution is similar to the previous problem, but here the frequency ν_0 associated with the energy $W_0 = 6 \cdot 10^{-19}$ J is $\nu_0 = 9 \times 10^{14}$ Hz, which is higher than $\nu_1 \approx 5.5 \times 10^{14}$ Hz. Since $\nu_0 > \nu_1$ the WIEN formula may be applied to estimate the fraction of absorbed energy (see Fig.4.11):

$$W_a \sim \frac{8\pi h}{c^3} \int_{\nu_0}^{\infty} \nu^3 \exp(-h\nu/k_BT)d\nu =$$

$$= \frac{8\pi k_BT}{c^3}\{\nu_0^3 + 3\frac{k_BT}{h}\nu_0^2 + 6\frac{(k_BT)^2}{h^2}\nu_0 + 6\frac{(k_BT)^3}{h^3}\} \exp(-h\nu_0/k_BT) \ ,$$

which results in a fraction of absorbed energy

$$\frac{W_a}{W} = \frac{15}{\pi^4}\left(\frac{W_0^3}{(k_BT)^3} + 3\frac{W_0^2}{(k_BT)^2} + 6\frac{W_0}{k_BT} + 6\right)\exp(-W_0/k_BT) \approx 0.035 \ .$$

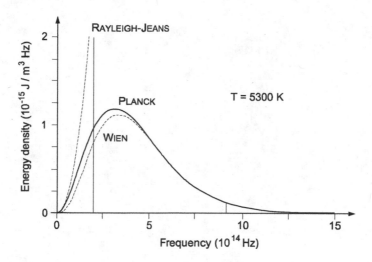

Figure 4.11. To problems 4.3 and 4.5: Comparison between the functions of RAYLEIGH-JEANS, PLANCK and WIEN. The frequencies treated in the problems are indicated by vertical lines.

4.5 According to the formula (4.35) the required ratio is given by the expression

$$\frac{B_m^n \rho}{A_m^n} = \frac{1}{\exp(w/k_B T) - 1} \quad ,$$

where the constants B_m^n, and A_m^n are connected with the transition from the excited level m to the lower level n. The factor $W/k_B T$ is approximately equal to 4, which gives $\exp(4) \approx 54$. We therefore estimate $B_m^n \rho / A_m^n \approx \exp(-4) \approx 2 \times 10^{-2}$.

4.6 We base this solution on the solution of the previous problem and write the ratio of the spontaneous to the induced transitions in the from

$$\frac{A_m^n}{B_m^n \rho} = \exp(W/k_B T) - 1 \geq 100 \quad .$$

Thus the following inequality is valid for the frequencies:

$$\nu \geq \frac{k_B T}{h} \ln(99) \quad .$$

Substitution of numerical values and calculation give $\nu \geq 3 \cdot 10^{13}$ Hz.

Chapter 5

SHOT NOISE

The intensity of light is measured by means of *photodetectors*. The response of a photodetector is a discrete process with random character due to the quantum character of the interaction between light radiation and the material of the photocathode. The term "random character" means that due to the same interaction a photoelectron may or may not be created. We assume that this process can be described only by the probability of photoelectron appearance when atoms of the photocathode are affected by an incident electromagnetic wave. In this chapter we are interested in the statistical properties of the photoeffect but not in statistical properties of the radiation. Thus, we assume monochromatic light radiation, which has no random parameters.

1.   Instantaneous intensity

The emission of photoelectrons or the phenomenon of the photoeffect is the basis of almost all intensity measurements in optics. Apart from technological aspects, measuring the intensity of a light field consists of the absorption of light energy by the electrons of a material. Sensors constructed to detect the light energy by converting it into the energy of moving electrons are called photodetectors. Usually, the transformation of the light energy takes place within a thin layer of a photosensitive material, called *photocathode*. Having absorbed a light quant, the photocathode emits an electron which is called photoelectron, just like in HALLWACHS' experiment considered in the previous chapter.

In accordance with quantum theory, the emission of a photoelectron under the action of the light field may take a few periods T of the oscillations of the electromagnetic field occurring with optical frequency (approximately 5.10^{14} Hz in the visible range), and we denote this time by Δt. The interval Δt is estimated to be 10^{-13} to 10^{-14} s. During this time interval, which we shall call the *elementary interval*, the light propagates a few wavelengths:

$$\Delta t = \frac{\text{a few wavelength } \lambda}{c}.$$

Let a plane monochromatic light wave with wavelength λ be incident normally on the plane surface of the photocathode of a photodetector. After interaction with an atom of the photocathode the electromagnetic field may impart its energy to the photosensitive layer and create a photoelectron. In accordance with quantum theory, such an interaction happens within an elementary volume, containing the atom. This elementary volume is equal, in order of magnitude, to λ^3. We surround the atom with an area of λ^2 normal to the propagation direction (Fig.5.1,a). We assume that the length of the cylinder is equal to a few wavelengths without loss of generality. Then the time needed for light propagation through the elementary volume is equal to a few periods of the monochromatic wave. If a photoelectron appears during light propagation through the elementary volume, we say that the photoelectron appeared within the elementary interval Δt.

Let us compute the energy of the light wave W_a within the elementary volume. For a given energy space density $\varepsilon_0 E^2$ of a monochromatic wave, and, thus an intensity $I = c\rho = c\varepsilon_0 E^2$, the energy in the elementary volume is

$$W_a = c\lambda^2\varepsilon_0 \int\limits_{t_0}^{t_0+\Delta t} E^2 dt = \varepsilon_0 c\lambda^2 E_0^2 \int\limits_{t_0}^{t_0+\Delta t} \cos^2(\omega t + \varphi_0) dt,$$

where ε_0 is the permittivity of vacuum; E_0 is the amplitude and φ_0 the initial phase of the wave; t_0 is the initial time. We may assume $t_0 = 0$ and $\varphi_0 = 0$, thus we rewrite the last expression:

$$W_a = \varepsilon_0 c\lambda^2 E_0^2 \int\limits_0^{\Delta t} \cos^2(\omega t) dt = c\Delta t\lambda^2\varepsilon_0 E_0^2 \left(\frac{1}{2} + \frac{\sin(2\omega\Delta t)}{2\omega\Delta t}\right).$$

Since the elementary interval is equal to a few periods of the wave, that means $\Delta t > 2\pi/\omega$, the function $\sin(2\omega\Delta t)/(2\omega\Delta t)$ should be nearly zero, and the energy under discussion takes the simple form $W_a = c\Delta t\lambda^2\varepsilon_0 E_0^2/2$. We conclude that the process of photoelectron creation reacts only on relatively slow variations of the intensity, thus on an intensity which is called the *instantaneous intensity I* as the energy density flux averaged over the elementary interval: $I = c\varepsilon_0 E_0^2/2$. We make use of the complex function E of electrical field of the wave for representation of the instantaneous intensity, to give:

$$I = c\varepsilon_0 EE^*/2.$$

In contrast to the classical picture of wave optics, the quantum mechanical explanation of the photoeffect showed us that always the energy $h\nu$ of one light quant is delivered to one photon. Thus, the wave must fill many times the elementary volume before one photoelectron arises.

In optics any measurement is based on the interaction between a light wave and matter. With the model of the photoeffect under discussion such an interaction results in the appearance of one photoelectron owing to absorption of one light quantum by an atom. Such a simple model is valid for a wide variety of photocathodes when absorbing thermal radiation, due to the extremely weak intensities of thermal sources. Since we assume that every time one photoelectron is created due to absorption of one light quantum, then no measurable quantity would depend on a time interval, being shorter than the elementary interval. In this connection the elementary interval is assumed to be the shortest time interval in this model of the photoeffect.

2.   Quantum efficiency

We consider a set of neighbouring atoms, each surrounded with the elementary volume, as shown in Fig. 5.1,b. With normal incidence of a plane monochromatic wave on the surface of the photocathode all the atoms are excited synchronously. But the number of photoelectrons n_e, appearing during every shortest interval, is always much smaller than the number of the atoms. Each elementary photoemission, i.e. the absorption of one light quant and the appearance of one photoelectron, is a random event due to the quantum nature of the photoeffect; such an event occurs with a certain probability. Most frequently, the energy of light is absorbed within the photocathode without setting free a photoelectron, and this energy is transformed into heat.

We represent the flux of the averaged energy density in terms of average density of quanta $\overline{n}$ of the monochromatic wave in the form: $I = c\rho = c\overline{n}h\nu$. The flux of photons (photons per second and unit area) is given by $I/(h\nu) = c\overline{n}$. The fact, that during each elementary interval the number of quanta per second and unit area is always larger than the number of photoelectrons per second and unit area, allows us to treat the photoemission in terms of averages. Thus we write

$$\overline{n_e} < c\overline{n} \quad .$$

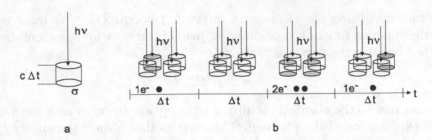

Figure 5.1. One elementary volume, surrounding an absorbing atom (a); a group of
elementary volumes, each considered within the interval Δt (b).

We introduce the *quantum efficiency* q of the photocathode:

$$\overline{n_e} = qc\overline{n} = \frac{\overline{n'_e}}{\Sigma \Delta t} \quad , \tag{5.1}$$

where Σ is the total base of the elementary volumes, Δt is again the
elementary interval, and $\overline{n'_e}$ is the averaged number of created photoelectrons.

We denote by j the average photocurrent density (A/cm^2). It follows
from the expression above that $j = e^- \overline{n'_e}/(s\Delta t)$, where e^- is the electron
charge. We get further

$$j = qe^- \frac{I}{h\nu} \quad . \tag{5.2}$$

The quantum efficiency q in (5.2) is a dimensionless quantity, which is
always less than unity. The formula (5.2) is a central law of photo-effect.
The inequality $q < 1$ testifies to the fact that a certain amount of light
energy, in average, may be converted into heat.

3.   The random experiment

The appearance of a photoelectron during the elementary interval is a
random event. There is need to take advantage of the ideas of the probability theory for treating such events. A set of rather simple regularities,
which result in probabilities for the appearance of photoelectrons, follow
from this treating. These regularities are so unsophisticated that most
of them may be expressed in obvious computer models. Because of this
fact our treatment of problems related to probability theory and mathematical statistics is accompanied by discussion of computer models. The
basis for a statistical model is often the requirement of a repeated use of
a *random experiment*. Usually, by a random experiment an imaginary
test is meant, producing outputs, which can be never predicted unam-
biguously. A peculiarity of statistics is that the random experiment can

be performing repeatedly under invariable conditions. We have a good chance of carrying out of such experiments by means of computers, and almost without exception a so-called *random generator* procedure is the heart of any random computer experiment.

3.1 The random number generator

The random generator is a computer procedure intended for generation of random numbers in series. We consider a linear congruent generator. Its operating principle is based on the use of integer numbers of 32 digit binary cells. Therefore the highest positive number of this kind is equal to $2^{31} - 1 = 2147483647$. In the first procedure step the random generator receives an initial integer number x_{in} ($0 < x_{in} \leq 2147483647$). The integer number resulting from the execution of this procedure is a new random value. Before the next running of the random generator the variable x_{in} is taking on the obtained value. To provide the return value to be within the interval [0,1) its value is divided by 2147483648.

The code of the random generator is presented in Appendix 5.A as **Rnd()** procedure. We should note that all procedures are given as pseudo-codes of C++ programming language.

3.2 The statistical trial

We consider a computer procedure, which we call the *statistical trial*, and which is an example of a random experiment. The code of the statistical trial is presented in Appendix 5A as procedure **Stat-Trial()**. In the first procedure step a variable x takes a value from the random generator; in the second procedure step the value of x is compared with a constant quantity denoted by p_e, with $p_e < 1$. If x is less than p_e, the procedure returns unity, otherwise it returns zero. We regard that the statistical trial may result either in event A_1, which takes the value 1, or in event A_2 with value 0.

Let the statistical trial be carried out M times, and let the total numbers of both events A_1 and A_2 be equal to n_1 and n_2, respectively. It is obvious that the relationship $n_1 + n_2 = M$ holds. It is convenient to define the *relative frequencies* of these events in terms of ratios: $\overline{m_1} = n_1/M$ for A_1 and $\overline{m_2} = n_2/M$ for A_2. Since $n_1 + n_2 = M$ the relative frequencies fulfill the normalization condition

$$\overline{m_1} + \overline{m_2} = 1.$$

The relative frequency is closely related to the probability that an event occurs. With $M \to \infty$ the relative frequency $\overline{m}$ is tending to the probability p :

$$p = \lim_{M \to \infty} \frac{n}{M} = \lim_{M \to \infty} \overline{m}.$$

Let us discuss as an example the modeling of the relative frequencies by means of the statistical trial for the following conditions: $M = 100,000$; $p_e = 0.25$. Using **Stat-Trial()** we find the following values: $\overline{m_1} = 0.25015$; $\overline{m_2} = 0.74985$. It is easy to verify that, in accord with the normalization condition, $\overline{m_1} + \overline{m_2} = 1$.

3.3   The uniform distribution of random value

As is evident from the example, the value of first relative frequency is approximately equal to the value of the parameter p_e, i.e. $\overline{m_1} \approx p_e$. One would expect that for $M \to \infty$ the absolute difference $|p_e - \overline{m_1}|$ is tending to zero. The quantity p_e is, by this hypothesis, the probability of event A_1, since $\overline{m_1}$ is the relative frequency of this event. The proposed hypothesis is not evident but follows from the fact, that the random values of x, formed by the random generator, have a so-called uniform distribution within the unit interval $[0, 1)$.

Let the unit interval be divided into K equal subintervals. We assign an event to each subinterval. For example, the event B_i means that the random magnitude x takes a value within i-th subinterval: $(i-1)/K \leq x < i/K$, where $i = 1, 2, ...K$. Let us assume that a reasonably long set of values of variable x was obtained to calculate relative frequencies of these events: $\overline{m_1}, ..., \overline{m_i}, ..., \overline{m_K}$. Each relative frequency $\overline{m_i}$ should be associated with a probability for obtaining event B_i:

$$P_i = P((i-1)/K \leq x < i/K) \ .$$

These probabilities P_i obey the normalization condition:

$$\sum_{i=1}^{i=K} P_i = 1 \ .$$

The probability density p_i is defined as P_i/Δ, where Δ is the length of the subinterval, in the case under consideration $\Delta = 1/K$. As is evident from the normalization condition for the probabilities, these probability densities obey a similar condition

$$\Delta \sum_{i=1}^{i=K} p_i = \frac{1}{K} \sum_{i=1}^{i=K} p_i = 1 \ .$$

If the random quantity x has a uniform distribution within the unit interval, all the probabilities will be equal to each other. Hence, with

$p_1 = p_2 = \ldots = p_K = p$, where p is the probability density of quantity x, from the last condition follows that $p = 1$. In turn, if $p = 1$, the probability P_i takes the following form: $P_i = p\Delta = \Delta$, i.e. its value is equal to the length of one subinterval. This is also true for a unit interval divided into two subintervals. Hence, the probability that the random value of quantity x occurs within the subinterval $[0, p_e]$ is equal to the length of this subinterval, which gives $P(0 \leq x \leq p_e) = p_e$.

4. Statistics of the number of photoelectrons

4.1 The counting time

As mentioned earlier, estimation of the elementary interval Δt gives $10^{-13} \div 10^{-14}$ s. In contrast, the limit frequency of photodetectors is some 10 GHz, corresponding to a resolution time of 10^{-10} s. Thus, in most of all practical measurements of the intensity of light we are working with intervals much longer than Δt. Because of this, the measurement of a photocurrent is often reduced to count the number of photoelectrons within a certain time interval T, which is called the *counting time*.

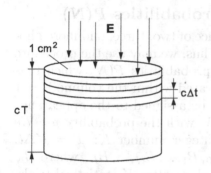

Figure 5.2 A volume of basis area 1 cm^2 in which a series of photoelectrons is generated during the counting time T.

Let us consider a cylindrical volume inside the body of a photocath-ode, assuming that a monochromatic light wave propagates normally to the base (area 1 cm^2) of a cylinder of length cT (Fig. 5.2). We divide this volume into discs with thickness $c\Delta t$, thus $\Delta V = 1$ cm$^2 \cdot c\Delta t$. We estimate the probability that one photoelectron is generated within this volume ΔV. Our prime interest within this book the description of thermal radiation within a limited spectral range. Such radiation stems from sources of limited illuminated areas, as in the case of interference or diffraction experiments. We shell see subsequently, that for such radiation the average amount of quanta, $\overline{n}_{\Delta V}$, within volumes of ΔV is quite small and never exceeds the value 10^{-3}. Even being very small, of course $\overline{n}_{\Delta V}$ is proportional to the light intensity delivered to the photocathode. The average amount of photoelectrons, $\overline{N}_{\Delta V}$ within this volume is given by $\overline{N}_{\Delta V} = q\overline{n}_{\Delta V} \approx q \cdot 10^{-3}$, where the quantum efficiency q is always smaller than 1. The magnitude $\overline{N}_{\Delta V}$ is at the same time the relative frequency for appearance of one photoelectron within the volume ΔV, since the light wave should, in average, pass throughout $1/\overline{N}_{\Delta V}$ volumes ΔV in order to generate one photoelectron. Therefore the probability p_e of such an event is estimated to be $\overline{N}_{\Delta V}$: $p_e \approx q \cdot 10^{-3}$. Since $p_e \ll 1$ we assume for the moment that we can neglect all events where two, three or more photoelectrons are generated within the elementary volume ΔV (there exist phenomena where such events have to be taken into account).

When the monochromatic wave is propagating through the volume V shown in Fig.5.2 ($V = 1 \text{ cm}^2 \cdot cT$), which is much larger, during the time T more than one photoelectron may be generated, and their number N is changing randomly. In other words, this number N is changing from one counting interval T to the next. In order to describe the physical process of the photoeffect, we have to calculate the probabilities $P(N)$, where N is an integer, for the appearance of N photoelectrons during a given counting time T.

4.2   A computer model of probabilities $P(N)$

We neglect such events as the appearance of two, three, and more photoelectrons within the elementary ΔV. Thus, we may use the procedure of the statistical trial for modeling the probabilities $P(N)$ which tells us how much electrons (N) may be generated within the volume $V = 1$ $\text{cm}^2 \cdot cT = \Delta V \cdot T/\Delta t$. Let the event A_1 be assigned to the appearance of a photoelectron within the volume ΔV with the probability p_e. We represent the counting time T by the integer number K: $T = K\Delta t$. Then we introduce a series of events: $B_0, B_1, ..., B_N, ..., B_K$, where B_N gives us the appearing of N events of kind A_1 after K statistical trials. For the calculation of the relative frequencies of these events we have to carry out repeatedly the following algorithm:

- The statistical trial is performed $K + 1$ times,

- The number N of events A_1 is computed

- The counter of the event B_N is increased by 1.

Let M be the total number of repetitions of this algorithm. Then the relative frequencies of events B are calculated as $\overline{m}_N = q_N/M$, where $N = 0, 1, ..., K$, and q_N is the counter of event B_N. Calculations were performed with three values of the probability p_e: 0.0025; 0.025 and 0.125. In all cases we used $M = 1,000,000$ and $K = 20$, giving thus $\overline{N} = 0.05$, 0.5 and 2.5, respectively. The code of this procedure is shown

Table 5.1. The first few relative frequencies $\overline{m}_N$ for three magnitudes of $\overline{N}$

N	$\overline{N} \approx 0.05$ $\overline{m}_N$	$\overline{N} \approx 0.5$ $\overline{m}_N$	$\overline{N} \approx 2.5$ $\overline{m}_N$
0	0.948725	0.587437	0.060806
1	0.050036	0.316860	0.181359
2	0.001214	0.080865	0.259525
3	0.000025	0.013162	0.234663
4	0.000001	0.001528	0.151159
5		0.000137	0.073421
6		0.000010	0.028006
7		0.000002	0.008406
8			0.002127
9			0.000438
10			0.000084
11			0.000007

in Appendix 5.B as **Poisson-Probabilities()**. The obtained values for the relative frequencies are presented in Table 5.1.

The average values of N represented in the first line of the table caption must fulfill also the condition

$$\overline{N} = \sum_{N=1}^{\infty} N\overline{m}_N.$$

which is the case. The obtained values were used for calculations of the dispersion of N by means of

$$\sigma_N = \sum_{N=0}^{\infty} (\overline{N} - N)^2 \overline{m}_N,$$

which gives 0.050079 (for $\overline{N} \approx 0.05$); 0.487444 ($\overline{N} \approx 0.5$) and 2.187703 ($\overline{N} \approx 2.5$). In all cases the dispersion is approximately equal to the appropriate value of $\overline{N}$. From the data presented in the table one can see that, if $\overline{N} < 1$, the relative frequencies are decreasing progressively with increasing N. For $\overline{N} \approx 2.5$ the relative frequencies have a maximum at $N = 2$. In all cases the relationship :

$$\overline{N} \approx K p_e = \frac{p_e}{\Delta t} T, \tag{5.3}$$

between p_e, $\overline{N}$ and K holds since we have assumed $K = T/\Delta t$. In other words, for a given value p_e and Δt the average number of N is directly proportional to the counting time.

4.3   The Poisson random process

The event A_1 associated with the appearance of one photoelectron may be represented as a point on a time axis. In the framework of our model of the photoeffect such an event happens during the shortest time interval Δt, hence, no measurable quantity dependents on a time interval which is shorter than Δt. Thus, each event of sort A_1 can be seen as a point on a time axis, and a series of such points is a point random process. Besides all types of point random processes it is of interest to consider a POISSON random process, as a point random process dealing with the formation of the photocurrent. This process possesses the following properties: First, only the event A_1 may take place within the elementary interval, thus, the probabilities for appearance of 2, 3, ..., points are assumed to be much smaller than the probability of event A_1. Second, any two points within neighboring intervals Δt appear independently of each other. Finally, for a given time interval T the power μ of the POISSON process, defined to be equal to

$$\mu = \overline{N}/T \quad , \tag{5.4}$$

is invariable. Here $\overline{N}$ is the average number of points within the time interval T. It may be regarded that these properties are in complete agreement with the properties of a time sequence of photoelectrons. The definition of the probability p_e for the volume ΔV is completely adequate to the first property of POISSON's process. According to the second property of POISSON's process, a photoelectron should be generated independent from the previous electron. When illuminating the photocathode with a plane monochromatic wave, having invariable instantaneous intensity, there are no reasons to assume that a correlation between the photoelectrons appears. Finally, let, as before, $\overline{N}_{\Delta V}$ be the average number of photoelectrons within the volume ΔV. Then for the volume $V = 1 \text{ cm}^2 \cdot cT$, the total number, denoted by $\overline{N}$, is given as $\overline{N} = \overline{N}_{\Delta V}(T/\Delta t)$, since $T/\Delta t$ is the total number of volumes ΔV inside the volume V. Hence, we can write the power of our POISSON process as

$$\mu = \overline{N}_{\Delta V}/\Delta t \quad , \tag{5.5}$$

and with $p_e = \overline{N}_{\Delta V}$, this magnitude becomes

$$\mu = p_e/\Delta t \quad . \tag{5.6}$$

$\overline{N}_{\Delta V}$ and p_e both are invariable quantities as long as the value of the instantaneous intensity is invariable, hence, the power μ is invariable, too.

Table 5.2. The first few POISSON probabilities for three magnitudes of $\overline{N}$.

N	$\overline{N} = 0.05$ $P(N)$	$\overline{N} = 0.5$ $P(N)$	$\overline{N} = 2.5$ $P(N)$
0	0.951229	0.606531	0.082085
1	0.047561	0.303265	0.205212
2	0.001189	0.075816	0.256516
3	0.000020	0.012636	0.213763
4		0.001580	0.133602
5		0.000158	0.066801
6		0.000001	0.027834
7			0.009941
8			0.003106
9			0.000863
10			0.000216
11			0.000043

In the case of a POISSON random process the probabilities $P(N)$ are subject to a POISSON distribution in a form

$$P(N) = \frac{\overline{N}^N}{N!} \exp(-\overline{N}),$$

where $\overline{N}$ obeys to eq. (5.4). We use now this POISSON distribution for the calculation of the average number and the dispersion of N. By definition, the quantity $\overline{N}$ has to be calculated as

$$\overline{N} = \sum_{N=0}^{\infty} \frac{\overline{N}^N}{N!} \exp(-\overline{N}) N = \overline{N} \exp(-\overline{N}) \sum_{N=0}^{\infty} \frac{\overline{N}^{N-1}}{(N-1)!} \quad .$$

We re-write the sum in the right-hand side in the form of a Taylor series:

$$\sum_{N=0}^{\infty} \frac{\overline{N}^{N-1}}{(N-1)!} = \sum_{K=0}^{\infty} \frac{\overline{N}^K}{K!} = \exp(\overline{N}) \quad ,$$

and get then $\overline{N}$. By definition the dispersion of N is given as

$$\sigma_N = \overline{(\overline{N} - N)^2} = \overline{N^2} - \left(\overline{N}\right)^2 \quad .$$

For the given value $\overline{N}$ the magnitude $\left(\overline{N}\right)^2$ is known. Let us calculate $\overline{N^2}$:

$$\overline{N^2} = \sum_{N=0}^{\infty} \frac{\overline{N}^N}{N!} \exp(-\overline{N}) N^2 = \exp(-\overline{N}) \sum_{N=1}^{\infty} \frac{\overline{N}^N}{N!} N^2 \quad .$$

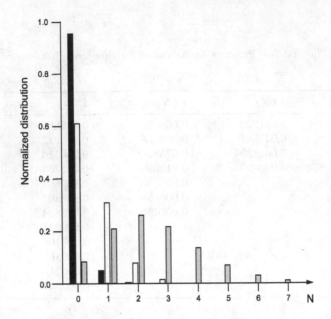

Figure 5.3. Three Poisson distributions for $\overline{N} = 0.05$ (black), for $\overline{N} = 0.5$ (white) and for $\overline{N} = 2.5$ (gray).

We re-arrange the right-hand side as follows:

$$\exp(-\overline{N}) \sum_{N=1}^{\infty} \frac{\overline{N}^N}{N!} N^2 = \exp(-\overline{N}) \sum_{K=0}^{\infty} \frac{\overline{N}^{K+1}}{(K+1)!} (K+1)^2 =$$

$$\overline{N} \exp(-\overline{N}) \left(\sum_{K=0}^{\infty} \frac{\overline{N}^K}{K!} K + \sum_{K=0}^{\infty} \frac{\overline{N}^K}{K!} \right) =$$

$$\overline{N} \exp(-\overline{N}) \left(\overline{N} \exp(\overline{N}) + \exp(\overline{N}) \right) = \left(\overline{N} \right)^2 + \overline{N} \quad .$$

Finally, the substitution of the obtained result into the formula for the dispersion gives

$$\sigma_{\overline{N}} = \overline{N}.$$

It can be seen that the average number of N determines completely the POISSON distribution.

For example, three POISSON distributions calculated for $\overline{N} = 0.05$, 0.5, and 2.5 are presented in Tab. 5.2. These data are in reasonably good agreement with the relative frequencies obtained with the computer simulation considered above (see Table 5.1). Three POISSON distributions are shown in Fig.5.3.

We call attention to a useful recurrence formula for computing a set of the POISSON probabilities:

$$P(N) = \frac{\overline{N}}{N} P(N-1), \tag{5.7}$$

which is valid for $N \geq 1$, and $P(0) = \exp(-\overline{N})$. This formula follows from the POISSON distribution.

We have used the concepts of a random variable and a random process. We remind that the output of the statistical trial is an example of a random variable, which may result in two events: A_1 and A_2. Here A_1 takes the value 1 with the probability p_e, and A_2 the value 0 with probability $1 - p_e$. Thus any random variable is characterized by the total number of its events together with their values and appropriate probabilities.

We turn to the concept of a random process for treating random events, where these events are described by means of functions. In the case of the random process which simulates the generation of photoelectrons, each elementary event may be represented on the time axis by a peaked function as δ-function. Such a function is assigned to the event A_1. With a POISSON random process for an event we should assign elementary peaked functions, distributed over a time interval, to a function of time. Thus, by a random process is meant a set of events together with their functions and probabilities assigned to these events. Such a process is used as a model for a physical phenomenon. In this connection this phenomenon is said to be subjected to a certain statistics. For example, appearance of a certain number of photoelectrons is subjected to POISSON statistics.

5. Detection of low intensities

5.1 Shot noise of a photodetector

The model used to describe the statistics of the number of photoelectrons shows that for a given average number $\overline{N}$ the number of observed photoelectrons varies randomly from one counting interval to the next. Such deviations around the mean value of N are called the noise of the photocurrent. The noise of the photocurrent has different physical reasons. In particular, the thermal motion of the electrons inside the photocathode is responsible for noise. For decreasing the thermal noise, the temperature of the photocathode is usually reduced. However, there exists a natural limit as specific noise is caused by the discrete nature of the photoeffect. The unavoidable noise due to the discrete and chaotic nature of the photoelectric emission is called the *shot noise*. The model of the observable amount of photoelectrons considered above allows us a mathematical representation of the shot noise. The shot noise is therefore subjected to the laws of POISSON statistics.

Besides emission of an electron due to interaction with a photon, sometimes electron emission happens if the thermal energy of an electron becomes large enough to overcome the detachment work of the photosensitive material. This thermal emission of electrons from the photocathode causes a so-called *dark current*, which is also present when no illumination of the photocathode takes place. The statistical properties of the electrons leading to dark current can also be well described in terms of POISSON statistics. However we shall take no special account of this type of noise. In other words, we will consider the shot noise to be the reason for fluctuations of the number of photocurrent pulses. Besides this unavoidable noise, additional noise is caused by thermal motion of electrons inside the elements of the outer electrical circuit of the photodetector unit.

5.2   Photomultiplier

For detecting low light intensities and even single light quanta $h\nu$, photomultipliers are usually used. A photomultiplier consists of a few simple elements: a photocathode for creating photoelectrons from light energy, and internal amplification stages (dynodes) for forming a current pulse even from a single photoelectron. Usually an outer electronic circuit is used to amplify these current pulses. If the incident intensity is large, the amplified current pulses may be averaged to a photocurrent, and the magnitude of this current is a measure for the instantaneous intensity. With low incident intensity, single amplified photocurrent pulses may be counted by an external counter. In this case, we say, the photomultiplier is working in the *photon-counting regime*, and we detect photon counting pulses. Such single amplified photon counting pulses are sometimes called photo-counts or counts for short.

Let us consider the photomultiplier, schematically shown in Fig.5.4, which is a so-called head-on type; various types with different layout and different properties of the photocathode are commercially available. The photocathode layer (sensitive area Σ)covered by a glass or a quartz window is mounted together with the dynodes inside an evacuated glass tube. The photosensitive layer of the cathode usually consists of material with a low value of the detachment work needed to set free an electron, for example it contains different kinds of alkali atoms, called "multialkali" type. The quantum efficiency of the photocathode is again denoted by q. The final electrode is called the anode. A voltage is applied between photocathode and anode, and each dynode is at a certain potential by means of a voltage divider. A light beam, passing through the

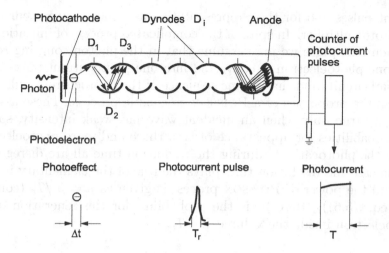

Figure 5.4. A photomultiplier. One photoelectron is initiated within the shortest interval Δt. The dinodes provide a current pulse of duration T_r. The output signal of the photomultiplier may be formed by accumulation of the current pulses over the counting time T.

entrance window, falls on the photocathode, which is at negative voltage with respect to the first dynode. A photoelectron, which is emitted from the photocathode due to the photoeffect, is accelerated by the electric field between photocathode and first dynode and hits the surface of the dynode with an energy in the order of 100 eV. Under the action of the accelerated photoelectron this dynode emits several secondary electrons, which are accelerating while traveling from the first dynode to the second one. Each of these electrons may initiate further secondary electrons on the surface of the second dynode, and so on. Finally, at the anode, a huge number of electrons arrive, which are detected as a *photocurrent pulse*. This pulse may contain up to 10^6 electrons and may be further amplified by the outer electric circuit of the photomultiplier.

5.3   Temporal resolution

As before, the initiation of one photoelectron should need as least the time Δt. But the transfer of electrons through the dynode system also needs also time. So, for the creation of one current pulse, a longer time is necessary, which we define as the *resolution time T_r* of the photomultiplier, since photocurrent pulses can not be distinguished if they arise within a shorter period. Nevertheless, the resolution time should be sufficiently shorter than the counting time T, which specifies the time span for counting single pulses or for an accumulation of the observed current pulses and for their representation as a mean output current of the photomultiplier. In spite of the complicated process of initiation of a photocurrent pulse we may assume that in the photon-counting regime only one photoelectron is converted into one photocurrent pulse in the external circuit, and that a time sequence of the photocurrent pulses will possess the properties of a POISSON stochastic process. These assumptions are true only when an incident wave has weak intensity, so that the probabilities for appearance of two, three and more photoelectrons inside the photocathode during the resolution time all are disregarded. In this case the resolution time plays the role of the elementary interval Δt, and the power of POISSON process is given as $\mu = p_e/T_r$ (compare with eq. (5.6)). Here p_e is the probability for the generation of one photoelectron inside the volume $\Sigma \cdot cT_r$.

5.4   A computer model of photocurrent pulses

Under the conditions considered above the number of counts, each recorded during the counting time T, becomes a measure of the incident instantaneous intensity. Since such a time sequence of counts shows the properties of a POISSON process, we have the possibility to simulate the photon counting regime of the photomultiplier.

For a given power of a POISSON distribution and a counting time T we assume that the average $\overline{N}$ is known. In the computer model the counting time T is divided into 16 intervals, each regarded to be equal to T_r, and the total time of observation is equal to $500T$.

Fig. 5.5 shows three series of obtained data presented as " time-records", each containing 500 points. The amplitudes of the photocurrent pulses within the counting intervals are shown by peaks of appropriate height. All "time-records" are simulated by means of a procedure, which is similar to that used for the demonstration of probabilities $P(N)$. The procedure **Photocurrent()** is presented in Appendix B. The "time-record" (a) was calculated for $\overline{N} = 0.025$, the second (b) for $\overline{N} = 0.25$ (b), and the third(c) for ($\overline{N} = 2.5$.

In Fig.5.5, the first "time-record" ($\overline{N} = 0.025$) is assumed to be found under extremely weak intensity. It consists only of single-electron pulses, their number is calculated to be 21 throughout 500 counting intervals. Thus the relative frequency corresponding to events of this sort is estimated to be $\overline{m}_1 = 0.042$. The relative frequency corresponding to the event that no counts occur within the counting interval T is estimated to be $\overline{m}_0 = 0.958$. The POISSON probabilities calculated are $P(0) = 0.975$ and $P(1) = 0.024$. The probability $P(2)$ to observe two counts ($N = 2$) within T is approximately equal to $3 \cdot 10^{-4}$, this means that two counts

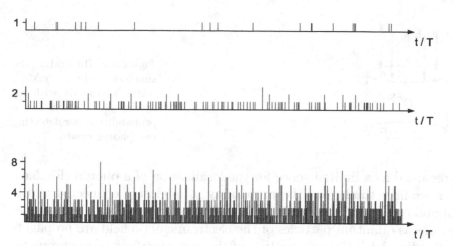

Figure 5.5. Data modeling the photocurrent in the photon counting regime

during interval T may appear only after $3 \cdot 10^3$ counting intervals, such events are absent in "time record" (a).

Within the second "time-record" (Fig.5.5,b) our calculations show that there are 378 counting intervals of the 500, where no counts exist. Therefore the relative frequency $\overline{m}_0$ of such events can be estimated to be about $\overline{m}_0 = 0.756$. Single-electron pulses are found within 110 counting intervals, which gives $\overline{m}_1 = 0.22$. The number of two-electron pulses is 22, thus $\overline{m}_2 = 0.022$. The appropriate POISSON probabilities are $P(0) = 0.779$, $P(1) = 0.195$, and $P(2) = 0.0243$. There is only one three-electron pulse among 500 counting intervals, thus $\overline{m}_3$ may be estimated to be equal to $\overline{m}_3 = 2 \cdot 10^{-3}$. The probability corresponding to such events is $P(3) = 2 \cdot 10^{-3}$. The third 'time-record" (Fig.5.5,c) is obtained at $\overline{N} = 2.5$ and does not contain any empty counting interval. There are a lot of single-electron pulses distributed over most of the counting intervals. We find a large amount of two- and three-electron pulses, and some four-electron pulses. A few eight-electron pulses are also present in this "time-record".

6. Poisson's statistics with the concept of photons

6.1 Probability waves

We have seen that the interaction between an atom and a monochromatic wave takes place within the elementary volume $\Delta V = c\Delta t \lambda^2$. To describe the appearance of a photoelectron within this volume, the probability p_e was introduced. In the context of a quantum treatment, where the monochromatic wave is described as flux of photons, this volume may

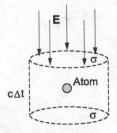

Figure 5.6 Illustrating the smallest volume $c\Delta t\lambda^2$, which surrounds with an atom, and for which the probability p_e for detecting one photon exists.

be regarded as a limited space for the localization of a photon (Fig.5.6). Thus we may think that the probability p_e is closely connected to the probability p_{ph} for the localization of the photon within this volume. Photons as quantum particles of the electromagnetic field are propagating together with the propagation of the corresponding monochromatic wave. Thus, we say that the probability p_{ph} is propagating together with this wave, and that the electromagnetic wave is transporting such a probability from one point to another. Therefore this wave fulfills the role of a *probability wave.*

In wave optics the volume $\Delta V \sim \lambda^3$ is regarded as a space point, whereas the quantity $p_{ph}/\Delta V$ is the probability density for photon localization at this point.

Let a remote point source emit a monochromatic wave of frequency ν, which passes though a volume of cylindrical shape of length l and of base Σ in the propagation direction (Fig.5.7). Because we assume that $l\Sigma \gg \Delta V$ one may find 0, 1, 2, 3 ... photons to be localized within the volume $l\Sigma$. With the given probability density $p_{ph}/\Delta V$ the average number of photons within this volume is equal to

$$\overline{N} = \frac{p_{ph}}{\Delta V}l\Sigma \ .\tag{5.8}$$

We assume that a few atoms are located at some points of the volume $l\Sigma$. Then the appearance of photoelectrons from these atoms are subject to POISSON statistics, where the probability p_e is a decisive factor. It is obvious that all regularities inherent in the photoeffect will take place if

Figure 5.7. A remote monochromatic point source S is emitting photons, which are passing throughout a volume $l\Sigma$ in such a way, that the amount of quanta detected within this volume is subject to POISSON's statistics.

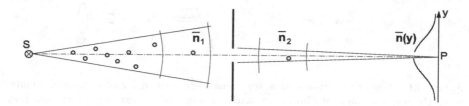

Figure 5.8. Variations of space density of photons $\bar{n}$ in the case of diffraction of monochromatic wave. A source S emits photons according with POISSON's statistics. The distributions of $\bar{n}$ in space are different in front of the aperture $\bar{n}_1$, and behind it $\bar{n}_2$, and in the vicinity of the screen for observation $\bar{n}(y)$.

we regard that emission of the photons from the light source is subject to POISSON statistics. Here the probability p_{ph} is a primary quantity, whereas the probability p_e is equal to $p_e = q p_{ph}$, where q is the quantum efficiency of the substance. Further, at every point of volume $l\Sigma$ the instantaneous intensity determines the probability density $p_{ph}/\Delta V$:

$$\frac{I}{h\nu} = c\frac{p_{ph}}{\Delta V} = c\bar{n},$$

where $\bar{n}$ is the space density of photons $(1/cm^3)$ at this point. The quantity $c\bar{n}$ $(1/(cm^2 s)$ is now the probability density flux. For a given counting time T there exists a volume $1cm^2 \cdot cT$, in which the probability for localization N quanta is equal to the appropriate POISSON probability $P(N)$. Such a treatment of the random nature of the photoeffect is an alternative way for the description of this phenomenon.

In such considerations the electromagnetic vectors $\mathbf{E}$ and $\mathbf{B}$ of the electromagnetic light wave can not be directly assigned to photons. Nevertheless, any propagation properties of the probability waves follow from MAXWELL's equations (J.C.Maxwell, 1831-1879). Concerning the phenomena of diffraction and interference, any space distribution of photons should obey intensity distribution laws established in wave optics. In this sense photons as quantum particles possess properties of electromagnetic waves. Thus, while interacting with atoms and electrons, photons show features peculiar to particles, and they are transporting quanta of energy, as well as quanta of momentum. Additionally, they propagate under the laws of wave optics. Emphasizing such a peculiar behavior of photons we say that light has a *dualistic nature*.

This treatment allows an alternative description of measurements in every respect. At first we discuss the wave properties of photons by the examples of interference and diffraction. Let a monochromatic wave from a remote point source fall on a circular aperture and, after diffrac-

Figure 5.9. Under conditions of a very small energy density of a monochromatic wave the mean amount of photons $\overline{N}$ within a space of laboratory dimensions may occur less than one.

tion, be spread in the space between the aperture and a screen for observation (Fig.5.9). Under conditions of FRAUNHOFER diffraction the probability density $\overline{n}$ of photons in the vicinity of the screen follows the intensity distribution, which corresponds to this type of diffraction. This distribution has a maximum at the center of the screen limited by the minima of first order. The maximal amount of quanta groups at the central part of the screen, whereas at points, corresponding to zeros of this distribution, no photons will appear. This propagation property of photons is independent on the magnitude of the instantaneous intensity. For this reason we may consider that under conditions of a sufficiently weak space density of the electromagnetic field the mean number $\overline{N}$ of photons in (5.8) may be less than 1 even if the volume of observation $l\Sigma$ occupies the whole laboratory space (Fig.5.9).

With the condition $\overline{N} < 1$ the largest POISSON probability is $P(0)$ followed by the probabilities $P(1) > P(2)$, $P(2) > P(3)$ and so on. Under such conditions the probability $P(1)$ is the magnitude which controls practically all propagation features of photons within this volume. In other words, such conditions are dealing with the propagation of single photons.

For example, for $\overline{N} = 0.05$ the first three probabilities are calculated to have the follows values: $P(0) = 0.95$, $P(1) = 0.0476$, and $P(2) \approx 0.0012$. Thus, we may regard only two events to be dominant. The first event is the absence of any photons with the probability $P(0) = 0.95$, and the second event is the presence of one single photon within this volume with probability $P(1) = 0.0476$. Nevertheless, in principal, there is a chance to detect any amount of photons, however POISSON probabilities associated with such events decrease progressively.

6.2   A computer model for space distributions of quanta

The POISSON statistics of photons of the monochromatic field allows the simulation of the statistical properties of the number of photons which may be localized (and detected) within a desired volume.

Let us consider YOUNG's double-slit experiment. Since the intensity distribution of the interference fringes is well known, and the space density of quanta $\overline{n}$ is proportional to the instantaneous intensity, we may write

$$\overline{n} = 2\overline{n}_0 \left(1 + \cos(k\Delta R)\right), \tag{5.9}$$

where $\overline{n}_0$ is the density of quanta of the incident wave, and ΔR is the path difference between two interfering waves. Let each point of the interference pattern be surrounded with a detecting volume of cylindrical shape and size $cT\lambda^2$. For a given magnitude of $\overline{n}$, the number N of photons within each volume can be calculated. The number N varies with the magnitude $\overline{n}$, resulting in a distribution of quanta localized over the interference pattern.

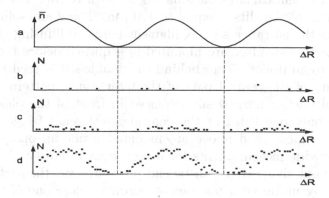

Figure 5.10. Data illustrating the interference effect as distribution of localized photons. The solid line shows the dependence of the probability density $\overline{n}$ on the path difference ΔR, which may be caused due to interference two monocrhromatic waves (a), the points below the solid line shows the dependence of localized quanta on ΔR, obtained for $\overline{N}_0 = 0.01$ (b), $\overline{N}_0 = 0.2$ (c), $\overline{N}_0 = 20$ (d).

The distribution (5.9) is shown in Fig.5.10,a by a solid line. Three distributions of N are also shown, each corresponding to a certain value of $\overline{N}_0 = cT\lambda^2\overline{n}_0$. These distributions were simulated by the algorithm **Photon's-distributions()** in Appendix 5.B. The vertical coordinates of the distributions are counted in N. The first distribution is obtained for $\overline{N}_0 = 0.01$. It can be seen that only a few photons are localized within one full interval of the path difference (Fig.5.10,b). The second distribution is obtained for $\overline{N}_0 = 0.2$. Here the amount of distributed quanta is larger, and they are grouped around the positions of intensity maxima (Fig.5.10,c). The third distribution is obtained for $\overline{N}_0 = 20$. This distribution reproduces roughly the original shape of the dependence $\overline{n}$ on ΔR (Fig.5.10,d). It should be noted that no photons are localized around zeros of the function (5.9) for all three distributions of N. Such simulations illustrate the fact that the interference effect will take place even if the mean number of $\overline{N}$ is much less than one, and when the conditions for the interference of single photons are valid.

7. Interference of single photons

7.1 The Young interferometer under weak intensity

The idea of interference of single photons can be demonstrated using YOUNG's double-slit interference scheme (Fig.5.11). The interferometer consists of an entrance slit of width 50 μm, and the double-slit is composed by two identical slits, each having a width 70 μm. The separation of the centers of the slits is equal to 120 μm. The double-slit is placed 60 cm from the entrance slit. A filament lamp, a diffusing glass plate, and an interference filter are mounted in sequence before the entrance slit. A photomultiplier 15 cm behind the double-slit is used to measure the intensity of the central part of the interference pattern. To select this central part, a narrow slit is placed in front of the photocathode which lets only one third of the central interference fringe pass. All the elements mentioned above are mounted inside an opaque tube in order to provide measurements for a very weak intensity. A thin plate with a straight sharp edge can be slid precisely over the surface of the double-slit by means of a micrometer screw to close one of the slits of the double-slit. This gives us the possibility to measure the intensity arriving at the photocathode if only one slit is open, and to compare it with the intensity under interference conditions, when both slits are open.

The photomultiplier works in the photon-counting regime. Thus, for a very weak intensity of the incident light the number of photocurrent pulses counted per second is a measure of the instantaneous intensity. The random time-sequence of photocurrent pulses represented by the number of pulses per second is averaged by the counter, and the resulting mean value of the count rate is a measure of the intensity. As a real photodetector, the photomultiplier also generates dark current pulses, which are produced even under full darkness of the photocathode. The used photomultiplier has a dark count rate of about $f \approx 100$ photocurrent pulses per second, and its resolution time is estimated to be about $T_r \approx 20$ ns. If one slit is closed the photocathode is illuminated by light from the central part of the diffraction pattern of the open slit. In this case, our measurements result in detection of about 200 pulses per second, but only $f_1 \approx 200 - f = 100$ among them caused by the illumi-

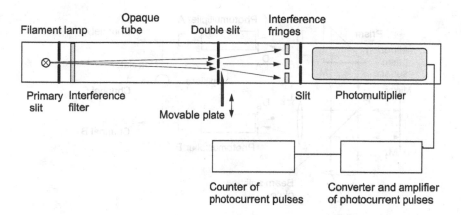

Figure 5.11. Setup for observation of an interference effect of single photons using YOUNG's double-slit interference scheme. The width of the primary slit is 50 μm; the widths of both slits of the double slit are 70 μm; their centers are separated 120 μm.

nation. With both slits open, the photocathode is illuminated with the intensity from the central part of the interference pattern, resulting in a count rate of about $f_2 \approx 480 - 100 = 380$ pulses per second. Thus, the ratio $f_2/f_1 = 3.8 \approx 4$ confirms that indeed an interference effect exists, otherwise a ratio $f_2/f_1 \approx 2$ would have been observed.

Let us estimate the mean number $\overline{N}$ of photons within the work space of the interferometer. The interference filter selects a narrow spectral range around the wavelength $\lambda = 643$ nm. For this wavelength the quantum efficiency of the photocathode is estimated to be $q = 0.015$. Thus, the mean number of photons incident on the photocathode may be estimated to be $380/q \approx 2.5 \cdot 10^4$ photons per second. The central part of the interference pattern consists of three bright fringes, but we shade 2/3 of the intensity. The full interference pattern is thus believed to be formed by $3 \cdot 3 \cdot 2.5 \cdot 10^4 \approx 2 \cdot 10^5$ photons per second. For the given distance $L = 75$ cm between the entrance slit and the photocathode the time interval τ needed for light passing through the space of our setup is about $\tau = L/c = 2.5 \cdot 10^{-9}$ sec. Hence, for the given amount of $2 \cdot 10^5$ photons per second the mean number of photons $\overline{N}$ may be estimated to be $\overline{N} = 2 \cdot 10^5 \cdot \tau = 5 \cdot 10^{-4}$. Therefore we see that the conditions for interference of single photons are fulfilled. Since the flux of quanta forming the interference pattern is about 10^5 quanta per second, the fringes can be well seen directly by eye, which allows the adjustment of the interferometer.

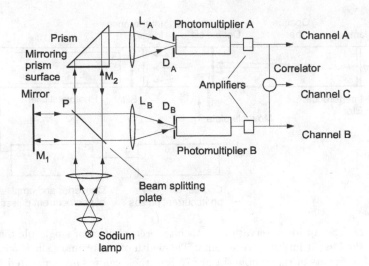

Figure 5.12. Setup for demonstration of the effect of interference of single photons with a MICHELSON interferometer.

7.2   Interference of single photons with a Michelson interferometer

Another interference experiment, where the conditions for the interference of single photons can easily be realized, uses MICHELSON's interference scheme. The outline of the experiment is shown in Fig.5.12. A light beam from a sodium lamp is collimated by a lens to form a nearly parallel beam which passes through the interferometer. It consists of a beam splitting plate P, a semi-transparent mirror M_1, and of the reflecting side M_2 of a prism. The reflectivity of the surfaces of the mirrors M_1 and M_2 is 50%, to give approximately equal intensities of the beams passing through the lenses L_A and L_B. The light path lengths between M_1 and P, and between M_2 and P, are both approximately equal to 5 cm. The mirror M_1 can be slid by a micrometer screw parallel to the incident beam to adjust the interferometer until a bright circle appears in the center of the interference pattern, as observed on the focal plane of lens L_B. The light beam from the central circle of the pattern passes thought a circular diaphragm D_B, and then falls on the photocathode of multiplier B. Another photomultiplier is mounted behind the prism to detect the intensity of the beam after partial transmission through mirror M_2. All the elements are mounted inside an opaque box to provide measurements under the conditions of weak intensities. Both photomultipliers operate in the photon-counting regime, and their photocurrent pulses are counted to form mean values of the counting rates f_A and f_B,

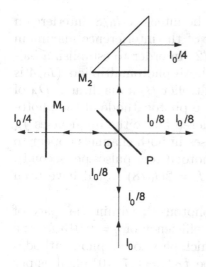

Figure 5.13 Intensity balance in the Michelson interference scheme with the semi-transparent mirrors M_1 and M_2.

which represent measures of the appropriate intensities. The photocurrent pulses of photomultiplier A detecting the intensity after the prism are counted in channel A; the pulses of photomultiplier B, which detects the intensity of the central interference fringe, are counted in channel B. In parallel, the photocurrent pulses are fed to the two entrances of a so-called correlator, which increases the value of channel C by 1 if two incoming photocurrent pulses reach the correlator within the resolution time T_r. The photomultipliers and the correlator have a resolution time of about $T_r = 20$ ns. Thus, apart from the two frequencies f_A and f_B, which represent the intensities A and B, the frequency f_C is a measure for coinciding photocurrent pulses and, therefore, for the multiplication of the photocurrent pulses given by both photomultipliers.

We consider a relationship between the intensities of the beams reaching the two photomultipliers. Let I_0 be the intensity of the original beam falling on the semi-transparent plate P. After the plate one half of the intensity, $I_0/2$, passes to mirror M_2, and the other half, again $I_0/2$, passes to M_1 (Fig.5.13). Since both mirrors M_1, M_2 are semi-transparent, they also divide the intensities equally. Hence, the intensity of the light beam passing through the prism towards multiplier A is equal to $I_0/4$. In a similar way, $I_0/4$ passes through the mirror M_1. The sum of these parts of the intensity is equal to $I_0/2$; thus, one half of the intensity of the incident beam can leave the interferometer.

Further, the intensity $I_0/4$ is reflected by the mirror M_1 back to P; hence, the intensity $I_0/8$ passes through P and the lens, and then falls on a diaphragm D_B in front of photomultiplier B. In the same way, $I_0/4$ is reflected by mirror M_2 towards P which also reflects $I_0/8$ to diaphragm

D_B. Thus, two light beams, each having the intensity $I_0/8$, interfere on the focal plane of lens L_B. The intensity of the interference maximum should then be $I = 2(I_0/8) \cdot (1+1) = I_0/2$. In order to establish a balance between the intensities detected by both photomultipliers ($I_0/4$ is detected by multiplier A, and $I_0/2$ by multiplier B) a diaphragm D_A of the area $S_A \approx 2S_B$ is placed in front of the photocathode of the photomultiplier A. Under the conditions considered above the frequencies f_A and f_B for counting the photocurrent pulses in both channels appear to be nearly equal: $f_A \approx f_B = 10^4$Hz (10^4 photocurrent pulses per second). Without interference, only the intensity $I = 2(I_0/8)$ would have been detected by multiplier B.

Let us estimate the mean number of photons $\overline{N}$ within the space of the interferometer. Assuming a quantum efficiency of $q = 0.015$ for the photocathodes, the amount of photons which pass to the photocathodes of both multipliers may be estimated to be $f_B/q = 6.7 \cdot 10^5$ photons per second. Since the paths between the mirrors M_1, M_2 and the plate P are both approximately 5 cm, the time τ needed for light traveling from P to M_1 and then back to P is about $\tau \approx 3 \cdot 10^{-9}$sec. For this reason the mean amount of photons $\overline{N}$ within the space of the interferometer may be estimated to be about $\overline{N} \approx \tau \cdot 6.7 \cdot 10^5 \approx 2 \cdot 10^{-3}$, thus the conditions for the interference of single photons are fulfilled.

7.3   Photons and classical mechanics

Historically, the interpretation of the phenomena of interference and diffraction with respect to quantum particles was often considered in *gedanken* experiments with single quanta. Such *gedanken* experiments were very useful in the understanding and establishment of principles of the quantum theory. For example, considering a photon as a particle, which possesses the smallest portion of light energy $h\nu$, we would like to understand how this individual particle would pass though two pinholes at the same time to form an interference pattern. If an answer to such a particular questions would exist, then the problems connected with propagation of photons could be solved. In classical mechanics the concept of a trajectory allows the determination of the coordinate and the velocity of a moving particle at any desired moment of time. Such a concept is based on a large amount of empirical observations which allowed the establishment of the basic principles of classical theoretical mechanics. But the statements of classical mechanics regarding the trajectory of photons are in contradiction to the interference effect.

In contrast to classical mechanics, quantum theory is based on measurements of another sort. Let us assume that a plane monochromatic wave with wavelength λ falls normally on an aperture (Fig.5.14). The

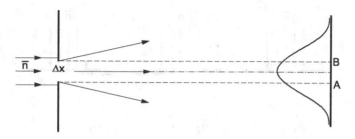

Figure 5.14. Measurements of the coordinate of a photon by means of an aperture of size Δx.

size of the aperture Δx can be changed to localize photons as well as possible. If such a localization is be done, we can regard photons as traveling along a trajectory drawn normally through the center of the aperture. Unfortunately, such localization is possible only for a sufficiently wide aperture, where the approximation of geometrical optics is valid and where the projections of the sides of the aperture may be drawn along light rays to points A and B of a screen for observation. Any classical particles will pass along such straight trajectories when we choose Δx as small as possible. But a decrease of Δx to localize photons of the incident wave more accurately will cause a divergence of the photons away from straight trajectories owing to diffraction. Since the angular dimensions θ of the principal diffraction maximum is limited by $\sin\theta = \pm\lambda/\Delta x$, a tangential component Δk_x of wave vector $\mathbf{k}$ will appear after the light passes through the aperture:

$$\Delta k_x = k\sin\theta = \frac{2\pi}{\Delta x} \quad . \tag{5.10}$$

When treating $\Delta k_x \hbar$ as the tangential component of the photon's momentum Δp_x, according with (5.10), we believe that a relationship exists between $\Delta k_x \hbar$ and Δx in a form:

$$\Delta p_x \Delta x \geq 2\pi\hbar \quad . \tag{5.11}$$

A relationship of this sort was first found by W.HEISENBERG (1901-1975) in 1927. The relationship (5.11) sets limits for a classical treatment of the propagation of quantum particles. It follows from (5.11) that the coordinate x and the momentum p_x can not have completely determined values at the same time: If x is determined, which means $\Delta x = 0$, then $\Delta p_x \to \infty$, and p_x will be undetermined. Nevertheless, we can find the positions of photons on the screen with a known distribution of the probability density $\overline{n}$. Thus, we have to keep the assumption that the

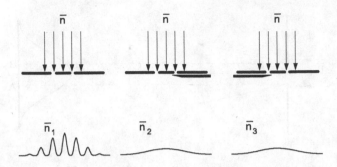

Figure 5.15. Three probabity density distributions for complete description of interference effect in the double-slit experiment.

propagation of photons occurs according to the space distribution of the probability density $\overline{n}$, which, in turn, results from a distribution of the instantaneous intensity.

We now discuss how the interference of photons in the double-slit experiment may be interpreted in terms of probabilities. Let a plane monochromatic wave which has a space probability density $\overline{n}$ pass normally through two identical pinholes. Then an interference pattern can be found on the screen of observation (Fig.5.15). As before, when discussing interference under single-photon conditions, we assume that the mean number of photons in the inner space of our setup is much less than unity. According to quantum mechanics any possible location of a photon on the screen should be predicted only by a probability distribution, which must occur in the space between the aperture and the screen. Fig.5.15 shows the three probability density distributions which can be found: n_1, where two slits are open; n_2, where only the left-hand slit is open, and n_3, where only the right-hand slit is open. These probability density distributions determine completely the propagation of the photons.

For the two last cases we can say definitely that the photon passed through a certain slit. In the first case, where n_1 takes the shape of interference fringes and where the interference effect takes place, we cannot determine through which slit the photon has passed. In the framework of quantum mechanics, any determination of the photon trajectory destroys the interference pattern. Thus the question-through which of the slits the photon is passing is incorrect. The distribution n_1 therefore gives the full description of photon's propagation in the case under consideration. We discuss the interference of single photons obtained with the MICHELSON interferometer considered above now in terms of probability distributions. Let us assume that a monochromatic wave with a probability density $\bar{n}_0$ falls on the semi-transparent plate P. Because the mirrors M_1 and M_2 are both semi-transparent (see Figs.5.12, 5.13), a balance between the probability densities exists inside and outside the interferometer (Fig.5.16). The probability densities of two of the waves, one $\bar{n}_1$ leaving the interferometer through M_1, and the other $\bar{n}_2$ leaving the interferometer through M_2, are both equal to $\bar{n}_0/4$. There are two further distributions of the probability density, associated with two interference patterns: one $\bar{n}_3$ is formed in the vicinity of the photomultiplier B, and the other $\bar{n}_4$ forms the interference pattern in the direction opposite to the incident wave. These distributions are mutually complementary to each other. It follows from the balance between the intensities considered above that these probability densities have to satisfy the following relationships:

$$\bar{n}_1 = \bar{n}_2 = \bar{n}_0/4 \quad ,$$

$$\bar{n}_3 = \frac{\bar{n}_0}{4}\left[1 + \cos(k\Delta R)\right] \quad ,$$

$$\bar{n}_4 = \frac{\bar{n}_0}{4}\left[1 - \cos(k\Delta R)\right] \quad , \tag{5.12}$$

where $k = 2\pi/\lambda$, and ΔR is the path difference between the interfering rays. In the framework of quantum mechanics eq. (5.12) provides completely the mean magnitudes for treating any problems with respect to propagation of photons through the space of the interferometer as well out from it. From the point of classical mechanics, each photon, even as an individual particle, is divided into two parts by the plate P in order to allow interference effects. In contrast, any questions of this sort are hidden in quantum mechanics, because for the case under discussion the space densities in (5.12) provide the full description of the problem. We note that the space probability distributions found with a monochromatic wave have to result in quantities given by POISSON statistics. In the experiment with the MICHELSON interferometer considered above, among other things, POISSON statistics must be evident as the statistical independence between the photocurrent pulses arising in both photomultipliers A and B. This statistical independence is proven by means of the signal of the correlator. Figure 5.17 illustrates two series of photocurrent pulses: one recorded in channel A and the other in channel B. If any pair of photocurrent pulses occurs within the time interval T_r, also a count of the correlator (channel C) appears . In other cases, the signal of the correlator is zero.

Let us assume f_A and f_B are the photocurrent pulse frequencies obtained for channels A and B. The total number of photocurrent pulses from channel A which would produce pulses of the correlator together

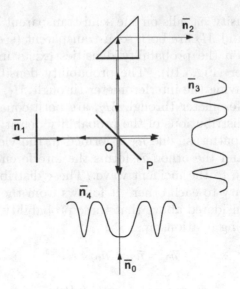

Figure 5.16. The four probability density distributions accompanied the MICHELSON interferometer with semi-transparent mirrors.

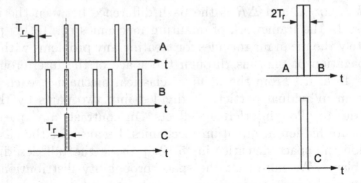

Figure 5.17. Illustrating the operating principle of correlator.

with any photocurrent pulse from channel B during one second, can be estimated to be $2f_A T_r$, where the factor 2 takes into account a partial overlap of the pulses from channels A and B. Since the number of pulses per second from channel B is just equal to f_B , the mean amount of pulses per second in the correlator is given as

$$f_C = 2T_r f_A f_B. \qquad (5.13)$$

For the given frequencies $f_A \approx f_B = 10^4$ Hz, and for $T_r = 2 \cdot 10^{-8}$ sec the value of f_C may be estimated to be $f_C = 2$ Hz . That is the magnitude measured in the experiment. This result shows us that the pulses counted in channels A and B are statistically independent.

SUMMARY

Light intensity measurements are based on the photoeffect. The elementary act of creating a photoelectron as a statistical event may be characterized by a probability p. This event needs a certain time span Δt. It was assumed that the quantity Δt exceeds several periods of oscillations of the monochromatic field of a light wave. Therefore a smoothing of the rapid oscillations of the field with the optical frequency takes place during creation of a photoelectron. The instantaneous intensity of the monochromatic light is a constant value. Therefore in order to describe the photodetection as a statistical process, only two values, p and Δt, are essential. This process is subjected to POISSON statistics. Principle parameters characterizing a photodetector, such as a quantum efficiency, temporal resolution, and counting time for the registration of weak light fluxes can be expressed in terms of the parameters p and Δt. The noise of a photodetector or the shot noise is also characterized by POISSON statistics.

On the methodological side it is reasonable to discuss the wave–particle light dualism based on the elementary concept of photons considered as quantum particles. The statistical model of photon registration coincides in fact with the model of photoelectron registration. The statistical models described above adequately simulate the results of real interference experiments for which conditions of single photon interference are realized. These conditions are analyzed for two examples which consider YOUNG and MICHELSON interferometers. The discussion of the results of these experiments is extremely important from the point of view of the methodology of statistical optics and the definition of its basic concepts.

PROBLEMS

5.1 A parallel monochromatic light beam propagates through a point B. The space density of quanta of the beam is $\overline{n} = 10^{-4}$ cm^{-3}. A beam splitter is placed into the beam at the angle 45° with respect to the propagation direction (Fig.5.18). What is the probability to detect a photon around point B inside the area $\Sigma = 0.1$ cm^2 if the counting time is $T = 10$ ns ? Does a correlation exist between any two photocurrent pulses observed at points A and B ?

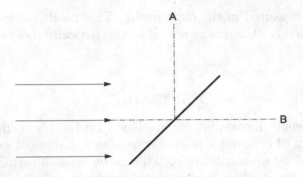

Figure 5.18.

5.2. The magnitude of the relative dispersion of a POISSON stochastic process, $\sqrt{\sigma_N}/N$, is a measure of shot noise when detecting a monochromatic wave. Estimate the relative dispersion for the previous problem. Consider that a photodetector is placed at point B (see Fig.5.18) for two cases: a) the beam splitter is removed, b) the beam splitter is present.

5.3. A monochromatic wave falls on two plane parallel plates of glass ($n = 1.5$), spaced in a way that a very thin parallel air film is formed, so that an interference pattern is obtained by the transmitted rays. What is the minimal flux of photons of the incident wave in order to distinguish the intensities at a maximum and a minimum of the interference fringes? What is the magnitude of the space density of photons associated with such a flux? The quantum efficiency of the photodetector is assumed to be $q = 0.01$; the sensitive area of the photocathode is $\Sigma = 1 \text{ mm}^2$; the counting time is $T = 500$ μs. The contrast sensitivity of the photodetector can be introduced as the ratio $V = (I_{\max} - I_{\min})/(I_{\max} + I_{\min})$.

5.4. The thermal radiation of the Sun is detected by a photodetector at the wavelength $\lambda = 555$ nm within the relative spectral range $\Delta\lambda/\lambda = 10^{-3}$ with the counting time $T = 1$ ms. Estimate the relative mean square fluctuation of the detected signal. The sensitive area of the photocathode is $\Sigma = 1 \text{ mm}^2$. The Sun is regarded as radiating as a black body at the temperature $T = 5300$ K; the angular diameter of the Sun may be regarded to be $\alpha \approx 10^{-2}$ rad.

SOLUTIONS

5.1. The intensity of the original beam is $I = c\bar{n}$. One half of this quantity is detected around point B. Since emission of quanta in a monochromatic field obeys POISSON statistics, the probability to detect a photon over the area Σ during the time interval T will be equal to

$$P(1) = \frac{c\bar{n}T\Sigma}{2} \exp(-c\bar{n}T\Sigma/2) \quad .$$

Substitution of the numerical data gives $P(1) \approx 1.5 \cdot 10^{-3}$ for this probability. There is no correlation between photocurrent pulses detected at points A and B, because the quanta reaching these points are independently emitted according to POISSON statistics.

5.2 The relative dispersion of a POISSON process $\sqrt{\sigma_N}/N$ may be expressed only by the quantity $\bar{N}$: $\sqrt{\sigma_N}/N = \sqrt{\bar{N}}/\bar{N} = 1/\sqrt{\bar{N}}$. If the mirror is absent the relative dispersion is $1/\sqrt{\bar{N}} = 1/\sqrt{c\bar{n}T\Sigma} \approx 18$. In the case that the mirror is present the magnitude of $\bar{n}$ decreases to half the value; therefore, the relative dispersion will increase by $\sqrt{2}$. Hence the relative dispersion will be equal to $18 \cdot \sqrt{2} \approx 25$.

5.3 We use the result for the intensity of an interference pattern for the transmitted rays, which we have found in Problem 1.8:

$$I \approx I_1 \left(1 + 2\mathcal{R} \cos \Delta\right) \quad ,$$

where $\mathcal{R} = (n-1)^2/(n+1)^2$, and I_1 is the intensity on the inner surface of the lower plate. By definition, the contrast of the pattern in the case under consideration is given as follows:

$$\frac{I_{\max} - I_{\min}}{I_{\max} + I_{\min}} = 2\mathcal{R} = 2\frac{(n-1)^2}{(n+1)^2} \quad .$$

We may assume that the main reason for the contrast sensitivity of the photodetector is the shot noise. The magnitude of the relative dispersion of this noise must be less than the quantity found above to distinguish such a pattern contrast from noise:

$$\frac{1}{\sqrt{\bar{N}}} \le 2\frac{(n-1)^2}{(n+1)^2} \quad .$$

From this inequality we get

$$\bar{N} \ge \frac{(n+1)^4}{4(n-1)^4} \quad .$$

For the space density of photons we can find the relation

$$\overline{n} \geq \frac{(n+1)^4}{4cqT\Sigma(n-1)^4} \; .$$

Finally, the magnitude of the flow of photons may be estimated to be

$$c\overline{n} \geq \frac{(n+1)^4}{4qT\Sigma(n-1)^4} \; .$$

Substitution of the numerical data gives $\overline{n} \geq 10^{-4}$ photons per cm^3 and $c\overline{n} \geq 10^7$ photons/(sm^2).

5.4. The relative mean square fluctuation of the detected signal is directly proportional to the magnitude $1/\sqrt{\overline{N}}$, where $\overline{N}$ is the mean amount of photons within a detecting volume, $cT\Sigma\overline{n}$. Here $\overline{n}$ is the space density of photons in the vicinity of the Earth's surface. For a given spectral density $\rho_1(\nu)$ of the Sun's radiation in the vicinity of the Earth $\overline{n}$ is represented as $\overline{n} = \rho_1(\nu)\Delta\nu/(h\nu)$. The magnitude of the spectral density decreases with the distance R_1 from the Sun, thereby, we may write the relationship:

$$\frac{\rho_1(\nu)}{\rho_0(\nu)} = \frac{R_0^2}{R_1^2} \; ,$$

where R_0 is the radius of the Sun. $\rho_0(\nu)$ is the spectral density of the Sun radiation in the vicinity of its surface. Assuming this radiation is emitted by a black body, we find for $\rho_1(\nu)$ the formula

$$\rho_1(\nu) = \rho_0(\nu)\frac{R_0^2}{R_1^2} = \frac{8\pi h\nu^3}{c^3}\frac{1}{\exp(h\nu/(k_BT))-1}\frac{R_0^2}{R_1^2} \; .$$

For the visible wavelength $\lambda = 555$ nm and $T = 5300$ K, the factor $\exp(h\nu/(k_BT)) \gg 1$. Since $\overline{n} = \rho_1(\nu)\Delta\nu/(h\nu)$, we can write

$$\overline{n} = \frac{R_0^2}{R_1^2}\frac{8\pi\nu^2}{c^3}\frac{\Delta\nu}{\exp(h\nu/(k_BT))} \; .$$

The angular diameter of the Sun $\alpha \approx 2\pi R_0/R_1$ allows the substitution $R_0^2/R^2 = \alpha^2/(4\pi^2)$. Since $\Delta\nu = c\Delta\lambda/\lambda^2$ we find

$$\overline{n} = \frac{\alpha^2}{\pi}\frac{2}{\lambda^3}\cdot\frac{\Delta\lambda}{\lambda}\cdot\frac{1}{\exp(h\nu/(k_BT))} \; .$$

The mean amount of photons within the detecting volume is therefore

$$\overline{N} = \overline{n}cT\Sigma = T\Sigma c\frac{\alpha^2}{\pi}\frac{2}{\lambda^3}\frac{\Delta\lambda}{\lambda}\frac{1}{\exp(h\nu/(k_BT))} \; .$$

Finally, the relative mean square fluctuation of the detected signal may be written as

$$\frac{1}{\sqrt{N}} = \frac{\lambda}{\alpha} \exp(hc/(2\lambda k_B T)) \sqrt{\frac{\lambda}{\Delta\lambda} \frac{\pi\lambda}{2cT\Sigma}} \quad .$$

Substitution of the numerical values gives $1/\sqrt{N} \approx 5 \cdot 10^{-5}$.

APPENDIX 5.A

Rnd():
```
x = xin ·    16807;
  if (x ≤ 2147483647)
xin = x;
  else {
x = x + 2147483648;
xin = x; }
  return x/2.147483648 · 10⁹;
```

Stat-trial(p):
```
  Rnd()
  if(x ≤ p)  return 1;
  else    return 0;
```

Rnd-events():
```
  xin = 1732; p = 0.25; Max = 100000;
  for(M = 0; M < Max; M + +){
  a =   Stat-trial(p)
  if   (a = 1)q1 = q1 + 1;
  else   q0 = q0 + 1;
  }
  m0 = q0/Max;
  m1 = q1/Max;
```

APPENDIX 5.B

Poisson-Probabilities():
```
  xin = 1732; p = 0.0025; Max = 1000000; Nmax = 20;
  for(M = 0; M < Max; M + +){
  N = 0;
  for(k = 0; k < Nmax; k + +){
```

```
        N = N+  Stat-trial(p)}
        q[N] = q[N] + 1;
        }
         for(k = 0; k < N_max; k + +)
        m[N] = q[N]/Max;
```

Photocurrent():
```
x_in = 17532; Max = 500; p = 0.025; N_max = 16;
    for(i = 0; i < Max; i + +){
    N = 0;
     for(k = 0; k < N_max; k + +){
    N = N+  Stat-trial(p);}
    q[i] = N;
    }
```

Photon's-distributions():
```
    x_in = 27351; N̄_0 = 0.01;
     for (i = 0; i < 100; i + +)
    ΔR = i − 50;
    I[i] = N̄_0 * (1 + cos(kΔR))
     for(i = 0; i<100; i + +){
    C = 0;
     for(k = 0; k < N_max; k + +){
    p = βI[i];
    C = C+  Stat-trial(p[i]);
    Count[i] = C;
    }
```

Chapter 6

WHITE GAUSSIAN LIGHT

The concept of measurements of the instantaneous intensity of a monochromatic wave developed in the previous chapter assumes that any real photodetector provides a smoothing of the intensity of the incident wave. Even in the case of an ideally detecting device a shortest interval Δt exists, which is needed to initiate photoelectrons. For this reason any detectable quantity can not depend directly on the carrier frequency of the incident wave but only on lower frequencies. A quasi-monochromatic wave possesses properties of such a smoothed field. Moreover, since any detectable magnitude is proportional to the magnitude of light energy, the statistical properties of a quasi-monochromatic wave must be developed from measurements of the instantaneous intensity or from other quantities treated in terms of the intensity.

1. Fluctuations

A quasi-monochromatic light wave may be created from natural light by means of an optical filter. Radiation of a certain spectral line, or a narrow group of such lines obtained from an optical line spectrum may also be regarded to be quasi-monochromatic. In each case the spectral range $\Delta \nu$ of the radiation is regarded to be much smaller than the carrier frequency ν_0, so that an inequality

$$\Delta\nu/\nu_0 \ll 1$$

can be treated as being a property of the quasi-monochromatic waves. For example, in the ideal case of smoothing of an original incident wave by detecting within the shortest time Δt, the spectral range $\Delta \nu$ may be estimated to be on the order of $(\Delta t)^{-1}$.

Natural light, for example the radiation from any thermal source, may be considered as consisting of a large amount of wave trains emitted by the atoms of the source. Under usual conditions the atoms of the source radiate due to spontaneous transitions, and those constitute the major portion of the radiated energy, whereas processes due to induced radiation are only observed as an insufficient small fraction among the total amount of the radiating processes. For these reasons, the amplitudes, initial phases, and frequencies of particular wave trains are all random quantities. This is still true for any narrow spectral range $\Delta\nu$, which is a measure of monochromaticity. Nevertheless, for a very small value of $\Delta\nu$ the oscillations of the electric field of the quasi-monochromatic wave may be well represented in terms of the slowly varying amplitude $A(t)$ and the phase $\Phi(t)$ as follows:

$$E(t) = A(t)\cos(2\pi\nu_0 t + \Phi(t)) \quad , \tag{6.1}$$

where ν_0 is the carrier frequency. Assuming the quasi-monochromatic wave to be linearly polarized, oscillations of the electric vector of such a wave are presented in Fig.6.1. It is seen that both functions, $A(t)$ and $\Phi(t)$, vary slowly with respect to the oscillations of the field, which occur with the period $1/\nu_0$.

Let us assume that a fast photodetector can reproduce such a slow dependence of the instantaneous intensity associated with the oscillations. We call the photodetector "fast" assuming that its resolution time satisfies the following condition:

$$T_r \leq 1/\Delta\nu \quad .$$

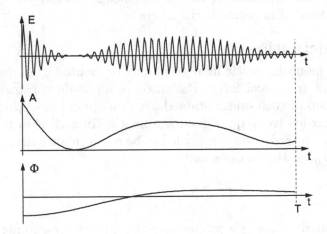

Figure 6.1. Oscillations of the electric vector of a quasi-monoctromatic wave; slow amplitude $A(t)$ and phase $\Phi(t)$ of the oscillations.

Here, we must assume that this quasi-monochromatic oscillations are obtained by means of optical filtration, so that $T_r \gg \Delta t$ is valid. If this is the case, the measured quantity may be proportional to the time dependence of the instantaneous intensity $I(t)$ for a given counting time T (Fig.6.2). For the interval T we may find the mean magnitude of the instantaneous intensity by integration of the measured quantity, for example in a form:

$$\overline{I(t)} = \frac{1}{T} \int_0^T I(t)dt \quad . \tag{6.2}$$

For any other interval T the value of $\overline{I(t)}$ will differ from that obtained during a previous interval due to the chaotic nature of the thermal radiation. Such variations of the measured quantity, the mean intensity in this case, are called *fluctuations of intensity*. The existence of the fluctuations of the instantaneous intensity is a basic property of the radiation of any thermal source. For a sufficiently short interval T the fluctuations of the instantaneous intensity may achieve values of the same order of magnitude as the mean value of this quantity.

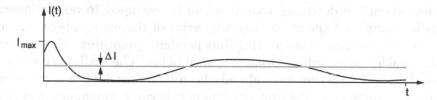

Figure 6.2. A time-realization of the instantaneous intensity for the period of observation t; the gray strip indicates a narrow interval ΔI around a certain magnitude of I.

1.1 The stationarity property

The existence of substantial fluctuations of the instantaneous intensity distinguishes quasi-monochromatic waves from others types of visible radiation such as laser light or the VAVILOV-CHERENKOV radiation. For this reason the statistical properties of any thermal visible radiation must be extracted from studies of the statistics of the fluctuations. It follows from (6.2) that with increasing counting interval we expect to find a constant value of $\overline{I(t)}$, provided that T extends to infinity: $T \to \infty$. Such a consideration is based on the intuitive assumption that the measurements of the instantaneous intensity of the observed radiation are performed under invariable external conditions. Conditions of this sort can be provided, for example, by measurements based on the model of black body radiation, where an energy balance between the radiated energy and the external heating of the light source exists.

In any laboratory experiment it is impossible to safeguard the invariability of measured quantities over a long period of time. Thus, the averaging procedure in eq. (6.2) causes some doubts upon the usefulness of such a description. Nevertheless, a statistical description method allows to overcome such difficulties, and provides adequate computations for the required mean magnitudes. Using this method and its requirements, the model of thermal radiation possesses a *stationarity property*. It turns out that with this model of the stationary radiation the computation of mean quantities is possible for any fixed moment of time. In this way there is no need for temporal averaging, where a time interval is tending to infinity. Moreover, the stationarity property allows the calculation of quantities related to intensity fluctuations, which is more important along with our treatment of thermal radiation.

Additionally the assumption is made that the operation regimes of the light source and the photodetector stay invariant. In the majority of cases, optical measurements are performed under invariable conditions, where the detecting radiation can by assumed to be *stationary*, which means that the statistical properties of the instantaneous intensity should be described by time-independent quantities. Further, we will consider only such stationary optical fields. The stationary properties of optical radiation have already been introduced when considering POISSON statistics in the previous chapter, because any monochromatic field was assumed as being stationary in the physical sense.

1.2   The probability density function

Under validity of the stationary condition, let the instantaneous intensity of a quasi-monochromatic wave be observed for a certain observation time t (Fig.6.2). If t is sufficiently long, the magnitude of the instantaneous intensity will take practically all possible values, which may be emitted from a source under certain conditions of radiation. We should find a maximal value I_{max} among others, because the magnitudes of the instantaneous intensity, obtained by smoothing over the finite time interval Δt can not exceed an extreme value. We also should find zero values, assuming t is long enough. Let the period of observation be represented by the shortest interval Δt as follows: $t = M\Delta t$, where $M \gg 1$. Then we subdivide the interval of possible variations of the intensity, from 0 to I_{max}, into $M + 1$ equal parts $\Delta I = I_{max}/M$, so that a particular value of the instantaneous intensity, specified by I_i, can be found within appropriate limits:

$$i\Delta I \leq I_i < (i+1)\Delta I \quad, \tag{6.3}$$

where $i = 0, ..., M$.

Now let the value I_i be fixed, then by going sequentially from one interval Δt to the next, an amount q_i of intervals, for which the inequality (6.3) holds, can be found under all M intervals. Thus, the relative frequency $\overline{m}_i$ to find the required value I_i, can be calculated:

$$\overline{m}_i = \frac{q_i}{M} .$$

In this way, by varying I_i, the relative frequencies for each value of the instantaneous intensity within the interval 0 to $I_{\max}$ can be found. It is clear that the relative frequencies $\overline{m}_i$ should satisfy the normalization condition

$$\sum_{i=0}^{M} \frac{q_i}{M} = \sum_{i=0}^{M} \overline{m}_i = 1 \quad,$$

since all possible events are taken into account when the index i runs from 0 to M.

We use the obtained relative frequencies for the calculation of probabilities. The probability that a value of I occurs within the interval (6.3) is given by $P(i\Delta I \leq I_i < (i+1)\Delta I)$. Let us assume that, with the requirement $M \gg 1$, P is approximately equal to the appropriate magnitude of the probability:

$$P(i\Delta I \leq I_i < (i+1)\Delta I) \approx \overline{m}_i.$$

Since the magnitude of the instantaneous intensity can not exceed a highest value, whereas the total amount M of intervals Δt is assumed to be as large as possible, the interval ΔI can adopt very small values. If this is the case, we may substitute for the probabilities a continuous function of the variable I, the probability density function $f(I)$:

$$P(i\Delta I \leq I_i < (i+1)\Delta I) = \int_{i\Delta I}^{(i+1)\Delta I} f(I)dI \tag{6.4}$$

For the probability density function the normalization condition is also true:

$$\int_{0}^{\infty} f(I)\,dI = 1. \tag{6.5}$$

In short, this function may also be called the *density of probability*. Although the highest value of the instantaneous intensity is limited by the value I_{max}, the upper limit of infinity is used here as a useful symbol for calculations.

For example, the so-called *statistical average* of the instantaneous intensity, denoted by $\langle I \rangle$, is represented in integral form as

$$\langle I \rangle = \int\limits_0^\infty I\,f(I)\,dI \; . \tag{6.6}$$

For a known density function we may calculate quantities which are characteristic for fluctuations. A measure of the fluctuations of the instantaneous intensity is the mean square deviation specified by $\sqrt{\sigma_I}$, where σ_I is the dispersion of the instantaneous intensity. The probability density function $f(I)$ allows evaluation of the dispersion as a mean square value of the fluctuations:

$$\sigma_I = \left\langle (I - \langle I \rangle)^2 \right\rangle = \int\limits_0^\infty (I - \langle I \rangle)^2 \, f(I)\,dI \quad . \tag{6.7}$$

We call attention to the fact that broken brackets are used for notation of magnitudes calculated by means of the probability density function, in contrast to an overline, which specifies a temporal averaging procedure.

Thus, a known probability density function allows us to evaluate such important statistical quantities as the mean of the intensity $\langle I \rangle$ (6.6), and the dispersion σ_I (6.7).

1.3   The statistical ensemble

The statistical description of an optical field emitted by any thermal source may be performed in terms of the mathematical model of a *random field*. Let us regard the instantaneous intensity $I_1(t)$ (or another optical magnitude), measured during a certain time interval, as a particular realization of such a random field. A priori this realization can be described by a certain probability density $f(I_1; t)$. During another measurement we should find a new value of the instantaneous intensity $I_2(t)$, which belongs to the same random field, and which may be observed with a probability density $f(I_2; t)$. In order to take into account all random values of the intensity, recorded during time intervals of the same duration, we assume that the random field associated with the optical field under consideration, is consisting of a family of functions: $I_1(t)$, $I_2(t)$, $I_3(t)$, ... These functions are called member functions of a *statistical ensemble*, where each function has a priority probability of $f(I; t)$. With such an approach the statistical description of each measurable quantity of the optical field can be represented in terms of a statistical average.

Unfortunately, there is no way to list all the probabilities as well as the functions of the random field associated with thermal radiation. But arguing with the physical nature of thermal radiation we can propose a few foundations for the statistical description of this sort of radiation. Then we shall take advantage of these foundations, without loss of generality, in order to derive all required regularities.

We consider now a way for the construction of the probability density function for the instantaneous intensity. Let

$$I_1(t), I_2(t), I_3(t), ..., I_k(t), ...$$

be the member functions of the statistical ensemble, each associated with a time record of the instantaneous intensity of the optical thermal radiation. We arrange these functions one over the other, having the time axes in common, as shown in Fig.6.3.

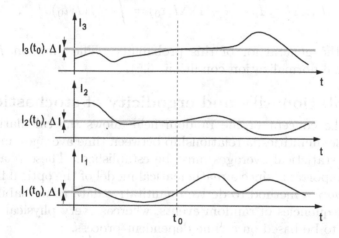

Figure 6.3. Three member functions of the statistical ensemble as the dependence of the instantaneous intensity on time. The notation $I(t_0), \Delta I$ indicates that for each function the interval between $I(t_0)$ and $I(t_0) + \Delta I$ is considered.

Each member function represents the random field as good as another one. Thus, crossing all the member functions at an arbitrary time moment t_0, we will find different values of the intensity, and - since the number of crossed functions is infinity - all possible intensity values. For this reason asuch a crossing of the statistical ensembe allows a complete representation of the probability density function for a given time moment.

We have seen that the upper limit of a value of the instantaneous intensity, I_{max} exists. When assuming a huge amount of member functions, we can divide the intensity range between 0 and I_{max} in intervals ΔI as small as desired. Our task is now to calculate the relative frequency of events that we find the intensity between I and $I + \Delta I$ for a fixed time moment t_0. First, as before, we divide the interval into M equal parts: $M = I_{max}/\Delta I$. Then, the probability to find the magnitude I within the interval $i\Delta I \leq I_i < (i+1)\Delta I$ can be represented in terms of the probability density function in integral form:

$$P(i\Delta I \leq I_i < (i+1)\Delta I; t_0) = \int_{i\Delta I}^{(i+1)\Delta I} f(I; t_0) dI \quad .$$

Here i is ranging from 0 to M; M may tend to infinity. The notation $f(I; t_0)$ emphasizes the crossing of the statistical ensemble at the time moment t_0. It follows from the last expression that we can write

$$P(i\Delta I \leq I_i < (i+1)\Delta I; t_0) = \int_0^{(I_0} f(I; t_0) dI \quad ,$$

which is the general for of this probability. The function $f(I; t_0)$ is fulfilling the normalization condition (6.5).

1.4   Stationarity and ergodicity of stochastic fields

Since the concept of the random field allows the calculation of all measurable quantities, a relationship between time averages and the appropriate statistical averages must be established. These relationships are very important, since a mathematical model of the optical field deals with a priori a method to derive quantitative data as probabilities and relative frequencies of random events, whereas every physical measurement has to be based on a time dependent process.

Often the stationary property of a random field is described using the properties of probability density functions of different orders. If, for an arbitrary member function $I(t)$, the first order probability density is denoted by $f(I; t)$ then the function $f(I; t)$ should not depend on time,

$$f(I; t) = f(I) \quad ,$$

when the random field is assumed to be stationary. For a stationary field, in a wider sense, the joined probability function of second order $f(I_1, I_2; t_1, t_2)$ should depend only on a time difference:

$$f(I_1, I_2; t_1, t_2) = f(I_1, I_2; t_2 - t_1) \quad .$$

We shall restrict our considerations to stationary fields described by probability functions as mentioned above. In particular, we will use that the probability density function $f(I, t_0)$ which describes the statistical ensemble does not depend on time: $f(I, t_0) = f(I)$.

The stationarity property of a random field alone does not guarantee that thermal optical radiation should show the appropriate statistical behavior of the mean magnitudes of the radiation. Since every measurement procedure is based on a time average, this random field must possess a property, which provides that a statistical mean calculated by the probability density functions will be equivalent to the appropriate mean found from a time averaging procedure. In order to satisfy the requirement above, the random field, assumed to be a statistical model of the radiation, must possess a so-called *ergodicity property*, and we say, the optical radiation is ergodic. This property belongs to a more limited class of stationary fields.

A random field we assume to be ergodic if an arbitrary member function takes values along the time axis with the same relative frequency as obtained by crossing all member functions of the statistical ensemble for arbitrary time moments. For example, with respect to the probability density function $f(I)$ the ergodicity property states that we should find the same function when going along any member function. We also should find the same function when crossing the ensemble at an arbitrary time point t_0.

Some member functions, each represented by one time-realization of the instantaneous intensity, are shown in Fig.6.4. For the arbitrary function I_1 the magnitude $\overline{I}$ is formed by averaging. The statistical value $\langle I \rangle$ results from all members of the ensemble.

For a given moment t_0 we assume that all possible values of the instantaneous intensity will be found with the same relative frequencies when going across the statistical ensemble and when progressing along the time axis of an arbitrary time realization. Let the amount of a limited set of member functions be $M^/$, and let us further treat a given value of the instantaneous intensity between I and $I + \Delta I$. By $q^/$ we denote the number of member functions, where, at a fixed time t_0, the intensity was found to be within the interval specified above. Then the relative frequency associated with the intensity being within this interval can be calculated as

$$\overline{m}(I; t_0) = \frac{q^/}{M^/} .$$

On the other hand, when traveling along the time axis for a given interval T, let us select such realizations which have an appropriate relative

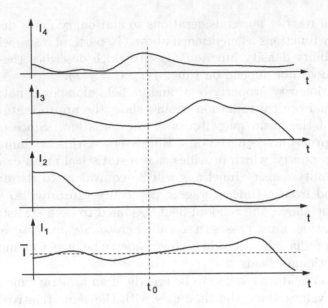

Figure 6.4. Statistical and time-averaging. Intensity realizations I_1, I_2, I_3, I_4 represent themselves as members of a statistical ensemble. For any arbitrary realization I_1 all relative frequencies associated with a probability distribution of the instantaneous intensity can be found as time averages as well as with going across the statistical ensemble at any moment t_0.

frequency $m_T(I)$, which can be represented, as before as

$$\overline{m_t}(I) = \frac{q}{M} \quad .$$

When the amount M' of member functions tends to infinity, the relative frequency $\overline{m}(I; t_0)$ will tend to a constant value, since q' will tend to infinity too; such a constant value becomes the probability for the instantaneous intensity $P(I, I + \Delta I; t_0)$. This probability does not depend on the moment in time: $P(I, I + \Delta I; t_0) = P(I, I + \Delta I)$, and it can be found for any other moment. The same probability can be achieved by increasing the time interval t in $\overline{m_t}(I)$. Finally, all the relative frequencies, as well as the probabilities found under the statistical and temporal procedures, have to be equivalent to each other for every magnitude of the intensity.

From the considerations above we follow that the ergodicity property indeed establishes the required relationship between a random field as a model of thermal radiation and a measurable quantity of the optical field. Nevertheless, ergodicity gives no solution to the problem of how one can find the relative frequencies as well the probabilities under discussion.

As in the case of black body radiation, the most plausible hypotheses related to statistical properties of optical thermal radiation follows from the physical nature of a radiating body, and from the conditions under which the radiation takes place.

In fact, all thermal sources emitting light under common conditions can be treated as being a colossal aggregate of radiating centers (atoms or molecules), each moving randomly under thermal energy and, in most cases, radiating independently from its neighbors. Further we will see that these two hypotheses are just enough to establish all substantial statistical laws of thermal radiation.

1.5   The concept of correlation

Since every macroscopic volume of any thermal source contains a huge amount of radiating atoms, the superimposed optical field emitted from this volume onto a point of observation has to be of chaotic nature. Neither the amplitude, nor the phase, nor the frequency, nor the polarization state of such a superimposed optical field can be assumed to be stable. However, all interference and diffraction phenomena prove the existence of stable interference patterns from thermal sources. The matter is that differential magnitudes related to amplitudes, phases and frequencies are responsible for any stable superposition of optical waves and not the quantities associated directly with amplitudes, or phases, or frequencies of such waves. This fact has to be evident in some quantities of a chaotic field, which contain differential magnitudes similar to the phase difference of the field found at two points, or at two moments of time. These quantities are stable under interference conditions and must have non zero values when calculating their statistical average.

Now we try to construct such a measurable quantity using the MICHEL-SON interference scheme as an example (Fig.1.30). For an arbitrary point of the radiating surface we specify the complex amplitude of a wave, reaching the plane of observation, by $E_1(t)$, where t is a moment associated with radiation of the wave at this point. In the same manner we specify the complex amplitude of the second interfering wave by $E_2(t-\tau)$, where $\tau = \Delta s/c$ (Δs is the optical path difference). We have seen in Chapter 1 that there is a requirement for observing interference, which can be expressed as the fact that the interference term $E_1(t)E_2^*(t - \tau)$ must be non zero. Since any photosensitive element performs averaging of the incoming optical signal, or, in fact, of every measurable quantity, we should consider the time-averaged magnitude

$$\overline{E_1(t)E_2^*(t - \tau)}$$

found for a long period of observation. Under stationary conditions, such a time-averaged value should not depend on time t, hence, this observable magnitudes becomes

$$\overline{E_1 E_2^*(\tau)} \quad . \tag{6.8}$$

Since the complex amplitude is represented by its phase, the parameter τ may be associated with a phase difference, which is invariant under the averaging during the time of observation. It is this invariable phase difference that provides the stability of the interference pattern. When such a case is realized, we say that there exists a *mutual correlation*, or simply correlation, between the fields E_1 and E_2. Existence of the correlation also implies that the random variables E_1 and E_2 are correlated quantities. All the interference and diffraction experiments with thermal sources considered in the previous chapters directly show, that, under appropriate conditions like small angular and linear dimensions of light sources and apertures, and large distances between the elements of an experimental arrangement, a mutual correlation of the fields can be successfully achieved.

A statistical treatment of the correlation between two fields can be illustrated by means of the member functions of a statistical ensemble with amplitudes $A_1(t)$, $A_2(t)$, ... (Fig.6.5). Two "vertical" section of the ensemble, one at time t, and the other at $t+\tau$, are needed to calculate the partial contribution to the statistical average by each member function in a form

$$A(t)A(t + \tau) \quad .$$

According to general rules, the magnitude of the statistical average should be represented in terms of a probability density function $f(A_1, A_2; t, t + \tau)$ of two variables:

$$\langle A(t)A(t+\tau)\rangle = \int\int A_1 A_2 f(A_1, A_2; t, t+\tau)dA_1 dA_2 \quad . \tag{6.9}$$

In general, in order to force that the statistical average $\langle E(t)E^*(t + \tau)\rangle$, which we call the *correlation function of field*, becomes completely equivalent to the time-average $\overline{E_1 E_2^*(\tau)}$, the random field E must have stationary and ergodicity properties.

We have already seen that in most cases of interference the condition of a narrow angular dimension of an interference pattern is fulfilled. Thus, the phase difference depends either on an angular difference, as in the cases of interference in parallel rays, or simply on the distance difference ΔR. Therefore we can believe that the correlation function of the field is, at least within a small space, independent on space coordinates. If

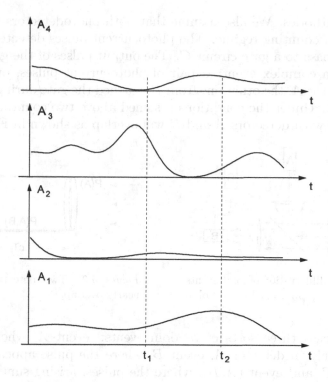

Figure 6.5. A statistical ensemble of field amplitudes $A(t)$. Statistical average of correlation function $\langle A(t_1)A(t_2)\rangle$ is formed with going across the statistical ensemble at two fixed moments t_1 and t_2.

this is the case, the field is said to be a *uniform optical field*. Then the correlation function of the uniform field takes the form

$$\langle E_1(R_1)E_2^*(R_2)\rangle = \int\int E_1(R_1)E_2^*(R_2)f(\Delta R)dR_1 dR_2, \qquad (6.10)$$

where E_1 and E_2 are two complex amplitudes at two points R_1 and R_2, respectively.

Let us discuss a criterion for correlation. It follows from general considerations of probability theory. We illustrate the appropriate probability treatment through the example of the operating principle of a correlator, similar to that considered in the previous chapter. A source of quasi-monochromatic light illuminates a beam splitting plate positioned at 45° with respect to the incident parallel beam (Fig.6.6). One half of the incident beam is reflected towards photodetector A, the other half of the beam is transmitted to detector B. Let us assume that the path difference between the beams is close to zero. Further, let both detectors have the same resolution time T_r and the same sensitive area

of photocathodes. We also assume that both photodetectors operate in
the photon-counting regime. The photocurrent pulses detected by both
detectors pass to a gate circuit C. The output pulses of the gate circuit
represent a complex event: a pair of photocurrent pulses, one coming
from detector A, the other one from B, reached the gate during the same
interval T_r. Under the conditions assumed above two counting volumes
associated with detectors A and B will overlap as shown in Fig.6.7.

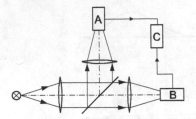

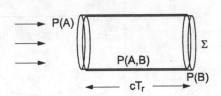

Figure 6.6. Illustration of joint events
with detecting photons by two photodetectors.

Figure 6.7. Two detecting volumes
nearly overlap.

We discuss three sorts of random events: event A, where a pulse
appears only in detector A; event B, where the pulse appears only in
detector B, and event (A, B), where the pulses, arising simultaneously
within the same period T_r, provoke the appearance of one output pulse
in C. After repetition of the measurements for a series containing M
periods T_r, let event (A, B) be found $q(A, B)$ times, event A be found
$q(A)$ times, and event B be found $q(B)$ times. The relative frequencies
of these events can be estimated to be

$$\overline{m}(A, B) = \frac{q(A, B)}{M} \quad , \qquad \overline{m}(A) = \frac{q(A)}{M} \quad , \text{ and } \quad \overline{m}(B) = \frac{q(B)}{M} \quad .$$

We introduce another type of events, so-called conventional events. For
example, the quantity $\overline{m}(A, B)$ may be represented by the quantity $\overline{m}(A)$
as follows:

$$\overline{m}(A, B) = \frac{q(A, B)}{q(A)} \frac{q(A)}{M} = \frac{q(A, B)}{q(A)} \overline{m}(A) \quad ,$$

where $q(A, B)/q(A)$ is the relative frequency of events where event B
occurs after event A has happened. In a similar way, $\overline{m}(A, B)$ may be
represented as

$$\overline{m}(A, B) = \frac{q(A, B)}{q(B)} \frac{q(B)}{M} = \frac{q(A, B)}{q(B)} \overline{m}(B) \quad ,$$

where $q(A, B)/q(B)$ is the relative frequency of events where event A
occurs after event B has happened. Fig.6.8 illustrates the calculations

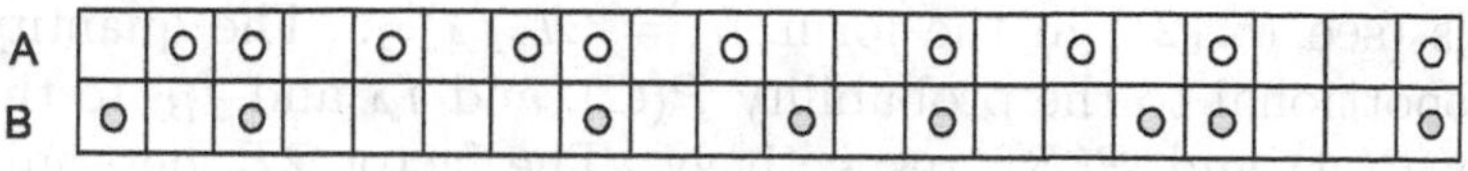

A		O	O		O		O	O		O			O		O		O			O
B	O		O				O			O		O			O	O				O

Figure 6.8. Two series of photocurrent pulses corresponding to detector A and detector B, respectively. It is assumed that $P(A) > P(B)$ is true here.

of the relative frequencies. 10 photocurrent pulses appeared from detector A, and 8 pulses from detector B, thus $\overline{m}(A) = 10/20$, and $\overline{m}(B) = 8/20$. There are 5 joint pulses in our 20 counting intervals, which gives $\overline{m}(A, B) = 5/20$. Further, we find $q(A, B)/q(B) = 5/8$ and $q(A, B)/q(A) = 5/10$.

Statistically, the probabilities for the events under consideration have to satisfy similar relationships

$$P(A, B) = P(B|A)P(A) \quad \text{and} \quad P(A, B) = P(A|B)P(B) \quad , \quad (6.11)$$

where the so-called joint probability $P(A, B)$ associated with $\overline{m}(A, B)$ may be represented either by means of the so-called conditional probability $P(B|A)$ and the marginal probability $P(A)$, or the conditional probability $P(A|B)$ and the second marginal probability $P(B)$. The conditional probabilities $P(B|A)$ and $P(A|B)$ are a measure for the correlation between two events A and B. If event B is statistically independent of any appearance of event A, then such events are called statistically independent, and there exists no correlation between these events. If this is the case, the conditional probability $P(B|A)$ is equal to the marginal one: $P(B|A) = P(B)$. At the same time $P(A|B) = P(A)$ is then valid. As follows from (6.10) the joint probability $P(A, B)$ of two mutually statistically independent events is equal to the product of two marginal probabilities:

$$P(A, B) = P(A)P(B) \quad . \quad\quad\quad (6.12)$$

In contrast, any correlation between two events increases the conditional probabilities to be larger than the appropriate marginal probabilities. Thus the joint probability is larger than the product of two marginal probabilities:

$$P(A|B) > P(A) \quad , \quad\quad P(B|A) > P(B) \quad , \quad\quad P(A, B) > P(A)P(B) \quad .$$

This treatment of the operating principle of a correlator is closely connected to that in the previous chapter when discussing the experiment on interference of single quanta in MICHELSON's scheme. The mean count rate of the correlator is the product of two count rates of both channels (see (5.12)) in the form $f_C = 2T_r f_A f_B$. The quantity f_C is just proportional to the probability $P(C)$, and f_A and f_B to the probabilities $P(A)$ and $P(B)$, respectively. The factor $2T_r$ determines the operating principle of the correlator. Therefore we state that under the conditions of this experiment the photocurrent pulses in both channels are mutually independent, suggesting that both photocurrent pulses are inherent to shot noise.

2.   The quadrature components

We consider a quasi-monochromatic wave with linear polarization propagating through a certain space point R. Then the oscillations of the electric vector at point R of the wave may be represented in terms of a slowly varying amplitude $A(t)$ and a slowly varying phase $\Phi(t)$:

$$E(t) = A(t)\cos(2\pi\nu_0 t - \Phi(t)) \quad , \tag{6.13}$$

where ν_0 is the carrier frequency. We expand the cosine function into two products:

$$\cos(2\pi\nu_0 t + \Phi(t)) = \cos(\Phi(t))\cos(2\pi\nu_0 t) + \sin(\Phi(t))\sin(2\pi\nu_0 t) \quad .$$

With the new functions

$$\zeta(t) = A(t)\cos(\Phi(t)) \quad , \tag{6.14}$$

and

$$\theta(t) = A(t)\sin(\Phi(t)) \quad , \tag{6.15}$$

the original oscillation function will take the form

$$E(t) = \zeta(t)\cos(2\pi\nu_0 t) + \theta(t)\sin(2\pi\nu_0 t) \quad . \tag{6.16}$$

The functions $\zeta(t)$ and $\theta(t)$ are called the *quadrature components* of the oscillation in (6.13) (Fig.6.9). The instantaneous intensity, defined by the amplitude as $I(t) = A(t)^2$, may be represented in terms of the quadrature components as

$$I(t) = A^2(t)\left[\cos^2(\Phi(t)) + \sin^2(\Phi(t))\right] = \zeta^2(t) + \theta^2(t) \quad . \tag{6.17}$$

Since the harmonic functions $\cos(2\pi\nu_0 t)$ and $\sin(2\pi\nu_0 t)$ are both determined functions of time, any stochastic properties of the quasi-monochromatic oscillations are caused by the quadrature components. Under the stationary property, any statistical average must be coincident with appropriate time averaging. For the same reason we may perform statistical averaging at any arbitrary moment, assuming an appropriate

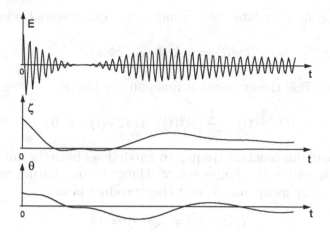

Figure 6.9. Quasi-monochromatic oscillations and the associated quadrature components.

probability density function to be known. For example, let us choose a time t_0 at which the term $\sin(2\pi\nu_0 t_0)$ in (6.16) is zero. Then we calculate the statistical average of the quantity $\zeta^2(t)\cos^2(2\pi\nu_0 t)$. At $t = t_0$, $\cos^2(2\pi\nu_0 t_0) = 1$, which gives $\langle\zeta^2\rangle$. Similarly, we can calculate the average $\langle\theta^2\rangle$ at another moment at which the term $\cos(2\pi\nu_0 t)$ in (6.16) is zero. Since both statistical averages represent the same quantity, and since choosing any moment in time will not influence the average of this value,

$$\langle\zeta^2\rangle = \langle\theta^2\rangle \tag{6.18}$$

must be true. Further, it follows from (6.17) and (6.18)

$$\langle I\rangle = \langle\zeta^2\rangle + \langle\theta^2\rangle \quad \text{and} \quad \langle I\rangle = 2\langle\zeta^2\rangle = 2\langle\zeta^2\rangle \ . \tag{6.19}$$

Using (6.18), we write $\langle\zeta^2\rangle - \langle\theta^2\rangle = 0$, and, using (6.14) and (6.15), we obtain

$$\langle A^2(t)\cos(2\Phi(t))\rangle = 0 \ .$$

Since $A^2(t) > 0$ for all time, the quantities $A^2(t)$ and $\cos(2\Phi(t))$ must be statistically independent:

$$\langle A^2(t)\cos(2\Phi(t))\rangle = \langle A^2\rangle\langle\cos(2\Phi)\rangle = 0 \ . \tag{6.20}$$

Because $\langle A^2\rangle = \langle I\rangle$, equality (6.20) shows that the statistical independence between the instantaneous intensity and the phase $\Phi(t)$ exists. The equality (6.20) should be valid if the phase is distributed uniformly within the limits $-\pi, \pi$.

Similarly, we calculate the product of the quadrature components

$$\zeta(t)\theta(t) = \frac{1}{2}A^2(t)\sin(2\Phi(t)) \quad ,$$

and then we find the correlation function $\langle\zeta(t)\theta(t)\rangle$:

$$\langle\zeta(t)\theta(t)\rangle = \frac{1}{2}\left\langle A^2(t)\right\rangle\langle\sin(2\Phi(t))\rangle = 0 \quad .$$

The fact that this function is equal to zero follows from the uniform phase distribution within the limits $-\pi, \pi$. Thus, the quadrature components are statistically independent, and their product is zero:

$$\langle\zeta(t)\theta(t)\rangle = \langle\zeta\rangle\langle\theta\rangle = 0 \quad .$$

Therefore both quadrature components average to zero.

By definition, the dispersion σ_ζ of the quantity $\zeta(t)$ is

$$\sigma_\zeta = \left\langle\left((\zeta - \langle\zeta\rangle)^2\right)\right\rangle = \left\langle\zeta^2\right\rangle - \langle\zeta\rangle^2 \quad .$$

With $\langle\zeta\rangle = 0$ this gives $\sigma_\zeta = \left\langle\zeta^2\right\rangle$. Similarly, for the dispersion of the quantity $\theta(t)$ we find $\sigma_\theta = \left\langle\theta^2\right\rangle$. Using (6.19) for the dispersions under consideration, the following relationship is true:

$$\sigma_\zeta = \sigma_\theta = \frac{1}{2}\langle I\rangle \quad . \tag{6.21}$$

2.1 Probability distributions of the quadrature components

In order to establish a probability law for the quadrature components we now use the stationary condition. We assume, that, at a certain moment and at a fixed space position, the complex amplitude of a quasi-monochromatic field results from the superposition of a large amount of partial contributions provided by the atoms of the light source.

In fact, such a superposition causes the random quantities $\zeta(t)$ and $\theta(t)$. For example, each partial contribution may be regarded as being a slow function in time: $a(t)\cos(\varphi(t))$, or $a(t)\sin(\varphi(t))$, because we may assume that any atom radiates a quasi-monochromatic wave train. Nevertheless, the most important fact used here is that such contributions must be mutually statistically independent, and each contribution should have a mean value of zero. Then, under the conditions mentioned above, and according to the general theorem of the probability theory, the sum of the contributions is a random quantity distributed according to a GAUSSian *probability law*.

As before we choose a moment in time to force the term $\theta(t)\sin(2\pi\nu_0 t)$ in (6.16) to be zero, and represent then the electric field $E(t)$ by the component $\zeta(t)$. Thus this quadrature component must be distributed according to a probability density function $f(\zeta)$ of GAUSsian shape:

$$f(\zeta) = \frac{1}{\sqrt{2\pi\sigma_\zeta}}\exp\left(-\frac{\zeta^2}{2\sigma_\zeta}\right) \qquad (6.22)$$

Similarly, for the probability density function related to the component θ, we may write

$$f(\theta) = \frac{1}{\sqrt{2\pi\sigma_\theta}}\exp\left(-\frac{\theta^2}{2\sigma_\theta}\right) \qquad (6.23)$$

Because the quadrature components are statistically independent, the joint probability distribution $f(\zeta,\theta)$ must have the form

$$f(\zeta,\theta) = f(\zeta)f(\theta) = \frac{1}{2\pi\sigma_\zeta}\exp\left(-\frac{\zeta^2+\theta^2}{2\sigma_\zeta}\right) , \qquad (6.24)$$

where $\sigma_\zeta = \sigma_\theta$. This is a two dimensional GAUSsian distribution on the (ζ,θ) plane.

Since the statistical averaging of one quadrature component, for example ζ, under the stationary condition will result in a quantity obtained with temporal averaging, the function $f(\zeta)$ may be treated geometrically. Let $\zeta(t)$ be found in a series of time samples, and let each value of $\zeta(t)$ be represented by one point on the ζ axis. After a long period of such trials the points will cover the axis with a density corresponding to the density function $f(\zeta)$, that is, with a GAUSsian density. If the second component θ is now represented by points on the θ-axis orthogonal to the ζ-axis, the joint distribution of the points over the (ζ,θ)-plane will then tend towards the function $f(\zeta,\theta)$ in (6.24) (Fig.6.10).

2.2 The probability distribution of the instantaneous intensity

The joint probability density for the quadrature components allows us to find the probability density of the instantaneous intensity, which may be calculated by means of the distribution (6.24) and the relation between the probability $dP(I)$ and the density function $f(I)$:

$$dP(I) = f(I)dI .$$

With the Cartesian system of the (ζ,θ)-plane considered above, the joint probability $dP(\zeta,\theta)$ is related with an element $d\zeta d\theta$ of the plane via the

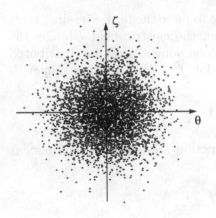

Figure 6.10 The GAUSSian distribution of 10,000 points, each having a random pair of the coordinates θ, ζ.

density function $f(\zeta, \theta)$:

$$dP(\zeta, \theta) = f(I; \zeta, \theta) d\zeta d\theta \quad . \tag{6.25}$$

The probability $dP(\zeta, \theta)$ for the appearance of a random point within the element

$$\zeta, \zeta + d\zeta \quad ; \qquad \theta, \theta + d\theta$$

may be calculated in polar coordinates centered at the origin of the Cartesian system. We introduce the radius $\rho = \sqrt{\zeta^2 + \theta^2}$ and the polar angle ϕ of this polar coordinate. Then, for an element at a distance ρ from the origin we write $\rho d\rho d\phi$, and we represent the probability on the right-hand side of (6.25) by new variables: $f(I; \rho, \phi)\rho d\rho d\phi$. Substitution of ρ for $\zeta^2 + \theta^2$ in $f(I; \zeta, \theta)$ provides a formula for $f(I; \rho, \phi)$:

$$f(I; \rho, \phi) = \frac{1}{2\pi\sigma_\zeta} \exp\left(-\frac{\rho^2}{2\sigma_\zeta}\right) \quad .$$

Thus, the elementary probability $f(I; \rho, \phi)\rho d\rho d\phi$ takes the form

$$\frac{1}{2\pi\sigma_\zeta} \exp\left(-\frac{\rho^2}{2\sigma_\zeta}\right) \rho d\rho d\phi \quad .$$

We see that this expression has no dependence on the polar angle, thus integration over ϕ between 0 and 2π gives the probability depending only on the variable ρ:

$$\frac{1}{\sigma_\zeta} \exp\left(-\frac{\rho^2}{2\sigma_\zeta}\right) \rho d\rho \quad .$$

Since $I = \zeta^2 + \theta^2 = \rho^2$, $\rho d\rho = dI/2$, and $2\sigma_\zeta = \langle I \rangle$, this expression may be represented in terms of the instantaneous intensity and its mean value as

$$\frac{1}{\sigma_\zeta} \exp\left(-\frac{\rho^2}{2\sigma_\zeta}\right) \rho d\rho = \frac{1}{\langle I \rangle} \exp\left(-\frac{I}{\langle I \rangle}\right) dI \quad.$$

The probability function $f(I)$ of the instantaneous intensity, therefore, takes an exponential form:

$$f(I) = \frac{1}{\langle I \rangle} \exp\left(-\frac{I}{\langle I \rangle}\right) \quad. \tag{6.26}$$

Let us verify calculations of the statistical average $\langle I \rangle$ by means of the probability function:

$$\langle I \rangle = \int_0^\infty I f(I) dI = \frac{1}{\langle I \rangle} \int_0^\infty I \exp\left(-\frac{I}{\langle I \rangle}\right) dI \quad.$$

With a new variable $x = I/\langle I \rangle$ the integral on the right-hand side takes the form:

$$\langle I \rangle \int_0^\infty x \exp\left(-x\right) dx = \langle I \rangle \left[1 - \int_0^\infty \exp\left(-x\right) dx\right] = \langle I \rangle \quad.$$

By definition, the dispersion σ_I of the instantaneous intensity is represented by

$$\sigma_I = \int_0^\infty (I - \langle I \rangle)^2 f(I) dI = \int_0^\infty \left(I^2 - 2I\langle I \rangle + \langle I \rangle^2\right) f(I) dI =$$

$$= \int_0^\infty I^2 f(I) dI - \langle I \rangle^2 \quad.$$

We substitute $f(I)$ from (6.26) and use again $x = I/\langle I \rangle$. For the dispersion σ_I, we then write

$$\sigma_I = \int_0^\infty I^2 f(I) dI - \langle I \rangle^2 = \langle I \rangle^2 \int_0^\infty x^2 \exp\left(-x\right) dx - \langle I \rangle^2 \quad.$$

Since

$$\frac{d}{dx}\left(x^2 \exp\left(-x\right)\right) = 2x \exp\left(-x\right) - x^2 \exp\left(-x\right) \quad,$$

we obtain

$$\int_0^\infty x^2 \exp\left(-x\right) dx = 2 \int_0^\infty x \exp\left(-x\right) dx - \int_0^\infty d\left(x^2 \exp\left(-x\right)\right) = 2.$$

It follows from the last result that the dispersion σ_I is given by

$$\sigma_I = \langle I \rangle^2 \quad, \tag{6.27}$$

and that the mean square fluctuation, calculated as $\sqrt{\sigma_I}$, is equal to the mean value of the instantaneous intensity: $\sqrt{\sigma_I} = \langle I \rangle$. In other words, the follows relationship is true:

$$\langle I^2 \rangle = 2 \langle I \rangle^2 \quad . \tag{6.28}$$

The probability distributions for the quadrature components and instantaneous intensity obtained under these considerations are valid in general, since we assumed the common properties of any thermal source. This means that these probability distributions are valid for the case of absolutely uncorrelated visible radiation, as *white light*. We note that white light treated in terms of the integration in (6.4) is considered during a very short counting interval $T \sim \Delta t$. It follows from (6.27) that the mean square fluctuation achieves the value of the mean intensity $\langle I \rangle$. Since the features of this radiation are completely determined by its quadrature components, which both have GAUSSian distributions, this radiation may also be called GAUSS*ian light*.

3.   Computer model for the Gaussian light

3.1   Method of polar coordinates

We have acertained that any measurable quantity related to stationary thermal radiation depends only on differences of time or coordinates. Such quantity is a slowly verying random function of time. In order to simulate such a function by a series of random values we need to calculate two time sequences of its quadrature components.

The random quantities θ and ζ obtained in a computer simulation must be statistically independent; they must have zero mean values and the same dispersions, and they must be described by the same GAUSSian distribution. The conditions listed above are satisfied when using the *polar coordinate* method. Thus, we consider the principle steps of the procedure **Polar-coordinates()**, the code shown in Appendix 6.A.

This routine generates two needed random magnitudes by means of the random generator in order to produce two new statistically independent quantities, denoted as θ and ζ, having zero mean value and a unit dispersion $\sigma = 1$. In order to use this routine for the generation of quantities distributed according to desired values σ_θ and σ_ζ of the dispersions, we have to multiply the results with appropriate factors.

3.2   Simulating distributions of quadrature components and intensity

Let the magnitude Max be the number of pairs θ, ζ which are produced when operating the routine **Polar-coordinates()**. Max constrains the accuracy of any relative frequency by the value $1/Max$. If $Max = 10^5$, then the minimal value of any relative frequency will be $1/Max = 10^{-5}$. Hence, events with probabilities less than 10^{-5} can not be found in this case. As an example we consider parameters of the probability function for the quantity θ. For a desired dispersion $\sigma_\zeta = \sigma_\theta = 1$ and for θ varying within the interval ± 5, and for a minimal step of the variable 0.2, the probability, that the magnitude of θ falls within the broad range between θ and $\theta + 0.2$ is estimated as

$$P(\theta) \approx \frac{0.2}{\sqrt{2\pi}} \exp(-\theta^2/2) \quad .$$

Let us estimate the probability at $\theta = 5$, that is at the end of the interval:

$$P(5) \approx \frac{0.2}{\sqrt{2\pi}} \exp(-25/2) \approx 3 \cdot 10^{-7}.$$

Any non zero relative frequency would occur if Max would have a value $1/P(5) \approx 2.5 \cdot 10^6$ or more. Hence, the choice $Max = 5 \cdot 10^5$ is just enough to observe all random values of θ distributed within the interval ± 5, and the relative frequencies associated with the distribution will be a satisfactory approximation for the probabilities. The estimations above are used in the procedure θ, ζ-**distributions()** presented in Appendix 6.A.

The relative frequencies of the quadrature components and the instantaneous intensity are calculated within the main cycle of the routine. The arrays $Freq_\theta[m]$ and $Freq_\zeta[m]$ are associated with the relative frequencies of the variables θ and ζ, respectively, where the index m specifies a subinterval according to the inequalities

$$-5 + m \cdot 0.2 > \theta \le -5 + (m+1) \cdot 0.2 \quad .$$

After each iteration of the **for** loop the appropriate terms of the arrays $Freq_\theta[m]$ and $Freq_\zeta[m]$ are incremented by the minimal value $1/Max$ as long as these inequalities are satisfied. Inside the **for** loop the array $Freq_I[m]$, associated with the relative frequencies of the instantaneous intensity, is also constructed. Here the inequalities

$$m \cdot 0.1 > \theta^2 + \zeta^2 \le (m+1) \cdot 0.1$$

are exploited to identify the appropriate subinterval of the instantaneous intensity. Data, approximating the distributions of the probability densities, are calculated upon completion of the main cycle; the arrays $f_\theta[m]$, $f_\zeta[m]$, $f_I[m]$ are used here. For example, an approximated value of the probability density of the quantity θ corresponding to the mth subinterval is evaluated to be equal to $Freq_\theta[m]/0.2$. The distributions of the probability densities of the quantities θ and ζ (arrays $f_\theta[m]$, $f_\zeta[m]$) run from -5 to 5 (Fig.6.11), and the distribution of the probability density of the instantaneous intensity (the array $f_I[m]$) from 0 to 5. (Fig.6.12). The distributions of θ and ζ have a GAUSSIAN shape.

We compare the maximal values of the θ- and ζ-distributions with that theoretically calculated from the GAUSSIAN distribution. The maximum of the functions (6.22),(6.23) is $1/(\sqrt{2\pi}) \approx 0.4$ at $\sigma_\theta = \sigma_\zeta = 1$. In the distributions under discussion the values corresponding to the centers of the distributions are both equal to 0.39. The maximal value of the probability density found for the instantaneous intensity is equal to 0.48,

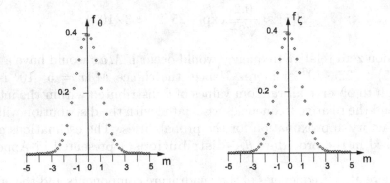

Figure 6.11. Distributions of the probability densities f_θ and f_ζ of the quantities θ and ζ, produced by means of the polar-coordinate method.

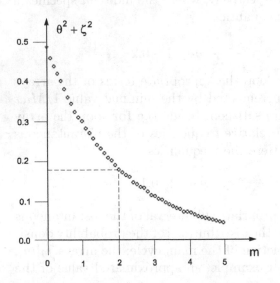

Figure 6.12 The distribution of the probability density of the instantaneous intensity $\theta^2 + \zeta^2$. The probability density obtained from a series of relative frequencies formed under the step of argument 0.1.

and close to the theoretical value 0.5. The value of the distribution corresponding to the subinterval $2 < \theta^2 + \zeta^2 \leq 2.1$ is equal to 0.189, which is in good agreement with the theoretical value $f_I(2) = 0.5 \exp(-1) \approx 0.185$ obtained from (6.27) under the assumption $\langle I \rangle = 2$.

While operating the main cycle, the variable $\overline{\theta\zeta}$ is formed as a sum of products $\theta\zeta$. This variable allows the verification of any statistical correlation between the magnitudes of θ and ζ. Upon completion of the main cycle the arithmetic mean $\overline{\theta\zeta}/Max$, or

$$\overline{\theta\zeta} = \frac{1}{Max} \sum_{k=0}^{Max-1} \theta_k \zeta_k \quad,$$

is calculated. The routine under consideration gives $\overline{\theta\zeta} = 0.01$.

SUMMARY

The fact, that detection of light is inevitable connected with smoothing of the measured quantities over at least some periods of the electromagnetic field, is the basis of our discussion of the emission of thermal light sources using statistical theory. With this respect, the properties of a quasi-monochromatic field possessing a relatively narrow spectral band is of particular interest.

Restriction to uniformity and stationarity of a radiation field is essential and is, indeed, realized for a large number of experimental situations. The conditions of stationarity and uniformity allow the construction of a statistical model of thermal visible radiation in terms of the statistical ensemble and the replacement of temporal averaging with statistical averaging. In other words, stationarity, uniformity, and ergodicity lead to the averaged values of statistical quantities as well as experimental observations, obtained by a time-averaging procedure.

The concept of a statistical ensemble to model a quasi-monochromatic field gives rise to a pictorial and comprehensive description of wave noise and intensity fluctuations. In this case the physical radiation field at a certain point of observation may be represented as the sum of two oscillations: a cosine and a sine wave, which are the quadrature components. In the case of a quasi-monochromatic light field the variations of the quadrature components occur chaotically over some macroscopic time interval. The stochastic variations of the quadrature components can be described statistically by a two-dimensional GAUSSIAN distribution.

APPENDIX 6.A

Polar-coordinates():
$x_{in} = 39152;$
$s = 1.0;$
   while $(s \geq 1.0)$
   {
   Rnd();
   $v_1 = 2x_1 - 1.0;$
   Rnd();
   $v_2 = 2x_2 - 1.0;$
   $s = v_1^2 + v_2^2;$
   }
    $\theta = v_1\sqrt{-2(\ln s)/s};$
    $\zeta = v_2\sqrt{-2(\ln s)/s};$

θ, ζ-**distributions**():
   $x_{in} = 18752;$
   for $(M = 0;\ M < Max;\ M + +)$
   {
   Polar-coordinates();
   $\overline{\theta\zeta} = \overline{\theta\zeta} + \theta * \zeta;$
   for $(m = 0;\ m < 50;\ m + +)$ {
   $u = -5.0 + 0.2 * m;$
   $v = -5.0 + 0.2 * (m + 1);$
   $s = \theta^2 + \zeta^2;$
   $r = 0.1 * m;$
   $z = 0.1 * (m + 1);$
    if $((\theta > u)\,\&\&\,(\theta <= v))$
   $Freq_\theta[m] = Freq_\theta[m] + 1.0/Max;$
    if $((\zeta > u)\,\&\&\,(\zeta <= v))$
   $Freq_\zeta[m] = Freq_\zeta[m] + 1.0/Max;$
    if $((s > r)\,\&\&\,(s <= z))$
   $Freq_I[m] \mathrel{+}= 1.0/Max;$ }
   }
   for $(m = 0;\ m < 50;\ m + +)$ {
   $f_\theta = Freq_\theta[m]/0.2;$
   $f_\zeta = Freq_\zeta[m]/0.2;$
   $f_I = Freq_I[m]/0.1;$ }

Chapter 7

CORRELATION OF LIGHT FIELDS

The statistical description of light fields emitted from thermal sources enables us to completely take into account all effects of source dimensions, the spectral composition of the radiation, and the geometrical characteristics of any interference phenomena. Again we assume that the light source consists of a huge number of elementary (atomic) sources, and in the present chapter we shall see, how effects of these sources can be found from simple considerations regarding their statistical properties. However, the existence of any correlation of the fields in a light beam is of interest in itself, because it is the field correlation that allows a stationary superposition of amplitudes at a point of observation to become observable. In other words, the observation of all diffraction and interference phenomena, produced by a superposition of light waves, is only possible since a field correlation exists.

Traditionally, investigation of the field correlation begins with studying the effects of the size and the spectrum of a light source on the contrast of the observed interference fringes. In order to demonstrate the relationship between field correlation and interference one can assume that two points, somewhere within a light beam, are both secondary sources, which produce interfering waves at a point of observation (Fig.7.1). With a strong correlation of the fields at these points we would obtain interference fringes with high contrast over a wide space of observation. Such a situation can be easily observed with a laser beam, which is characterized by a very high coherence. If certain necessary requirements for a high field correlation are not fulfilled, e.g. for radiation emitted by a thermal source, an interference pattern with a low contrast may be observed.

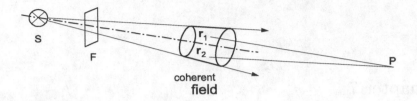

Figure 7.1. Relationship between field correlation and interference. Source S and the optical filter F provide a coherent field existing within a remote volume. Incident rays from points r_1 and r_2 should test the correlation of the field by means of interference around point P.

1.   Visibility and complex degree of coherence

In order to illustrate the relationship between field correlation and interference phenomena we consider two sorts of interference schemes: one introduced by MICHELSON and the other by YOUNG.

In MICHELSON's interference scheme a nearly parallel beam of quasi-monochromatic light is used for illuminating the interferometer. A semi-transparent plane parallel plate acts as a beam splitter and generates two identical light beams. The interference pattern obtained by recombining these parallel rays (two-beam interference) is located at infinity. Here, two secondary sources radiating interfering rays are located somewhere on the beam axis. For example, place the secondary sources at points r_1 and r_2, as shown in Fig.7.2,b. The path difference is mainly determined by the geometrical path difference Δs between r_1 and r_2. We can represent Δs by the time interval $\tau = t_2 - t_1$ the light wave needs to travel from point r_1 to point r_2:

$$\Delta s = c\tau = c(t_2 - t_1) \quad .$$

From these secondary sources a stable interference pattern may appear somewhere in space if the complex amplitudes of the light field at points r_1 and r_2 provides a non zero statistical average in the form of the correlation function (see (6.13)):

$$\langle EE^* \rangle = \int \int E(t_1)E^*(t_2)f(t_2 - t_1)dt_1dt_2 \quad ,$$

where the joint probability function $f(t_2 - t_1)$ depends only on the interval $\tau = t_2 - t_1$. Such a dependence on τ is true under the stationary property of the optical field, as discussed in Chapter 6. The requirements above are also true for all other interference schemes of this sort, like a plane parallel glass plate, a LUMMER-GEHRCKE plate, a FABRY-PEROT interferometer, and others.

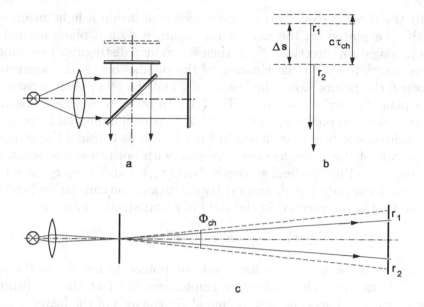

Figure 7.2. In a MICHELSON interferometer with an extended source, two nearly parallel beams are separated by $\Delta s = c\tau$ (a). This is just equivalent to two secondary sources at points $\mathbf{r_1}$ and $\mathbf{r_2}$ located at the axis of the beam (b). Any interference from these points exists as long $\Delta s < c\tau_{ch}$ is valid. In a YOUNG interferometer, two points $\mathbf{r_1}$ and $\mathbf{r_2}$ are located at a plane normal to light propagation; any interference from $\mathbf{r_1}$ and $\mathbf{r_2}$ exists as long as these points are inside a region limited by an angle Φ_{ch} (c).

Now we discuss YOUNG's interference scheme, to which also other interferometers are related, like LLOYD mirrors, FRESNEL mirrors, FRESNEL bi-prisms, and others. In all these schemes the contrast of the fringes depends on the dimensions of the light source, as well on the separation between the secondary sources. Here, the secondary sources are located within a plane normal to the light propagation at points r_1 and r_2 (Fig.7.2,c). The path difference of waves emitted by these secondary sources is determined by their separation and the correlation function will take the form

$$\langle EE^* \rangle = \int \int E(r_1)E^*(r_2)f(r_2 - r_1)dr_1 dr_2 \quad ,$$

where the joint probability function $f(r_2 - r_1)$ depends only on the difference in distance $\Delta r = r_2 - r_1$, if the condition of an uniform optical field is assumed to be true (see (6.22)). An interference pattern with certain contrast will be observed somewhere behind the plane of the sources, if the statistical average $\langle EE^* \rangle$ has a non zero value.

In the general case, two points somewhere inside a light beam can neither be placed on the axis of the beam, nor .on a plane normal to the propagation direction. Nevertheless, we may distinguish two dimensions associated with the location of the points: one is the separation between the points along the beam axis, and the other their separation in a plane normal to this axis (Fig.7.3). In order to describe this case we consider two points $\mathbf{r}_1$ and $\mathbf{r}_2$ within a space illuminated by a quasi-monochromatic beam, as shown in Fig.7.3. Let us consider the complex amplitude of the beam as slowly varying with both time and space coordinates, and let the field strengths be $E(\mathbf{r}_1, t_1)$ and $E(\mathbf{r}_2, t_2)$ at a unit distance from positions $\mathbf{r}_1$ and $\mathbf{r}_2$. By definition, the correlation function of the field is represented in the form of a statistical average as

$$\Gamma_{12}(\tau) = \frac{c\varepsilon_0}{2} \langle E(\mathbf{r}_1, t_1) E^*(\mathbf{r}_2, t_2) \rangle \quad , \qquad (7.1)$$

where the subscript $_{12}$ specifies that our points do not lie on the axis of the beam, and the variable τ emphasizes the fact that a distance $\Delta s = c\tau$ exists between these points along the axis of the beam.

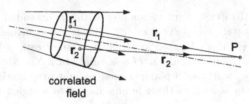

correlated
field

Figure 7.3.   To interference from two points inside a partially coherent field.

For a given wavelength λ, we consider the points $\mathbf{r}_1$ and $\mathbf{r}_2$ to emit spherical waves to form interference at a point of observation P. The amplitudes at point P are then given as

$$E_1 = \frac{E(\mathbf{r}_1, t_1)}{r_1} \exp(ikr_1) \quad \text{and} \quad E_2 = \frac{E(\mathbf{r}_2, t_2)}{r_2} \exp(ikr_2) \quad ,$$

where r_1 is the distance from point $\mathbf{r}_1$ to P, and r_2 from point $\mathbf{r}_2$ to P. Since $\mathbf{r}_1$ and $\mathbf{r}_2$ are located within a space illuminated by a quasi-monochromatic wave, we assume that the polarization states of E_1 and E_2 are the same. The superposition of fields E_1 and E_2 can be written as

$$E_1 + E_2 = \frac{E(\mathbf{r}_1, t_1)}{r_1} \exp(ikr_1) + \frac{E(\mathbf{r}_2, t_2)}{r_2} \exp(ikr_2) \quad .$$

The instantaneous intensity at P, calculated as before, then has the form

$$I_P \sim (E_1 + E_2)(E_1 + E_2)^* = E_1 E_1^* + E_1^* E_2 + E_1 E_2^* + E_2 E_2^* \quad . \qquad (7.2)$$

It is seen that the first and fourth term on the right-hand side of (7.2) become

$$E_1 E_1^* \sim \frac{I_1}{r_1^2} \quad \text{and} \quad E_2 E_2^* \sim \frac{I_2}{r_2^2} \quad , \tag{7.3}$$

where I_1 is the instantaneous intensity at the first radiating point $\mathbf{r}_1$, and I_2 is the instantaneous intensity at the second point $\mathbf{r}_2$. Thus, statistical averaging of the instantaneous intensity at P gives

$$\langle I_P \rangle = \frac{\langle I_1 \rangle}{r_1^2} + \frac{c\varepsilon_0}{2} \langle E_1^* E_2 \rangle + \frac{c\varepsilon_0}{2} \langle E_2^* E_1 \rangle + \frac{\langle I_2 \rangle}{r_2^2} \quad . \tag{7.4}$$

$\langle E_1^* E_2 \rangle$ becomes

$$\langle E_1^* E_2 \rangle = \frac{\langle E^*(\mathbf{r}_1, t_1) E(\mathbf{r}_2, t_2) \rangle}{r_1 r_2} \exp(ik(r_2 - r_1)) \quad ,$$

and, in similar manner, $\langle E_1^* E_2 \rangle$ becomes

$$\langle E_1 E_2^* \rangle = \frac{\langle E(\mathbf{r}_1, t_1) E^*(\mathbf{r}_2, t_2) \rangle}{r_1 r_2} \exp(-ik(r_2 - r_1)) \quad .$$

We represent the average on the left-hand side of the first equality using the correlation function $\Gamma_{12}(\tau)$ in (7.1):

$$\frac{c\varepsilon_0}{2} \langle E_1^* E_2 \rangle = \frac{\Gamma_{12}(\tau)}{r_1 r_2} \exp(ik(r_2 - r_1)) \quad , \tag{7.5}$$

whereas we apply the complex conjugate of $\Gamma_{12}(\tau)$ for the average $\langle E(\mathbf{r}_1, t_1) E^*(\mathbf{r}_2, t_2) \rangle$ on the left-hand side of the second equality:

$$\frac{c\varepsilon_0}{2} \langle E_1 E_2^* \rangle = \frac{\Gamma_{12}^*(\tau)}{r_1 r_2} \exp(-ik(r_2 - r_1)) \quad . \tag{7.6}$$

Hence, the average intensity at the point of observation becomes

$$\langle I_P \rangle = \frac{\langle I_1 \rangle}{r_1^2} + \frac{1}{r_1 r_2} \Gamma_{12}(\tau) \exp(ik(r_2 - r_1)) +$$

$$+ \frac{1}{r_1 r_2} \Gamma_{12}^*(\tau) \exp(-ik(r_2 - r_1)) + \frac{\langle I_2 \rangle}{r_2^2} \quad . \tag{7.7}$$

It follows from (7.7) that interference at P will exist if the correlation function $\Gamma_{12}(\tau)$ has a non-zero value. In other words, for a given quasi-monochromatic beam, illuminating a space where two points play the role of secondary sources, a superposition of waves emitted from these sources will take a stable form if the correlation function calculated for these sources has a non-zero value.

As any complex function, the correlation function $\Gamma_{12}(\tau)$ can be represented by its modulus $|\Gamma_{12}(\tau)|$ and its argument $arg(\Gamma_{12}(\tau))$:

$$\Gamma_{12}(\tau) = |\Gamma_{12}(\tau)| \exp(i\, arg(\Gamma_{12}(\tau))) \quad . \qquad (7.8)$$

In turn,

$$\Gamma_{12}^*(\tau) = |\Gamma_{12}(\tau)| \exp(-i\, arg(\Gamma_{12}(\tau)))$$

is true, thus the sum of second and third terms in (7.7) can be rewritten in the form

$$2\frac{|\Gamma_{12}(\tau)|}{r_1 r_2} \cos(arg(\Gamma_{12}(\tau)) + k(r_2 - r_1)) \quad .$$

Finally, using $\delta = r_2 - r_1$, from (7.7) together with the sum found above, we find for the average intensity at P

$$\langle I_P \rangle = \frac{\langle I_1 \rangle}{r_1^2} + 2\frac{|\Gamma_{12}(\tau)|}{r_1 r_2} \cos(arg(\Gamma_{12}(\tau)) + k\delta) + \frac{\langle I_2 \rangle}{r_2^2} \quad , \qquad (7.9)$$

where δ is the geometrical path difference between the interfering rays, with assuming that the waves propagate in vacuum.

Let us assume that when varying the path difference δ, a set of interference fringes is observed around P. The averaged intensity will then have minimal values $\langle I_P \rangle_{\min}$ as well as maximal values $\langle I_P \rangle_{\max}$. According to (7.9) the minimal value is observed when

$$\cos(arg(\Gamma_{12}(\tau)) + k\delta) = -1 \quad ,$$

and the maximal value when

$$\cos(arg(\Gamma_{12}(\tau)) + k\delta) = 1 \quad .$$

Therefore, we can represent $\langle I_P \rangle_{\min}$ as

$$\langle I_P \rangle_{\min} = \frac{\langle I_1 \rangle}{r_1^2} - 2\frac{|\Gamma_{12}(\tau)|}{r_1 r_2} + \frac{\langle I_2 \rangle}{r_2^2} \quad , \qquad (7.10)$$

and $\langle I_P \rangle_{\max}$ as

$$\langle I_P \rangle_{\max} = \frac{\langle I_1 \rangle}{r_1^2} + 2\frac{|\Gamma_{12}(\tau)|}{r_1 r_2} + \frac{\langle I_2 \rangle}{r_2^2} \quad . \qquad (7.11)$$

MICHELSON was the first who introduced a *visibility function* for the interference fringes in order to represent the contrast of an interference pattern. According to his definition, for the visibility function V we write

$$V = \frac{\langle I_P \rangle_{\max} - \langle I_P \rangle_{\min}}{\langle I_P \rangle_{\max} + \langle I_P \rangle_{\min}} \quad . \qquad (7.12)$$

Substitution for $\langle I_P \rangle_{min}$ and $\langle I_P \rangle_{max}$ from (7.10) and (7.11) gives

$$V = \frac{|\Gamma_{12}(\tau)|}{r_1 r_2} \frac{2}{\langle I_1 \rangle / r_1^2 + \langle I_2 \rangle / r_2^2} .$$

For many circumstances the distance between r_1 and r_2 can be insignificant, thus one can regard $r_1 \approx r_2$. Hence the expression for V may be simplified to

$$V = \frac{2|\Gamma_{12}(\tau)|}{\langle I_1 \rangle + \langle I_2 \rangle} .$$

If we assume that the averaged intensities of the secondary sources are equal to each other, the visibility function can be represented in terms of a normalized form of the correlation function:

$$V = \frac{2|\Gamma_{12}(\tau)|}{\langle I_1 \rangle + \langle I_2 \rangle} = \frac{|\Gamma_{12}(\tau)|}{\langle I \rangle} = |\gamma_{12}(\tau)| , \tag{7.13}$$

where $\gamma_{12}(\tau)$ is called the *complex degree of coherence*.

Taking into account the approximations above, we represent the average intensity of the interference in terms of $\gamma_{12}(\tau)$ as

$$\langle I_P \rangle = \frac{2 \langle I \rangle}{r^2} [1 + |\gamma_{12}(\tau)| \cos(arg(\Gamma_{12}(\tau)) + k\delta)] , \tag{7.14}$$

where r specifies the distance from point $\mathbf{r}_1$ to point P, as well as from $\mathbf{r}_2$ to point P. It can be seen that the averaged intensity of the interference pattern has an oscillating form provided by the factor $\cos(arg(\Gamma_{12}(\tau)) + k\delta)$, while the modulus of the complex degree of coherence remains a non-zero magnitude.

In fact, the complex degree of coherence can vary between zero and unity:

$$0 \leqslant |\gamma_{12}(\tau)| < 1 .$$

Completely incoherent light is associated with $|\gamma_{12}(\tau)| = 0$. In this case we assume that the light is absolute white, that means, all spectral harmonics are represented equally in its spectrum. In the case of a white spectrum, light is believed to be emitted by absolutely independent sources. A model of completely incoherent light will be utilized later to establish important properties of partially coherent light. With partially coherent light, we mean a field where $0 < |\gamma_{12}(\tau)| < 1$. Most types of typical thermal sources emit partially coherent light. The hypothetical case of $|\gamma_{12}(\tau)| = 1$ is associated with light which is emitted without the important statistical feature of fluctuations. No light sources are known today, for which fluctuations are completely absent.

2.   General form of the correlation of fields

Correlation of light fields possesses features assigned to different directions with respect to light propagation – parallel to the light beam axis and perpendicular to it. Mathematically, the directions under discussion were taken into consideration by the function $\Gamma_{12}(\tau)$, as mentioned before. Distinction between the axial and normal arguments of the correlation function $|\Gamma_{12}(\tau)|$ is caused by the shape of the light source, and by its spectrum; thus, all features of correlation can be achieved by choosing proper shape, size and spectrum of a light source.

In order to establish rules for obtaining the desired properties of the correlation function, we have to make assumptions concerning the source, which emits the light beam. Let a plane radiating surface of an original source be divided into emitting elements. We make the following assumptions concerning these elements:

1 The mean amount of elements on a unit surface of the source is a constant value, which means, that the surface density of the source is invariable (until assumed otherwise).

2 Irrespective of an element's position on the surface, the mean flux of radiation from every element within a frequency interval ν and $\nu + \Delta\nu$ has to be the same.

3 Statistical independence exists between all the elements.

The assumptions above allow the representation of the correlation function $\Gamma_{12}(\tau)$ in terms of a surface integral over the plane of the original surface, and in an integral over the frequencies contained in the spectrum of the source. Let the z-axis of a Cartesian system be directed along the beam propagation and ξ, η be the coordinates of the radiating surface element. We draw a line r_{11} from one element (ξ_1, η_1) to point r_1 somewhere in a remote space, and another line r_{12} from the same element to an another point r_2, close to r_1, as shown in Fig.7.4. Then the field emitted from the element under consideration has a spherical wave front and spreads to points r_1 and r_2. We can describe it using the complex functions

$$E_{11} = \frac{e(\xi_1, \eta_1)}{r_{11}} \exp(i\psi_{11}) \quad \text{and} \quad E_{12} = \frac{e(\xi_1, \eta_1)}{r_{12}} \exp(i\psi_{12}) \quad , \quad (7.15)$$

where E_{11} and E_{12} are the fields received from the element at the points r_1 and r_2, respectively. $e(\xi_1, \eta_1)$ is the complex amplitude, related to a certain spectral interval between ν and $\nu + \Delta\nu$, defined at unit distance from the element. ψ_{11} and ψ_{12} are phase increments due to light

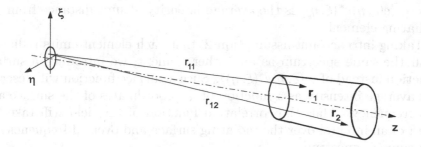

Figure 7.4. Illustrating the contribution from one element of a radiating surface to the correlation function.

propagation from the element to points r_1 and r_2, respectively. With an arbitrary position of another element (ξ_m, η_m) the fields of its spherical wave at the same pair of points can be represented in a similar manner as

$$E_{m1} = \frac{e(\xi_m, \eta_m)}{r_{m1}} \exp(i\psi_{m1}) \quad , \qquad E_{m2} = \frac{e(\xi_m, \eta_m)}{r_{m2}} \exp(i\psi_{m2}) \quad ,$$

$$(7.16)$$

where r_{m1}, r_{m2} are the distances to points r_1, r_2, ψ_{m1}, ψ_{m2} are the phase increments, and $e(\xi_m, \eta_m)$ is as before the complex amplitude associated with this element. According with (7.15) and (7.16) the superimposed fields at points r_1 and r_2 caused by these illuminating elements have the form

$$E(r_1) = E_{11} + E_{m1} \quad \text{and} \qquad E(r_2) = E_{12} + E_{m2} \quad .$$

Let us implement statistical averaging of the product $E(r_1)E^*(r_2)$. This will give us a correlation function as the sum

$$\langle E(r_1)E^*(r_2) \rangle = \langle E_{11}E_{12}^* \rangle + \langle E_{11}E_{m2}^* \rangle + \langle E_{m1}E_{12}^* \rangle + \langle E_{m1}E_{m2}^* \rangle \quad .$$

When substituting E_{11} and E_{m2}^* by the expressions in (7.15), (7.16), the second term becomes

$$\langle E_{11}E_{m2}^* \rangle = \frac{\langle e(\xi_1, \eta_1)e^*(\xi_m, \eta_m) \rangle}{r_{11}r_{m2}} \exp(i(\psi_{11} - \psi_{m2})) \quad ,$$

but $\langle e(\xi_1, \eta_1)e^*(\xi_m, \eta_m) \rangle = 0$, due to the statistical independency of the elements (assumption 3 above); hence, $\langle E_{11}E_{m2}^* \rangle = 0$. For the same reason the third term of the sum must also be equal to zero. Hence, any contribution to the correlation function has the form

$$\frac{\langle e(\xi, \eta)e^*(\xi, \eta) \rangle}{r_1 r_2} \exp(i(\psi_1 - \psi_2)) \quad , \qquad (7.17)$$

where $\langle e(\xi, \eta)e^*(\xi, \eta)\rangle$ is the average intensity at unit distance from the radiating element.

Taking into account assumption 2, that each element emits radiation with the same spectrum as any other element, we introduce a surface function instead of $\langle e(\xi, \eta)e^*(\xi, \eta)\rangle$. Such a surface function will describe the average intensity as a function of the coordinates of the surface and the frequency. Thus the correlation function of the field will take the form of an integral over the radiating surface and over all frequencies of the source's spectrum:

$$\Gamma_{12}(\tau) = \int_\xi \int_\eta \int_\nu \frac{G(\xi, \eta, \nu)}{r_1 r_2} \exp(i(\psi_1 - \psi_2)) d\xi d\eta d\nu, \qquad (7.18)$$

where $G(\xi, \eta, \nu)$ is the surface density function of the average intensity. It is seen now that a correlation of the fields at points $\mathbf{r}_1$ and $\mathbf{r}_2$ can exist in any case, even if an original source shows no correlation (this facts is specified by $G(\xi, \eta, \nu) = const$), provided that the radiation of the source is either limited in space or in its spectrum.

3.   Spatial correlation of the field

Using the general integral form of the correlation function $\Gamma_{12}(\tau)$ in (7.18) we now derive an expression for the correlation function associated with interference schemes of the YOUNG-type. Such a correlation function is called the *spatial correlation function of a light field*. In fact, the integral (7.18) allows calculations under different geometrical parameters of observation. Nevertheless, the approximation of FRAUNHOFER diffraction becomes most significant and useful for practical applications, since all interference experiments considered in the previous chapters were arranged under this approximation.

We assume that a light beam is generated from a restricted area (ξ, η) of the source. The spatial distribution of the average intensity on the surface of the source is uniform. Thus, the surface density of the average intensity depends only on the frequency. Our task is to find the correlation of the field on a remote plane (x, y) normal to the beam axis. With a Cartesian system on plane (x, y) let the coordinates of points $\mathbf{r}_1$ and $\mathbf{r}_2$ be x_1, y_1 and x_2, y_2, respectively. Under the FRAUNHOFER approximation in the form of (2.26) and for a certain frequency ν, the phase difference $\psi_1 - \psi_2$ can be represented in terms of the separations $x_1 - x_2$ and $y_1 - y_2$ as follows:

$$\psi_1 - \psi_2 = k(R_1 - R_2) \approx$$

$$= k\left(\xi\frac{x_2 - x_1}{z_0} + \eta\frac{y_2 - y_1}{z_0} + \frac{x_2^2 - x_1^2}{2z_0} + \frac{y_2^2 - y_1^2}{2z_0}\right) \quad, \qquad (7.19)$$

where $k = 2\pi\nu/c$, and z_0 is the distance between the planes (ξ, η) and (x, y). In additional, let us choose points $\mathbf{r}_1$ and $\mathbf{r}_2$ symmetrically with respect to the beam axis, so that the conditions

$$x_1 = -x_2 \quad \text{and} \quad y_1 = -y_2$$

will be true (Fig.7.5). It then follows from (7.19) that the phase difference $\psi_1 - \psi_2$ becomes linearly dependent on the coordinate differences $x_2 - x_1$ and $y_2 - y_1$:

$$\psi_1 - \psi_2 = k\xi\frac{x_2 - x_1}{z_0} + k\eta\frac{y_2 - y_1}{z_0} \quad . \tag{7.20}$$

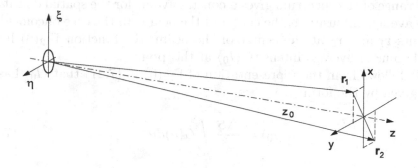

Figure 7.5. For the case of spatial correlation under the approximation of FRAUN-HOFER diffraction two points $\mathbf{r}_1$, $\mathbf{r}_1$ lie symmetrically on a plane normal to the beam propagation: $x_1 = -x_2$, $y_1 = -y_2$.

Let us estimate the effect of a certain spectral composition on the phase difference, provided that the radiation is considered to be quasi-monochromatic. This means that the effective spectral width of the radiation is much smaller than the carrier frequency: $\Delta\nu \ll \nu_0$. Since $k = 2\pi\nu/c$, this fact can be represented as

$$k + \Delta k = 2\pi\nu_0/c + 2\pi\Delta\nu/c,$$

where Δk can be regarded as the maximal variation of k. Under these requirements the phase difference in (7.20) is a sum of four terms:

$$\psi_1 - \psi_2 = k_0\xi\frac{\Delta x}{z_0} + k_0\eta\frac{\Delta y}{z_0} + \Delta k\xi\frac{\Delta x}{z_0} + \Delta k\eta\frac{\Delta y}{z_0} \quad .$$

Under the quasi-monochromatic condition the terms containing k_0 are much larger than those dependent on Δk. Therefore the third and fourth items, containing $\Delta k \Delta x$ or $\Delta k \Delta y$, respectively, can be omitted. This implies that the correlation function is determined by geometrical parameters, rather than by the spectral composition of the light. Emphasizing this fact, we specify the spatial correlation function by $\Gamma_{12}(0)$, assuming $\tau = 0$. Thus, for the spatial correlation function, we find the integral form

$$\Gamma_{12}(0) = \frac{1}{z_0^2} \int_\xi \int_\eta \exp\left[ik_0\xi\frac{\Delta x}{z_0} + ik_0\eta\frac{\Delta y}{z_0}\right] d\xi d\eta \int_\nu G(\nu)d\nu \quad , \quad (7.21)$$

where $G(\nu)$ describes the spectral composition of the light beam, and where $k_0 = 2\pi\nu_0/c$ is specified only by the carrier frequency ν_0. The integration of the spectral density of the average intensity $G(\nu)$ over the full range of the spectrum gives a constant value for the spatial density of the average intensity at the center of the beam. In the case of coinciding points $\mathbf{r}_1$ and $\mathbf{r}_2$ at the center of the beam, the function $\Gamma_{12}(0)$ has a real value of average intensity $\langle I_0 \rangle$ at this point.

It follows from the representation of $\Gamma_{12}(0)$ in (7.21) that $\langle I_0 \rangle$ has to be given by the form

$$\langle I_0 \rangle = \frac{\Delta S}{z_0^2} \int_\nu G(\nu)d\nu \quad , \quad (7.22)$$

where the integral over ν gives an average spectral density of intensity, and the integration over the coordinates of the source at $\Delta x = \Delta y = 0$ gives the value ΔS, which is equal to the area of the source. This relationship allows us to determine $\Gamma_{12}(0)$ as the surface integral

$$\Gamma_{12}(0) = \frac{\langle I_0 \rangle}{\Delta S} \int_\xi \int_\eta \exp\left[ik_0\xi\frac{\Delta x}{z_0} + ik_0\eta\frac{\Delta y}{z_0}\right] d\xi d\eta \quad . \quad (7.23)$$

When placing a splitting element on the plane (x, y), such as a double slit or FRESNEL's mirrors, as an element which provides two interfering rays, we can form an interference pattern somewhere behind plane (x, y). Under proper geometrical parameters of the beam splitter, the contrast of the pattern at a certain distance from the splitter is determined by the modulus of $\Gamma_{12}(0)$. The fact that $\Gamma_{12}(0)$ mainly depends on geometrical conditions results in colored interference fringes.

The spatial correlation is a special case of the correlation of light fields, where the correlation function of the field $\Gamma_{12}(0)$ is mainly determined by the shape and the dimensions of the radiating surface and by the mutual position of the points $\mathbf{r}_1$ and $\mathbf{r}_2$, but only marginally by the spectral distribution of the average intensity.

3.1   Typical cases of spatial correlation

The integral in (7.23) allows us to directly derive the correlation function $\Gamma_{12}(0)$ for some of the most typical shapes of radiating surfaces.

3.1.1   Rectangular aperture and slit

We consider an aperture of rectangular shape. Let the long side b and the short side a of the rectangle be positioned along the vertical axis η and the horizontal axis ξ of a Cartesian reference system on the radiating surface, and let the origin O be located at the axis of a light beam (Fig.7.6). Then the two-dimensional integral in (7.23) will take the form of two integrals, one dependent on variable ξ, and the other on variable η :

$$\int_{-b/2}^{b/2} \int_{-a/2}^{a/2} \exp(ik_0[\xi\Delta x + \eta\Delta y]/z_0)d\xi\,d\eta =$$

$$= \int_{-a/2}^{a/2} \exp(i(k_0\xi\Delta x)/z_0)d\xi \int_{-b/2}^{b/2} \exp(i(k_0\eta\Delta y)/z_0)d\eta \ .$$

Both integrals are already known (see (2.29), (2.30)):

$$\int_{-a/2}^{a/2} \exp(i(k_0\xi\Delta x)/z_0)d\xi = \frac{\sin(k_0 a\Delta x/(2z_0))}{k_0 a\Delta x/(2z_0)} \quad \text{and}$$

$$\int_{-b/2}^{b/2} \exp(i(k_0\eta\Delta y)/z_0)d\eta = \frac{\sin(k_0 b\Delta y/(2z_0))}{k_0 b\Delta y/(2z_0)} \ .$$

Hence, the correlation function takes the form

$$\Gamma_{12}(0) = \langle I_0\rangle \left(\frac{\sin(k_0 a\Delta x/(2z_0))}{k_0 a\Delta x/(2z_0)}\right) \left(\frac{\sin(k_0 b\Delta y/(2z_0))}{k_0 b\Delta y/(2z_0)}\right) \ , \quad (7.24)$$

where $\langle I_0\rangle$ is the intensity at point O.

The spatial correlation function is real and is represented by a two-dimensional distribution of the variables $\delta_x = (k_0 a\Delta x)/(2z_0)$ and $\delta_y = (k_0 b\Delta y)/(2z_0)$. The principle maximum of the distribution is located within a rectangular of sides Δx_{ch} and Δy_{ch}, which follow from the conditions $\delta_x = \pi$, $\delta_y = \pi$:

$$\Delta x_{ch} = z_0 \frac{\lambda_0}{a} \ , \qquad \Delta y_{ch} = z_0 \frac{\lambda_0}{b} \ . \quad (7.25)$$

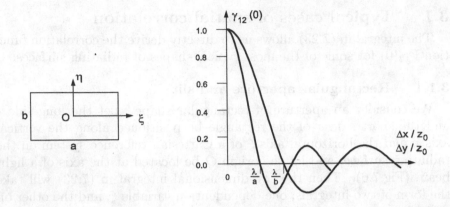

Figure 7.6. A regtangular aperture and the correlation functions of the field corresponding to two orthogonal dimensions of the aperture.

Here, the magnitudes Δx_{ch} and Δy_{ch} specify the boundary of the so-called *area of coherence* which is the area where the function $\Gamma_{12}(0)$ has significant values. For comparison, the FRAUNHOFER diffraction pattern formed by diffraction of a plane monochromatic wave of wavelength λ_0 on the same rectangle aperture will have its first zeros at $x = \pm z_0\lambda_0/a$ and at $y = \pm z_0\lambda_0/b$. Hence, the $x-$ and $y-$dimensions of the central maximum are twice as large when compared to the case of the correlation function under consideration.

If the short side a of the rectangle is much smaller than the long side, one can regard the aperture to be a narrow slit of width a. In the limiting case of $b \sim z_0$, where the distance z_0 is treated as being an extremely large linear dimension, we obtain $\Delta y_{min} \sim \lambda_0$ as an estimation of the critical size of the correlation function. Such an estimation means that the spatial correlation in the direction parallel to the longer side of the slit ranges only over about a distance λ_0. The short width a of the slit is responsible for the extension of the correlation in the direction orthogonal to the longer side. The correlation function is shown in Fig.7.7.

The spacing of the minima and of the maxima of the correlation function is equal to $\Delta x = m\lambda z_0/a$, where $m = 1, 2, \dots$. Thus the correlation function of a slit is a real function:

$$\Gamma_{12}(0) = \langle I_0\rangle \frac{\sin(ka\Delta x/(2z_0))}{ka\Delta x/(2z_0)} \quad . \tag{7.26}$$

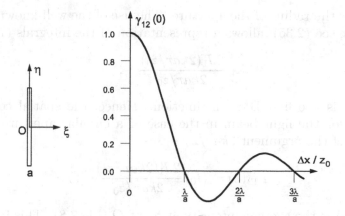

Figure 7.7. A narrow vertical slit of width a and the correlation function of the field associated with the slit.

3.1.2 A circular aperture

In a similar way, we analyze the correlation function for a circular aperture. Here, it is appropriate to use polar coordinates. Let (ρ, θ) be the polar coordinates of a point of the aperture (Fig.7.8), represented in terms of the coordinates ξ, η in the form

$$\rho \cos \theta = \xi \quad , \qquad \rho \sin \theta = \eta \quad . \tag{7.27}$$

Now let (r, φ) be polar coordinates of a point in the plane x, y :

$$r \cos \phi = x \quad , \qquad r \sin \phi = y \quad . \tag{7.28}$$

By choosing two points (x_1, y_1) and (x_2, y_2) symmetrical with respect to the beam axis, we find, using (7.27) and (7.28), the relationship

$$\Delta x = 2r \cos \phi \quad , \qquad \Delta y = 2r \sin \phi \quad . \tag{7.29}$$

For the polar coordinates (7.27), (7.28) and with $\Delta x, \Delta y$ from (7.29), the phase of the integrand in (7.23) becomes

$$k(\xi \Delta x + \eta \Delta y)/z_0 = 2kr\rho \cos(\theta - \phi)/z_0 \quad .$$

Hence, the two-dimensional integral in (7.23) becomes

$$\int_0^a \int_0^{2\pi} \exp(i2kr\rho \cos(\theta - \varphi)/z_0)\rho \, d\rho \, d\theta \quad , \tag{7.30}$$

where a is the radius of the aperture. The use of the well-known BESSEL functions, see (2.35), allows a representation of the integrals (7.30) as

$$\frac{2J_1(2kar/z_0)}{2kar/z_0} \quad,$$

where J_1 is the first BESSEL function. Hence, the spatial correlation function for the light beam in the case of a circular aperture is a real function of the argument $2kar/z_0$:

$$\Gamma_{12}(0) = \langle I_0 \rangle \frac{2J_1(2kar/z_0)}{2kar/z_0} \quad, \tag{7.31}$$

where $\langle I_0 \rangle$ is the average intensity at point O (Fig.7.8). This function is real due to the axial symmetry of the aperture. The complex degree of spatial coherence is also real and can be represented by the expression

$$\gamma_{12}(0) = \frac{2J_1(2kar/z_0)}{2kar/z_0} \quad. \tag{7.32}$$

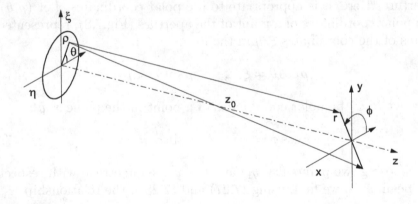

Figure 7.8. Polar coordinates to calculate the field correlation from a circular aperture.

This is a function of the dimensionless argument $\delta = 2kar/z_0$. The position of the first minimum is given by the value $\delta = 1.22\pi$. For the diameter of the central circle, which should be associated with the coherent area produced by the circular aperture, we get the relation

$$d_{ch} = 2r_{ch} = 0.61 \cdot z_0 \frac{\lambda_0}{a} = 1.22 \cdot z_0 \frac{\lambda_0}{D} \quad, \tag{7.33}$$

where $D = 2a$, d_{ch} is the diameter and r_{ch} is the *radius of coherence*. This result is similar to the formula $x_0 = 0.61 \cdot z_0\lambda_0/a$ discussed in

Chapter 2 in connection with FRAUNHOFER diffraction from a circular aperture. The value d_{ch} determines the limits of the region within which the correlation function has the principle maximum. This fact is specified by the subscript $_{ch}$. For the same radius a and the same optical carrier frequency ν_0, the radius of the central disk r_{ch} of the coherent area is numerically one half of the radius r_0 of AIRY's disk (see (2.36)).

The normalized correlation function from a radiating disk of diameter $D = 2a$ is shown in Fig.7.9 as a function of the dimensionless parameter $2r/z_0$. The first two minima occur at $2r/z_0 = 1.22\lambda/D$ and $2.23\lambda/D$.

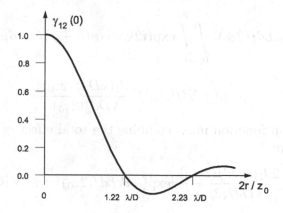

Figure 7.9. The correlation function of the field from a circular aperture of diameter D.

3.1.3 Two radiating disks

Now we consider the spatial correlation from a source in the form of two identical radiating discs. Let D be the diameters of disks, and L the separation of their centres O_1 and O_2. Let the origin of a Cartesian system $\xi^{(1)}, \eta^{(1)}$ be placed at the middle of the line O_1O_2, and let (ρ, φ) be the polar coordinates of a point on the first disk. The coordinates of both systems then satisfy the relations

$$\xi^{(1)} = -L/2 + \rho\cos\theta \quad \text{and} \quad \eta^{(1)} = \rho\sin\theta \ . \tag{7.34}$$

For a given distance $d = 2r$ between points (x_1, y_1) and (x_2, y_2), the integration over the surface of the first disk can be performed:

$$\exp(-ikLd/(2z_0)) \int_0^a \int_0^{2\pi} \exp(i2k\rho r\cos(\theta - \varphi)/z_0)\,d\rho\,d\theta =$$

$$= \exp(-ikLd/(2z_0))\frac{2J_1(kDd/(2z_0))}{kDd/(2z_0)} \quad . \tag{7.35}$$

In a similar way, for an arbitrary point $\xi^{(2)}, \eta^{(2)}$ of the second disk the following relationships are true:

$$\xi^{(2)} = L/2 + \rho\cos\theta \quad \text{and} \quad \eta^{(2)} = \rho\sin\theta \quad . \tag{7.36}$$

Thus, the integration over the surface of second disk gives a similar expression:

$$\exp(ikLd/(2z_0)) \int_0^a \int_0^{2\pi} \exp(i2k\rho r \cos(\theta - \varphi)/z_0)\, d\rho\, d\theta =$$

$$= \exp(ikLd/(2z_0))\frac{2J_1(kDd/(2z_0))}{kDd/(2z_0)} \quad . \tag{7.37}$$

The correlation function must combine the total effect of both disks in a real function:

$$\Gamma_{12}(0) = \langle I_0 \rangle \frac{2J_1(kDd/(2z_0))}{kDd/(2z_0)} [\exp(-ikLd/(2z_0)) + \exp(ikLd/(2z_0))] =$$

$$= \langle I_0 \rangle \frac{2J_1(kDd/(2z_0))}{kDd/(2z_0)} \cos[kLd/(2z_0)] \quad , \tag{7.38}$$

where $\langle I_0 \rangle$ is the average intensity at the axis of the beam caused separately by each disk. For an illustration, the function

$$(2J_1(\delta)/\delta)\cos(1.64\delta)$$

of the phase δ is shown in Fig.7.10. With increasing δ, it oscillates with gradually diminishing amplitude. The spatial frequency of its oscillations is directly proportional to the separation L, and the function $2J_1(\delta)/\delta$ is the envelope of the oscillations.

3.2   Demonstration of spatial correlation

3.2.1   Visibility of fringes obtained with the Young double–slit interferometer

We have seen that the visibility function V as a measure of the contrast of the interference pattern is directly connected with the normalized correlation function of the light field, the so-called degree of coherence (see (7.13)). In order to demonstrate the effect of geometrical parameters on the contrast of interference fringes we analyze results obtained

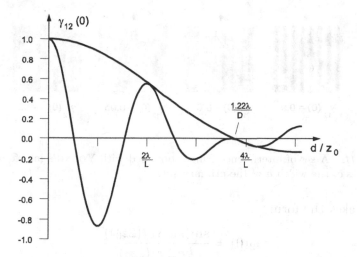

Figure 7.10. The correlation function of the field from two identical radiating disks (oscillating line) and the analogous function produced by each disk.

for some typical interference schemes. We emphasize once more that in such a simple scheme as YOUNG's double-slit interferometer the spatial degree of coherence is a real function. This fact allows a simple analysis of the observed interference fringes.

All principle features of the behavior of the contrast of the fringes with the geometrical parameters of YOUNG's interferometer can be demonstrated by means of the experimental equipment considered in Chapter 1 (Fig.1.2). A set of interference fringes, each corresponding to a certain width a of the primary slit, is shown in Fig.7.11. In the first picture the visibility is 0.9 and it decreases with increasing the width a, according to different values of the function $\gamma_{12}(0)$. Then $\gamma_{12}(0)$ becomes nearly zero (0.05), and further it has a negative value. With a positive value of $\gamma_{12}(0)$ the central interference fringe is bright, and all fringes of even orders are bright, too. In the case of a negative value of $\gamma_{12}(0)$ the central fringe is dark as well all other fringes of even orders, whereas now all fringes of odd orders are bright.

We should emphasize that in the experimental scheme under discussion (Fig.1.2) the spatial correlation of the field caused by the primary slit will keep its value in the region between objective O_1 and the double slit, because here the light propagates as a parallel ray. Thus, the spatial correlation function formed by the slit is submitted to (7.26); in turn,

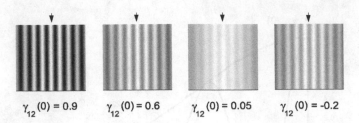

$$\gamma_{12}(0) = 0.9 \qquad \gamma_{12}(0) = 0.6 \qquad \gamma_{12}(0) = 0.05 \qquad \gamma_{12}(0) = -0.2$$

Figure 7.11. A set of interference fringes obtained with YOUNG's interferometer for four values of the width a of the primary slit.

$\gamma_{12}(0)$ takes the form

$$\gamma_{12}(0) = \frac{\sin(ka\Delta x/(2z_0))}{ka\Delta x/(2z_0)} \quad ,$$

where a is the width of the primary slit, $z_0 = f = 10$ cm is the focal length of objective O_1. In the case under discussion $\Delta x = b = 0.8$ mm, where b is the distance between the centers of the double-slit (consisting of slits which are 0.05 mm wide). Since the degree of coherence is a real function, its argument must be zero. Hence, it follows from (7.14) that the average intensity of the interference fringes varies according to

$$\langle I_P \rangle = 2 \langle I \rangle \left[1 + \gamma_{12}(0) \cdot \cos(k\Delta R) \right] \quad . \tag{7.39}$$

Let us estimate the width of the slit which provides a zero value of $\gamma_{12}(0)$. According to (7.25) the first zero of the degree of coherence $\gamma_{12}(0)$ appears at $b = \Delta x_{ch}$; hence, $b = z_0\lambda_0/a$ has to be true. Thus, for the wavelength $\lambda_0 = 580$ nm, a has to be equal to $z_0\lambda_0/b \approx 0.07$ mm. The value $\gamma_{12}(0) \approx 0.9$ is obtained for $a = 0.03$ mm.

3.2.2   The Michelson stellar interferometer

The idea that the width of the primary source affects the visibility of interference fringes in an interference scheme of the YOUNG-type was used by MICHELSON for measurements of the angular dimensions of such remote astronomical objects as stars and double-stars. The basic arrangement, which is called the MICHELSON *stellar interferometer*, is shown in Fig.7.12. Since any angular dimension of a star under investigation is an invariable magnitude, the visibility of the fringes is changed with the MICHELSON stellar interferometer by varying the spatial separation between two interfering rays. Two apertures S_1 and S_2 limit the telescope objective. These apertures emit two partially coherent beams, which interfere within the focal plane of the objective. Two mobile mirrors M_3 and M_4 play the role of the double-slit of the YOUNG

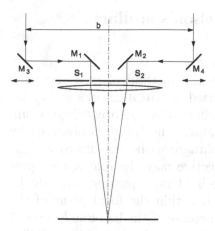

Figure 7.12 The Michelson stellar interferometer. Mirrors M_3 and M_4 are movable in order to change the baseline b.

interferometer. The distance between the centers of the mirrors can be varied. In such a way, the parameter Δx, which is called baseline b of the interferometer here, is varied, which affects the visibility of the fringes. After reflection on the mirrors M_3 and M_4 the two rays are reflected on the fixed mirrors M_1 and M_2, pass through the apertures S_1 and S_2, and are then focused by the telescope objective to form an interference pattern.

By modeling a star with a radiating disk of angular diameter ψ the measurements of the visibility should show that the size of the baseline b providing the first minimum of the visibility has to be equal to $b = d_{ch} = 1.22 \ \lambda_0/\psi$, which follows from (7.33). Here, $\psi = D/z_0$, where D is the diameter of the star, and z_0 is the distance from the star to the stellar interferometer. Thus, for the angular diameter, we get the expression:

$$\theta = 1.22 \times \lambda/d_{ch} \quad . \tag{7.40}$$

Beteigeuze (α Orionis) was the first star for which the angular diameter was measured successfully. Its value was found to be about 0.047 arc seconds. The angular dimension could be measured only for a few of other stars. All of them are giant stars, many times larger than the Sun. Typical difficulties inherent to such measurements are connected to the disturbing effects of atmospheric turbulence. Another difficulty is the weak intensity over the focal plane of the objective, which restricts any investigations to bright stars.

3.3   A labor model of Michelson's stellar interferometer

3.3.1   A "star"

We consider a setup to demonstrate the method of measuring the angular dimensions of remote objects used in MICHELSON's stellar interferometer. A small slit, which is illuminated by the bright light beam of a mercury lamp, plays the role of a star. The "star" is observed by means of a telescope (Fig.7.13). A diaphragm with two slits S_1 and S_2 positioned in front of the telescope objective play the role of the apertures S_1 and S_2 shown in Fig.7.12. The light rays, passing through the slits, give rise to an interference pattern within the focal plane of the objective. In order to demonstrate the increase of the baseline b, a set of double-slit diaphragms with different distances b is used. The visibility of the fringes should decrease with increasing separation b between the slits.

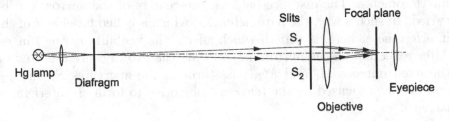

Figure 7.13.   A model of the stellar interferometer.

Interference patterns obtained with this setup are shown in Fig.7.14. The pictures are obtained for five diaphragms, where the separation b between the slits is varied from $b = 0.5$ mm to $b = 2.5$ mm. We see that the visibility decreases with the increasing value of b. The angular size θ of the "star" can be found for a certain baseline, where the fringes just disappear (as in Fig.7.14 at $b = 2.5$ mm). The "diameter" of the "star" is 0.5 mm, and the distance from the "star" to the objective is $z_0 = 2$ m. Thus, we find for the angular dimension $\theta = 0.5/2000 = 2.5 \times 10^{-4}$ rad. We call attention to a major problem when measuring the angular size of a "star": The spacing of the fringes decreases in parallel with the visibility. Thus the disappearance of the fringes can be estimated only roughly.

3.3.2   A "double star"

In a similar way, measurements of the spatial structure of a "double star" can be performed with our model of MICHELSON's stellar inter-

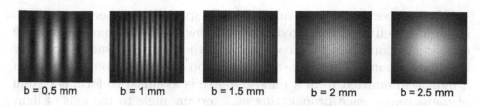

b = 0.5 mm   b = 1 mm   b = 1.5 mm   b = 2 mm   b = 2.5 mm

Figure 7.14. A set of pictures observed with our model setup when varying the separation b between two slits in front of the telescope objective.

b = 0.5 mm   b = 1 mm   b = 2 mm

Figure 7.15. A set of pictures observed by means of a model of stellar interferometer in the case of a "double star". A particular picture obtained at a certain separation b between two slits in front of the telescopic objective.

ferometer. This time, two small identical slits, each having a width $d = 0.12$ mm, play the role of a double star. The separation a between the centers of the slits is 0.5 mm. Light from the "double star" passes to the telescope as above. Three diaphragms with different baselines b between the slits S_1, S_2 are used for the analysis of the interference patterns. If b is small enough (0.5 mm), distinct interference fringes are observed (Fig.7.15). With a second diaphragm with a larger separation of the slits ($b = 1$ mm) the interference fringes disappear. Further, with a third diaphragm ($b = 2$ mm) the interference fringes appear again, but the central fringe is now dark as well as all fringes of even orders ($\pm 2, \pm 4, \pm 6, ...$), whereas all fringes of odd orders are bright. The visibility of this pattern seems to be nearly the same as in the case of the smallest baseline used, $b = 0.5$ mm. The visibility function, calculated for the case under consideration, is shown in Fig.7.10.

We call attention to the fact that no extra optical filters are used to improve the quasi-monochromaticity of the light beam used in the present demonstration. The quasi-monochromaticity of the line spectrum of the mercury lamp is sufficient. Moreover, the spectral composition of the light has little influence on the spatial correlation compared to the effects of the geometrical parameters.

We have seen that even an absolutely spatially incoherent source is enough to deliver the desired results. Now we make some qualitative conclusions concerning the correlation of the field from such a spatially incoherent source. Under the approximation of FRAUNHOFER diffraction a light field emitted from a remote source is assumed to form a system of plane waves, each propagating at a certain angle to the axis of light beam. In the case of diffraction, non-diffracted rays pass along this axis. An analysis of the diffraction image allows us to ascertain the shape of the source and its spatial features, provided that the full spatial spectrum reaches the observer. For example, we can distinguish between sources of rectangular and circular shape. Let us assume that our observation is restricted to the first few spatial harmonics; then the shape of the original source stays undetermined to a certain degree. In the limiting case of only one system of plane waves, each propagating at zero diffraction angle, there is no chance to reconstruct the original shape. If this is the case, we say the light source is a point source.

As above, similar considerations will be true if we discuss the spatial correlation instead of the diffraction field. We can be learn from the examples shown with the model setup of MICHELSON's stellar interferometer that the full spatial spectrum provides the correct reconstruction of the source shape. Thus, we must expand the baseline b sufficiently to investigate the required variations of the visibility function. In reality, every elementary source on the original radiating surface emits light just as any other. Within the space of observation there is no chance to distinguish waves from different sources until the mutual positions of the radiating elements are at our disposal. It is the mutual position of these elements that causes the decrease in correlation, since every element emits light independently. The elements located on the periphery of the light source give rise to destructive contributions to the correlation function with respect to contributions from central elements. In turn, all elements of the central part of the source provide a nearly invariable spatial correlation. If this is the case, we speak of a remote point source, or of a light field formed by one parallel beam.

4.   Temporal correlation of the light field

Now we consider another special case of correlation of a field propagating as a nearly parallel beam. We can assume that such a beam is emitted from a remote source so that, for the space of observation, the approximation of FRAUNHOFER diffraction is true. Hence, the shape and the dimensions of the source will not affect the correlation at points r_1 and r_2, if these points lie on the axis of the parallel beam (Fig.7.16). Let z_1 and z_2 be the coordinates of these points. Using the approxima-

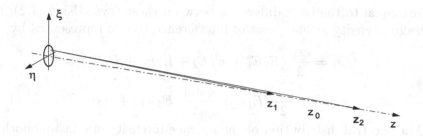

Figure 7.16. Two points on the beam axis in the case of temporal correlation of the field.

tion of FRAUNHOFER diffraction, the phase difference in (7.18) takes the simple form

$$\psi_1 - \psi_2 = k(r_1 - r_2) = 2\pi\nu(z_1/c - z_2/c) \quad . \tag{7.41}$$

The quantities z_1/c and z_2/c are the time intervals needed for the light to travel from any point of the radiating surface to the points z_1 and to z_2, respectively. In turn $\tau = (z_2 - z_1)/c$ is the time delay associated with the path difference $z_2 - z_1$.

The representation of the field correlation in (7.18) now has the form of one integral over the spectral distribution of the source, and another integral over the coordinates of the surface of the source:

$$\Gamma_{11}(\tau) = \frac{1}{z_1 z_2} \int_\xi \int_\eta d\xi d\eta \int_\nu G(\nu) \exp(i2\pi\nu\tau) d\nu \quad ,$$

The subscript $_{11}$ indicates that both points $\mathbf{r}_1$ and $\mathbf{r}_2$ are located on the z-axis. We assume further that z_0 is located somewhere between points z_1 and z_2, thus we can use $z_1 z_2 = z_0^2$. The integration over the surface of the source gives the area ΔS of the source. Thus, for the correlation function we get

$$\Gamma_{11}(\tau) = \frac{\Delta S}{z_0^2} \int_\nu G(\nu) \exp(i2\pi\nu\tau) d\nu \quad . \tag{7.42}$$

This function is called the *temporal correlation function of the light field*.

We emphasize that with a quasi-monochromatic wave of carrier frequency ν_0 it is reasonable to describe the temporal correlation function as a slowly varying function of the frequency difference $\nu - \nu_0$, instead of as a function of the original frequency ν. To find such a function we discuss interference arising from a superposition of two rays radiated from points z_1 and z_2. Since the interfering rays propagate along one direction within a nearly parallel beam, the path difference $\Delta s = c\tau$ above is just equal to the path difference between these rays. Using (7.2), the average intensity at one point of interference can be represented by

$$\langle I_P \rangle = \frac{c\varepsilon_0}{2} \langle E_1 E_1^* + E_1^* E_2 + E_1 E_2^* + E_2 E_2^* \rangle =$$

$$= \langle I_1 \rangle + \frac{c\varepsilon_0}{2} \langle E_1 E_2^* \rangle + \frac{c\varepsilon_0}{2} \langle E_2 E_1^* \rangle + \langle I_2 \rangle \quad . \qquad (7.43)$$

The spectral distribution of the average intensity of quasi-monochromatic light is usually treated in terms of a function of the detuning Ω of the circular frequency ω from the circular frequency of the carrier, ω_0:

$$\Omega = \omega - \omega_0 = 2\pi(\nu - \nu_0) \quad .$$

Let us introduce a function $G(\Omega)$ specifying the spectral distribution around the carrier frequency ν_0. With the new variable Ω the integrand in (7.42) can be written as

$$G(\nu) \exp(i2\pi\nu\tau) = \exp(-i2\pi\nu_0\tau) G(\Omega) \exp(i\Omega\tau) \quad . \qquad (7.44)$$

Substituting the right-hand side of (7.42) by the left-hand side of (7.44), the temporal correlation function takes the form of an integral over the new spectral distribution:

$$\Gamma_{11}(\tau) = \exp(i2\pi\nu_0\tau) \frac{\Delta S}{z_0^2} \int_\Omega G(\Omega) \exp(i\Omega\tau) d\Omega \quad . \qquad (7.45)$$

We introduce the correlation function $\Gamma'_{11}(\tau)$, which does not have the exponential term $\exp(i2\pi\nu_0\tau)$, and, therefore, takes the form

$$\Gamma'_{11}(\tau) = \frac{\Delta S}{z_0^2} \int_\Omega G(\Omega) \exp(i\Omega\tau) d\Omega \quad . \qquad (7.46)$$

Here the subscript $_{11}$ specifies that both points lie on the beam axis. The variable τ specifies the path difference between these points. Thus, the average intensity in (7.43) takes the form of a sum:

$$\langle I_p \rangle = \langle I_1 \rangle + \Gamma'_{11}(\tau) \exp(i2\pi\nu_0\tau) + \Gamma'^*_{11}(\tau) \exp(-i2\pi\nu_0\tau) + \langle I_2 \rangle =$$

$$= \langle I_1 \rangle + 2 \left| \Gamma'_{11}(\tau) \right| \cos(\Theta + 2\pi\nu_0\tau) + \langle I_2 \rangle \quad , \qquad (7.47)$$

where $\Theta = arg(\Gamma_{11}(\tau))$ is the argument of $\Gamma_{11}(\tau)$, which, in general, has to be assigned to a complex function. As before it is the term $2 \left| \Gamma_{11}(\tau) \right| \cos(\Theta + 2\pi\nu_0\tau)$ which can provide any interference and, for that reason, it is called the interference term.

It can be seen that such a temporal correlation function $\Gamma_{11}(\tau)$ has a real value of an average intensity at $\tau = 0$. We denote $\langle I_0 \rangle = \Gamma_{11}(0)$ and represent $\langle I_0 \rangle$ in terms of the spectral distribution $G(\Omega)$ in the following way:

$$\langle I_0 \rangle = \frac{\Delta S}{z_0^2} \int_\Omega G(\Omega) d\Omega \quad . \tag{7.48}$$

We can use this result to substitute $\Delta S / z_0^2$ in (7.46), and we get

$$\Gamma'_{11}(\tau) = \langle I_0 \rangle \frac{\int_\Omega G(\Omega) \exp(i\Omega\tau) d\Omega}{\int_\Omega G(\Omega) d\Omega} \quad .$$

Let $G(\Omega)_{\max}$ be the maximal value of the distribution $G(\Omega)$, then $G(\Omega)$ can be represented by a normalized function in the form $C(\Omega) = G(\Omega)/G(\Omega)_{\max}$. The use of such a dimensionless function gives

$$\frac{\int_\Omega G(\Omega) \exp(i\Omega\tau) d\Omega}{\int_\Omega G(\Omega) d\Omega} = \frac{\int_\Omega C(\Omega) \exp(i\Omega\tau) d\Omega}{\int_\Omega C(\Omega) d\Omega} \quad ,$$

where the integral $\int_\Omega C(\Omega) d\Omega$ is assigned to the area below the distribution, which can be seen as the effective width $\Delta\Omega_{eff}$ of the spectral distribution:

$$\Delta\Omega_{eff} = \int_\Omega C(\Omega) d\Omega \quad . \tag{7.49}$$

Thus, the temporal correlation becomes

$$\Gamma'_{11}(\tau) = \langle I_0 \rangle \frac{\int_\Omega C(\Omega) \exp(i\Omega\tau) d\Omega}{\Delta\Omega_{eff}} \quad , \tag{7.50}$$

where $C(\Omega)$ specifies the spectral contour of the radiation.

This special case of correlation of the field at two points, where the statistical correlation of the fields is mainly determined by the spectral distribution of the intensity, but depends only weakly on the dimensions and on the shape of the light source, is called temporal correlation.

4.1   Typical cases of temporal correlation

Now we analyze a few typical cases of temporal correlation. Since the essential part of any temporal correlation function is determined by the spectral contour of the radiation, given by the integral

$$\int_\Omega C(\Omega) \exp(i\Omega\tau) d\Omega \quad ,$$

we will discuss different cases of the function $C(\Omega)$, which can be associated with some important spectral distributions.

4.1.1 The Lorentzian spectrum

Let us consider the emission of atoms at rest. Radiation then occurs with the natural width of each line (determined by the life time of the excited atomic level, compare Chapter 5), and the lines of the emission spectrum have also a LORENTZian shape:

$$C(\Omega) = \frac{\gamma^2}{\gamma^2 + \Omega^2} \quad , \tag{7.51}$$

where γ is a constant which specifies the natural width of the spectral line. The function $\gamma^2/(\gamma^2 + \Omega^2)$ in the interval $\Omega \gtrless 0$ is shown in Fig.7.17,a; $C(\Omega) = 0.5$ at $\Omega = \gamma$ by definition. Integration of the LORENTZian curve gives an effective width $\Delta\Omega_{eff}$ below the curve:

$$\Delta\Omega_{eff} = \int_0^\infty \frac{\gamma^2}{\gamma^2 + \Omega^2} d\Omega = \gamma\pi/2.$$

We emphasize that $C(\Omega)$ is an even function, hence, the imaginary part of the integral in (7.50) must have zero value. We also restrict our calculation to the specific case of $\Omega \geq 0$ and $\tau \geq 0$, since any others cases can be calculated in a similar way. Under the restrictions above, the integral in (7.50) becomes

$$\int_0^{+\infty} \frac{\gamma^2}{\gamma^2 + \Omega^2} \cos(\Omega\tau) d\Omega = (\gamma\pi/2) \exp(-\gamma\tau).$$

For the temporal correlation function, therefore, the following expression is true:

$$\Gamma'_{11}(\tau) = \langle I_0 \rangle \exp(-\gamma\tau) \quad , \text{ for } \quad \tau \geq 0 \quad \text{and} \quad \gamma > 0 \quad . \tag{7.52}$$

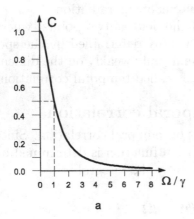 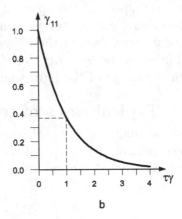

Figure 7.17. A LORENTZian spectral profile (a) and the temporal correlation fuction of the field associated with this spectral curve (b).

Since $\Gamma'_{11}(\tau)$ is a real function, it follows from the definition of the complex degree of coherence that, for the LORENTZian contour, the temporal degree of coherence has the simple form

$$\gamma_{11}(\tau) = \exp(-\gamma\tau). \qquad (7.53)$$

The function $\gamma_{11}(\tau)$, corresponding to the LORENTZian spectrum, is shown in Fig.7.17,b.

Even we have discussed the emission of atoms at rest, a LORENTZian profile can be observed under very special conditions, where the line broadening due to so-called homogeneous mechanisms is larger than the thermal (DOPPLER-) broadening (see next paragraph). Under such conditions, the shape of the lines is approximately a LORENTZian curve with a large value of the half width γ.

4.1.2 The Doppler spectrum

In low pressure discharges the radiating atoms move with a MAXWELLian velocity distribution. In the reference frame of the laboratory the frequencies of the emitting light wave undergoes a DOPPLER shift (C.DOPPLER (1803-1853)). Such a frequency shift is directly proportional to the velocity of the atom in the laboratory reference frame. Because the projections of the atomic velocities along the line of sight are distributed according to a GAUSSian curve, a so-called DOPPLER distribution of frequencies exists in the laboratory frame. Thus the spectral contour takes the shape of a GAUSSian curve:

$$C(\Omega) = \exp(-\Omega^2/\gamma_D^2) \quad , \qquad (7.54)$$

where γ_D is a constant associated with the effective width of such a DOPPLER profile. In low pressure discharges, the DOPPLER width is approximately 100 times larger than the natural line width. As before, $C(\Omega)$ is an even function, and the area below the curve can be calculated by integration over positive magnitudes of Ω. This integration gives the effective width

$$\Delta\Omega_{eff} = \int_0^\infty \exp(-\Omega^2/\gamma_D^2)d\Omega = \gamma_D\sqrt{\frac{\pi}{2}} \quad .$$

For the integral part of the correlation function we can now write

$$\int_0^\infty \exp(-\Omega^2/\gamma_D^2)\cos(\Omega\tau)d\Omega = \gamma_D\sqrt{\frac{\pi}{2}}\exp(-\gamma_D^2\tau^2/4) \quad . \qquad (7.55)$$

Thus, the temporal correlation function formed by a line with a DOPPLER shape must be written as

$$\Gamma'_{11}(\tau) = \langle I_0 \rangle \exp(-\gamma_D^2 \tau^2/4) \quad . \tag{7.56}$$

In turn, the complex degree of coherence has the shape of a GAUSSIAN curve over the variable τ :

$$\gamma_{11}(\tau) = \exp(-\gamma_D^2 \tau^2/4) \quad . \tag{7.57}$$

Contour $C(\Omega)$ and function $\gamma_{11}(\tau)$ are shown in Fig.7.18,a,b.

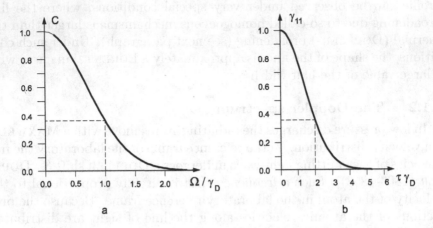

Figure 7.18. A DOPPLER spectral contour and the temporal correlation function of field associated with this contour.

4.1.3   A double spectral line

Let us discuss the correlation effect caused by two identical spectral lines, which both have a DOPPLER contour. It is clear that here two carrier frequencies have to be taken into account: ν_1 and ν_2, with $\nu_1 < \nu_2$. We interpret the frequency difference between the centers of the lines as a positive constant value $2\Delta\Omega_0 = 2\pi(\nu_2 - \nu_1)$. The low frequency variable Ω is calculated, as before, from the central frequency:

$$\Omega = 2\pi(\nu_1 + \nu_2)/2 - 2\pi\nu.$$

Thus, the center of the low frequency spectral line is located at $\Omega = -\Delta\Omega_0$, whereas the center of the high frequency spectral line is at $\Omega = -\Delta\Omega_0$. The total spectral profile is the sum of both contours:

$$C(\Omega) = \exp[-(\Omega + \Delta\Omega_0))^2/\gamma_D^2] + \exp[-(\Omega - \Delta\Omega_0))^2/\gamma_D^2] \quad ,$$

where γ_D specifies the widths of both contours. For the integral in (7.50), we find

$$\int \left(\exp[i\Omega\tau - (\Omega + \Delta\Omega_0)/\gamma_D)^2] + \exp[i\Omega\tau - (\Omega - \Delta\Omega_0)/\gamma_D)^2] \right) d\Omega \quad .$$

We restrict our consideration to the simple case of narrow lines, where the frequency difference between the centers of the lines is larger than the width of the line profiles, $2\Delta\Omega_0 > 2\gamma_D$ (Fig.7.19,a). Under this assumption the integration over each contour can be performed separately. For example, integration over the lower frequency line profile gives

$$\int \exp[i\Omega\tau - (\Omega + \Delta\Omega_0)/\gamma_D)^2]d\Omega =$$

$$= \exp(-i\Delta\Omega_0\tau) \int \exp[-(\tilde{\Omega}/\gamma_D)^2]d\tilde{\Omega} \quad ,$$

where $\tilde{\Omega} = \Omega + \Delta\Omega_0$ is a new variable. It is clear that the integration over $\tilde{\Omega}$ must give the function $\exp(-\tau^2\gamma_D^2/4)$, as before. Therefore, the contribution of the first spectral line to the correlation function has to take the form $\exp(-\tau^2\gamma_D^2/4)\exp(-i\Delta\Omega_0\tau)$. In a similar way we find $\exp(-\tau^2\gamma_D^2/4)\exp(i\Delta\Omega_0\tau)$ for the second spectral line. The sum of the terms gives us the correlation function

$$\Gamma'_{11}(\tau) = \langle I_0 \rangle \exp(-\gamma_D^2\tau^2/4)\left(\exp(-i\Delta\Omega_0\tau) + \exp(i\Delta\Omega_0\tau)\right) =$$

$$= 2\langle I_0 \rangle \exp(-\gamma_D^2\tau^2/4)\cos(\Delta\Omega_0\tau) \quad , \tag{7.58}$$

where $\langle I_0 \rangle$ is the average intensity from one spectral line at the axis of the beam.

It follows from (7.58) that for the temporal degree of coherence we find

$$\gamma_{11}(\tau) = \exp(-\gamma_D^2\tau^2/4)\cos(\Delta\Omega_0\tau) \quad . \tag{7.59}$$

Two DOPPLER broadened spectral lines and the corresponding temporal degree of coherence $\gamma_{11}(\tau)$ are shown in Fig.7.19. For both spectral lines vertical dashed lines are drawn through the points: $2\pi\nu/\gamma_D = \pm 1$. The normalized intensity value at these points is e^{-1}. Thus, for the case of two separated GAUSSIAN lines, the degree of temporal coherence has exponential envelopes $\exp(-\gamma_D^2\tau^2/4)$, and a harmonic oscillating factor $\cos(\Delta\Omega_0\tau)$. The oscillations shown in Fig.7.19 are caused by the value of the phase, and the period is $2\gamma_D/(\nu_2 - \nu_1)$.

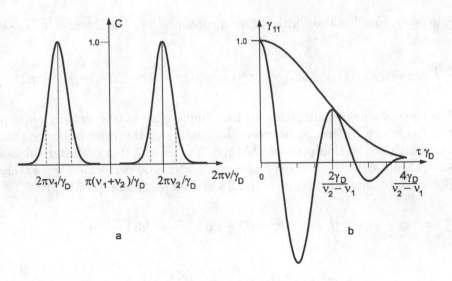

Figure 7.19. Two identical lines with DOPPLER profile (a) and the temporal correlation function of the field associated with this doublet (oscillating line) (b). The correlation function caused by each line is specified by one half of the GAUSSIAN shaped line.

4.1.4   Visibility of the yellow double-line of mercury

Experiments with a MICHELSON interferometer illuminated by yellow light from a mercury lamp (selected by a so-called Hg-yellow filter) gives a compelling evidence to the existence of the compound spectrum of the mercury lamp. Under lower pressure conditions, two yellow lines ($\lambda_1 = 577.0$ nm, $\lambda_2 = 579.1$ nm) are emitted. By changing the path difference between the two interfering rays in the MICHELSON interferometer, the visibility of interference fringes must show periodical decreases and increases following the temporal degree of coherence inherent to the double-line spectrum.

The scheme of the experiment is shown in Fig.7.20. A light beam from a low pressure mercury lamp passes through the optical filter. The yellow rays are formed by the first objective into a nearly parallel beam, which passes into the MICHELSON interferometer. One mirror is movable by a specific system which provides a very smooth and accurate motion of the mirror. The system consists of two cylinders and pistons; the space between the pistons is filled with glycerine, as shown in Fig.7.20. The mobile mirror M_2 is mounted on the first piston, and the second piston can be pushed precisely by a micrometer screw. Because glycerine has high viscosity, this system allows smooth movement of M_2 with high precision. When shifting M_2 progressively in one direction, one

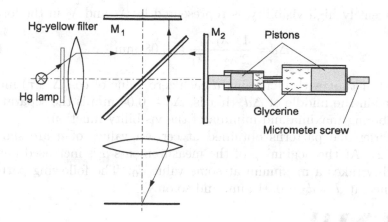

Figure 7.20. Setup for demonstrating visibility variations with a MICHELSON inter-ferometer illuminated by the yellow double-line of mercury.

observes that variations of the interference pattern typical to a double-line spectrum appear.

With an appropriate adjustment of the interferometer, motion of the mobile mirror shows that the visibility remains high although the mirror is moved several millimeters, but the visibility in the central part of the interference pattern decreases periodically each 0.08 mm. This observa-tion shows that the envelope of the visibility decreases slowly with the parameter τ (see Fig.7.19), in contrast to rather fast decreases and in-creases due to the existence of two spectral lines. It also tells us that the components of the double-line should be regarded as non-overlapping.

In order to make estimations of the parameters of our experiment, let us consider the relationship between the observed facts, the path difference, and the wavelengths. According to (7.59) such a periodic variation of the visibility is caused by a cosine function $\cos(\Delta\Omega_0\tau)$, where $\Delta\Omega_0 = \pi(\nu_2 - \nu_1)$, and $\tau = 2d/c$ (d is the shift of the mobile mirror). $\nu_1 = c/\lambda_1$, $\nu_2 = c/\lambda_2$. Where $\cos(\Delta\Omega_0\tau)$ has its extreme values 1, -1, the visibility has also maximal and minimal values. If this is the case, the argument of $\cos(\Delta\Omega_0\tau)$ satisfies the requirement:

$$\Delta\Omega_0\tau = 2\pi d \left[\frac{1}{\lambda_1} - \frac{1}{\lambda_2}\right] = \pi|m| \quad , \quad \text{for} \quad m = 0, \pm 1, \pm 2, \dots \quad ,$$

where m is the order of the extreme value of the periodical part of the vis-ibility. The spacing Δd between two values of $\cos(\Delta\Omega_0\tau)$, corresponding to extremely high visibility, is represented by λ_1 and λ_2 in the form:

$$\Delta d = \frac{1}{2}\frac{\lambda_1\lambda_2}{\lambda_2 - \lambda_1} \approx 0.08 \text{ mm} \quad . \tag{7.60}$$

It is clear that with increasing d each zero-value of $\cos(\Delta\Omega_0\tau)$ must be located in the middle of Δd, that is, $\Delta d \approx 0.04$ mm is the separation of neighboring maxima and minima of the visibility function.

Interference patterns obtained at certain values of d are shown in Fig.7.21. At the beginning of the measurements d is increased until the visibility takes a maximum at some value d_0. The following pattern is obtained at $d = d_0 + 0.04$ mm, and so on.

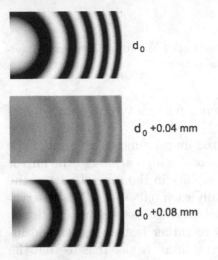

d_0

d_0 +0.04 mm

d_0 +0.08 mm

Figure 7.21 Interference patterns observed with a MICHELSON interferometer when increasing the position d of the mobile mirror. The interferometer is illuminated by two yellow lines of mercury ($\lambda_1 = 577.0$ nm, $\lambda_2 = 579.1$ nm).

Let us estimate the amount of interference fringes between neighboring maxima and minima. For the central part of the interference pattern a shift of $\Delta d = \lambda/2$ is needed to change the order of interference by one. Since each interference fringe has to be associated with a certain order of interference, the total number M of fringes between neighboring maxima and minima can be estimated as $0.04/(\lambda/2) \approx 150$. Using (7.60) we get the same estimation represented in a general form:

$$M = \frac{1}{4}\frac{\overline{\lambda}^2}{\lambda_2 - \lambda_1} \cdot \frac{2}{\overline{\lambda}} = \frac{\overline{\lambda}}{2\Delta\lambda} \quad ,$$

where $\overline{\lambda} = (\lambda_2 + \lambda_1)/2$, $\Delta\lambda = (\lambda_2 - \lambda_1)$. $\overline{\lambda}/\Delta\lambda$ is the degree of monochromaticity introduced to describe a quasi-monochromatic wave, so the amount of fringes between neighboring maxima and minima of the visibility function has to be numerically equal to one half of the degree of monochromaticity.

5. The spatial mode

Now we consider the general case of field correlation from a remote source, where two points r_1 and r_2 have arbitrary positions x_1, y_1, z_1 and x_2, y_2, z_2, respectively (Fig.7.22). Again we assume that all required geometrical parameters satisfy the FRAUNHOFER approximation.

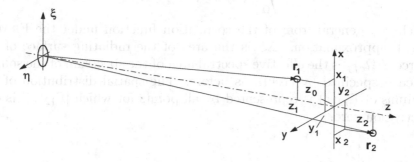

Figure 7.22. Mutual position of two points with respect to the beam axis in the case of the general spatial mode.

On the basis of the assumptions about the properties of the light source we conclude that the correlation function has to take the form of the product of two integrals for the general case. The first one is the integration over the surface of the source, and the second is the integration over the spectral shape of the source. According to the integrals we have considered above, we can assume that $\Gamma_{12}(\tau)$ should have the form

$$\Gamma_{12}(\tau) = \frac{1}{z_0^2} \int_\xi \int_\eta \exp\left[ik_0\xi\frac{\Delta x}{z_0} + ik_0\eta\frac{\Delta y}{z_0} \right] d\xi \, d\eta \cdot$$

$$\cdot \int_\nu G(\nu)\exp(i2\pi\nu\tau)d\nu = \Gamma_{12}(0)\cdot\Gamma_{11}(\tau), \qquad (7.61)$$

where z_0 is the distance between the source and the space of observation, and $G(\nu)$ describes the spectral distribution of the source. With the introduction of the frequency difference $\Omega = 2\pi(\nu_0 - \nu)$ and the dimensionless function $C(\Omega)$, the integral over the frequency variable should have a form similar to (7.50). In order to do this, we introduce a maximal value in the spectral distribution, which is denoted by $\langle I(\nu)_{\max}\rangle$. Hence, the integral over the frequency variable is given as

$$\int_\nu G(\nu)\exp(i2\pi\nu\tau)d\nu = \cdot$$

$$= \langle I(\nu)_{\max}\rangle \exp(i2\pi\nu_0\tau)\cdot \int_\Omega C(\Omega)\exp(i\Omega\tau)d\Omega \quad.$$

In turn we can write

$$\Gamma_{12}(\tau) = \frac{\langle I(\nu)_{\max} \rangle}{z_0^2 \Delta S \Delta\Omega_{eff}} \exp(i2\pi\nu_0\tau) \int_\xi \int_\eta \exp\left[ik_0\xi\frac{\Delta x}{z_0} + ik_0\eta\frac{\Delta y}{z_0}\right] d\xi d\eta \cdot$$

$$\cdot \int_\Omega C(\Omega)\exp(i\Omega\tau)d\Omega \quad , \tag{7.62}$$

which is a general from of the correlation function under the FRAUN-HOFER approximation. ΔS is the area of the radiating surface of the source, $\Delta\Omega_{eff}$ is the effective spectral area of the spectrum of the source, which is specified by (7.49), as before. The spatial distribution of the modulus of $\Gamma_{12}(\tau)$ is represented by all points for which $|\Gamma_{12}(\tau)|$ is distinctly non-zero.

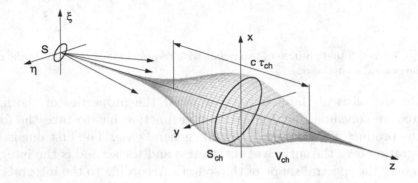

Figure 7.23. The spatial distribution of the function $\Gamma_{12}(\tau)$ for a circular source, which has a DOPPLER spectral contour. The coherence volume is given by the product: $V_{ch} = c\tau_{ch}S_{ch}$.

For example, a substantial part of $|\Gamma_{12}(\tau)|$ can take a form as shown in Fig.7.23 in the case of circular shape of the radiating surface, and with a GAUSSIAN shape of the spectral distribution. Such a spatial distribution of $|\Gamma_{12}(\tau)|$ extends roughly over a range of $c\tau_{ch}$ along the propagation direction, where τ_{ch} is the so-called *coherence time*. Assuming $\Delta\Omega_{eff}$ to be the effective width of the spectral distribution, the value of the coherence time can be estimated by

$$\tau_{ch}\Delta\Omega_{eff} \sim 2\pi \quad .$$

Let us denote by r_{ch} the radius of the maximal cross section of the distribution shown in Fig.7.23, and let us call r_{ch} the radius of coherence. In the case under discussion the radius of coherence satisfies Eq. (7.33), where $\Delta d_{ch} = 2r_{ch}$ is the diameter of the coherence distribution of the radiating disk. It follows from (7.33), that

$$r_{ch}\frac{D}{\lambda_0 z_0} = 1 \; . \qquad (7.63)$$

Such a spatial distribution is called the *spatial mode*, and its volume is called the *coherence volume*. In the particular case under discussion the coherence volume has to be $V_{ch} = c\tau_{ch}S_{ch}$, where the coherence area S_{ch} is the area of the maximal cross section of the spatial mode; here $S_{ch} = \pi r_{ch}^2$.

In the general case the dimensions of the coherent volume follow from relations which are typical for any arguments of FOURIER integrals, since the surface integral and the spectral integral in (7.62) both have the form of FOURIER integrals. The relations under questions are

$$k_0\xi_{\max}\frac{\Delta x_{ch}}{z_0} \geq 2\pi \quad , \qquad k_0\eta_{\max}\frac{\Delta y_{ch}}{z_0} \geq 2\pi \quad , \qquad (7.64)$$

$$\tau_{ch}\Delta\Omega_{eff} \geq 2\pi \quad . \qquad (7.65)$$

Here, $\xi_{\max}$ and $\eta_{\max}$ specify the dimensions of the source. Δx_{ch} and Δy_{ch} are assumed to be the dimensions of the maximal cross section of the coherence volume. Parameter $\Delta\Omega_{eff}$ denotes the effective width of source's spectrum. Inequalities (7.64) and (7.65) are called the *coherence conditions*, which allow estimation of the coherent volume in the arbitrary case of any partially coherent light beam. The inequality (7.65) represents the so-called *temporal coherence condition*, whereas the inequalities (7.64) are named the *conditions of spatial coherence*. For a given distance between the source and the plane of observation of the correlation, the following statement is true: The smaller the linear dimensions $\xi_{\max}$ and $\eta_{\max}$ are, the larger the coherence dimensions Δx_{ch} and Δy_{ch} will be.

It is appropriate to mention here, that, for a given wavelength λ_0 associated with the carrier frequency, angular variables are useful to make estimations of the geometrical parameters of spatial coherence. For example, by introducing the angular dimensions of the cross-section of the coherence volume, $\Phi_{ch}^{(x)} = \Delta x_{ch}/z_0$, $\Phi_{ch}^{(y)} = \Delta y_{ch}/z_0$, the inequalities (7.64) become

$$\Phi_{ch}^{(x)} \geq \lambda_0/\xi_{\max} \quad \text{and} \quad \Phi_{ch}^{(y)} \geq \lambda_0/\eta_{\max} \; . \qquad (7.66)$$

Opposite, with the angular dimensions of the source, $\Psi_x = \xi_{\max}/z_0$ and $\Psi_y = \eta_{\max}/z_0$, the linear dimensions of the cross-section of the coherence volume should satisfy the inequalities

$$\Delta x_{ch} \geq \lambda_0/\Psi_x \quad \text{and} \quad \Delta y_{ch} \geq \lambda_0/\Psi_y \; . \qquad (7.67)$$

Roughly speaking, the conditions (7.67) emphasize the fact that a spatial limit of the order of magnitude of λ exists, which restricts any area of spatial coherence to a value of λ^2. This follows from (7.67) for $\Psi_x \to 2\pi$ and $\Psi_y \to 2\pi$; and such a limited case must be associated with an extremely extended source. As follows from (7.66), in the other limiting case of a "point source", where $\xi_{max} \to \lambda_0$ and $\eta_{max} \to \lambda_0$, $\Phi_{ch}^{(x)}$ and $\Phi_{ch}^{(y)}$ will both tend to 2π. This can be interpreted as an entirely opened wave front from a "point source". In other words, such a "point source" emits a quasi-monochromatic wave with a spatial coherence of extremely high quality ($|\gamma_{12}(0)| \approx 1$).

5.1   Coherence conditions and interference

The coherence conditions allow us a qualitative analysis of typical cases of interference. In every interference scheme of the MICHELSON type, the spatial mode is represented by a cylindrical volume of length $c\tau_{ch}$ and a radius which is small compared to $c\tau_{ch}$, which arises after a beam splitting element (Fig.7.24,a). For example, the semi-transparent plate of a MICHELSON interferometer or the reflecting surfaces of a FABRY-PEROT interferometer can play the role of such a beam splitter. In the case of any interference scheme of the YOUNG type, the spatial mode takes the shape of a flat disk, because, here, the coherent length $c\tau_{ch}$ is relatively short compared to the radius of the maximal cross-section of the coherent volume. For example, the coherence volume formed by radiation from a star can have a diameter of tens of meters, whereas $c\tau_{ch}$ may range only over several micrometers (Fig.7.24,b).

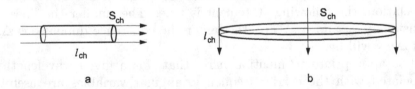

Figure 7.24. Two typical cases of the coherent volume: a thin cylindrical volume belongs to the MICHELSON type of interference (a); a flat disc coherence volume, e.g. from a star, belongs to the YOUNG type (b).

Let us now consider how we can observe interference fringes from the light of an extended source. In a simple case, a thin plane parallel glass plate can provide interference when illuminating the plate with a quasi-monochromatic light beam from an extended source, as shown in Fig.7.25,a. For a given distance z_0 between the radiating surface of the source and the plate, the coherence volume somewhere close to the plate will look like a needle-shaped cylinder. When both angular dimensions of the source tends towards 2π, according to (7.66) we can regard both dimensions of the coherence area to be about λ_0. Hence, two points on the upper surface of the plate must be considered as incoherent sources, because these points belong to different coherent volumes. Secondary waves emitted by such points can not interfere. Thus, only one way exists to form interference, namely, when the interfering waves origin from one coherence volume: one wave is created due to reflection of the incident wave on the outer surface, and the other wave due to refraction and reflection on the lower surface (Fig.7.25a). Here, interference fringes of equal inclination will appear, if the optical path difference Δs between the interfering rays is smaller than the length $c\tau_{ch}$ of the coherence volume. Therefore, such an extended source can provide interference only in the case of a thin plane parallel plate, and the interference fringes are located at infinity.

In the case of a source with limited angular dimensions, the coherence area can have a relatively large size on the upper surface of the plate, covering points 1 and 2 in Fig.7.25,b. Here, two arbitrary points within this area can play the role of partially coherent secondary sources, because these points belong to one spatial mode. It follows from our considerations of interference on plates (see Chapter 1), that interference fringes of an arbitrary location can be observed under the conditions mentioned above. For example, fringes of equal thickness may appear due to two waves: one is reflected at point 2 of the outer surface, and the other originates from point 1 and leaves the plate at point 2 after refraction and reflection at the other side of the plate. Nevertheless, the temporal coherence condition $\Delta s < c\tau_{ch}$ must also be fulfilled here.

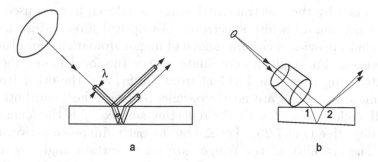

a   b

Figure 7.25. Needle-shaped coherence volume associated with interference from a plane parallel plate using an extended source (a). Typical coherence volume with fringes of equal thickness located close to the surface of a non-parallel plate (b).

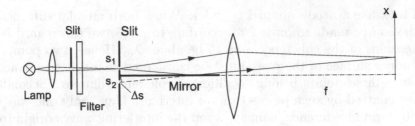

Figure 7.26. Setup for observation of fringes with a LLOYD mirror using a halogen lamp and a filter.

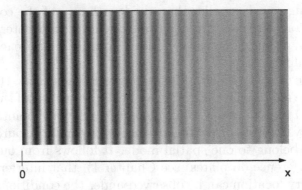

Figure 7.27. Fringes with the Lloyd mirror with a wider spectral distribution of the source.

Let us now discuss an interference experiment using a LLOYD mirror, where geometrical parameters, as well as the spectral distribution of the source, affect the visibility of the fringes. In order to observe the effect caused by the spectral distribution, a halogen lamp is used which emits a continuous white spectrum. An optical filter is inserted into the beam to provide a certain degree of monochromaticity for observing interference. For a given coordinate x over the focal plane, there are two interfering rays: one incident from slit S_1, and the other from the slit's image S_2. Both are nearly parallel to each other, and both make a small angle $\alpha = x/f$ with the reflecting surface. f is the focal length of the objective (Fig.7.26). Let Δs be the path difference between these rays. The visibility of the fringes formed at certain angles α may be dependent on the spectral distribution after the optical filter F, since the path difference Δs can become comparable to the coherence length $l_{ch} = c\tau_{ch}$. If this is the case, the visibility must be decrease dramatically. Such a case is shown in Fig.7.27, where fringes of higher orders vanish.

We now derive an expression for the average intensity of the interference pattern. We emphasize that the degree of spatial coherence should become real, just like the function $\gamma_{12}(0)$, corresponding to a slit. Since all geometrical parameters affecting the visibility are assumed to be invariable, $\gamma_{12}(0)$ has a constant value. Due to diffraction on slit S_1 (its width is d) the brightness of the peripheral fringes decreases. This effect is described by the factor $(\sin^2 u)/u^2$, where $u = \pi dx/(\lambda f)$. Let us assume that the optical filter cuts out a rectangular spectral region around the central frequency ν_0 within the limits $-\Delta\Omega/2$, $+\Delta\Omega/2$. The temporal degree of coherence will then take the real form

$$\gamma_{11}(\tau) = \frac{\sin(\Delta\Omega\tau)}{\Delta\Omega\tau} \ . \tag{7.68}$$

Taking into account the considerations made before, we find the intensity at any point of the pattern as

$$\langle I \rangle = 2 \langle I_0 \rangle \frac{\sin^2(u)}{u^2} \left[1 - \gamma_{12}(0)\gamma_{11}(\tau)\cos(2\pi\Delta s/\lambda_0)\right] \ , \tag{7.69}$$

where the sign "$-$" specifies visibility in the case of the LLOYD interference scheme. Because we assume $\gamma_{12}(0) = const.$, any variations of the visibility over the pattern must be due to variation of Δs. Since $\Delta s = \alpha b/2$, the time delay τ, which depends on α and x, satisfies the relationship

$$c\tau = \Delta = \alpha b/2 = \frac{xb}{2f} \ . \tag{7.70}$$

It follows from (7.68) and (7.70) that the visibility of the fringes of low orders is mainly determined by the spatial degree of coherence, since $\gamma_{11}(\tau) = \sin(\Delta\Omega\tau)/(\Delta\Omega\tau) \approx 1$ at $\tau \approx 0$. At the periphery, where $\Delta s \sim c\tau_{12}$, the visibility decreases to zero. It seen in Fig.7.27 that in spite of the decreasing brightness due to the diffraction effect, a set of peripheral fringes exists with nearly zero contrast (around an order of $m_{\max} = 16$). Thus, the coherence length represented in terms of $m_{\max}$ and λ_0 is given as

$$c\tau_{ch} \sim m_{\max}\lambda_0 \ , \tag{7.71}$$

because the fringe spacing corresponds to a path difference of λ_0. Through substitution of numerical values, we estimate τ_{ch} to be about $\tau_{ch} = 16/\nu_0 \approx 3.2 \cdot 10^{-14}$ s, and the effective spectral width of the filter to be about $\Delta\nu_{eff} = 1/\tau_{ch} \approx 3 \cdot 10^{13}$ Hz. The expression (7.71) gives an estimation of $m_{\max}$, since the existence of fringes of low contrast at the right-hand side of the picture shows that the temporal degree of coherence has nearly-zero values at $\tau > \tau_{ch}$. This situation is similar to the appearance of low contrast fringes in the YOUNG interference scheme at $\gamma_{12}(0) < 0$ in Fig.7.11.

5.2   The spatial mode of a black body source

Let us assume that black body radiation of the temperature T passes through a small circular aperture of area σ in one wall of the black body (Fig.7.28). The radiation becomes quasi-monochromatic after passing an optical filter with effective spectral width $\Delta\Omega_{eff}$. For the given values λ_0 and σ, let z_0 be the distance which provides validity of the FRAUNHOFER diffraction approximation. It would be of interest to find the amount of quanta within the coherence volume under the parameters given above.

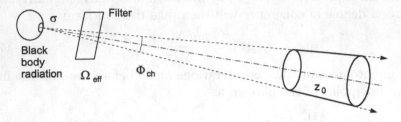

Figure 7.28.   A spatial mode from a black body and an optical filter.

According to PLANK's law (4.15), the energy density $\rho(T)$ close to ν_0 has the form

$$\rho(\nu_0) = \frac{8\pi h\nu_0^3}{c^3}\overline{n} \quad , \qquad (7.72)$$

where $\overline{n} = 1/(\exp(h\nu_0/k_BT) - 1)$. The amount of light energy passing through the aperture and the optical filter per unit time has to be equal to $c\sigma\rho(\nu_0)\Delta\nu_{eff}$, where

$$\Delta\nu_{eff} = \Delta\Omega_{eff}/(2\pi) = 1/\tau_{ch} \quad .$$

Hence, the amount of energy during the coherence time becomes

$$c\sigma\rho(\nu_0) \quad .$$

Since any black body radiation shows spatial anisotropy, the light flux emitted by the aperture spreads over the solid angle of 2π. The solid angle associated with coherence area S_{ch} of the coherence volume can be estimated by

$$\Psi_{ch}^2 = \frac{S_{ch}}{z_0^2} = \frac{S_{ch}}{z_0^2} = \frac{S_{ch}\lambda_0^2}{4\sigma} \quad .$$

Thus, the amount of light energy considered here is given as

$$\sigma c\rho(\nu_0)\frac{\Psi_{ch}^2}{2\pi} = \frac{1}{8\pi}c\lambda_0^2\rho(\nu_0) = \frac{c^3}{8\pi\nu_0^2}\rho(\nu_0) \quad .$$

Substitution for $\rho(\nu_0)$ by its magnitude from the right-hand side in (7.72) gives the expression

$$U_{ch} = h\nu_0\overline{n} = \frac{h\nu_0}{\exp(h\nu_0/k_BT) - 1} , \qquad (7.73)$$

where U_{ch} specifies the average light energy within the coherence volume at the carrier frequency ν_0 over the given frequency range $\Delta\Omega_{eff}$. Thus, we achieved the fundamental result that - for black body radiation - the average amount of quanta within the coherence volume, and within the spatial mode as well, must be exactly equal to the average amount of quanta $\overline{n} = 1/(\exp(h\nu_0/k_BT) - 1)$ at a given carrier frequency ν_0.

We have to emphasize that in any interference and diffraction phenomena, the existence of any stable pattern of fringes must be caused by quasi-monochromatic waves incident from one spatial mode. A superposition of two waves, one wave emitted from one spatial mode, and the other wave from another spatial mode, will produce a statistical average of the fields of zero. For this reason, the average number of light quanta which actually form the interference pattern will always be within the limits of $\overline{n}$ for any thermal source. We have seen in Chapter 4 that for black body radiation in the visible spectral range $\overline{n} \ll 1$ is valid due to the relatively low temperatures of heated bodies. Thus, the situation of interference of single quanta is believed to be typical for all interference conditions.

In order to describe the amount of quanta within a spatial mode of non-equilibrium thermal radiation the *degree of degeneracy* δ is used. For all spectral lines the degree of degeneracy is within the limits of 10^{-3}, where $\delta \approx 10^{-3}$ belongs to the most coherent spectral lines. The monochromaticity degree $\nu_0/\Delta\nu$ of such spectral lines can reach values of 10^8 which corresponds to magnitudes of the coherent length of $l_{ch} = c\tau_{ch} \approx 1$ m. In any case such a low value of the degree of degeneracy causes a weak intensity in most experimental schemes for observation of interference. The experiments on the interference of single quanta considered in Chapter 5 exemplify such a weak intensity from thermal sources.

6. Computer simulation of field correlation

The GAUSSian noise of the quadrature components associated with optical fields from thermal sources is mainly responsible for all the features of field correlation of such radiation. This idea provides us with the basis for the computing of field correlation.

We discuss here the simulation of spatial and temporal correlation functions for a few typical simple cases. We restrict our considerations to a slit, a disk, a DOPPLER contour, and a double spectral line, thus, to cases of real correlation functions. This allows us to propose a simple method for computing the quadrature components of the correlated field. First, the distribution of the quadrature components θ, ζ of the source is calculated by means of the polar coordinate method (see Chapter 6). Such a distribution is determined either by the geometrical shape or by the spectral contour of the source. Nevertheless, in all cases the quadrature components θ_c, ζ_c can be computed by means of only cosine functions:

$$\theta_{c,i} = \sum_k \theta_k \cos \phi_{ik} \quad \text{and} \quad \zeta_{c,i} = \sum_k \zeta_k \cos \phi_{ik} \quad , \qquad (7.74)$$

where the i-th term of the distribution of the components θ_c, ζ_c is a cosine expansion of the distribution of the source, θ_k, ζ_k. Here, the phase ϕ_{ik} links the appropriate magnitudes of the source's distribution and the distribution of the correlated field. The summations in (7.74) are similar to the integral representations applied above to evaluate correlation functions.

6.1   A slit

Algorithm **CFSlit()** allows the computation of a model of a correlated field and the appropriate correlation function for a narrow slit (Appendix 7.A). The variables $\theta_O[k]$, $\zeta_O[k]$ are associated with quadrature components of a field emitted by the slit, where integer magnitude k specifies the coordinate of a point within the slit. In a similar manner the variables $\theta_C[m]$, $\zeta_C[m]$ have to be assigned to a correlated field, where the point's coordinate is specified by the integer m. One random realization of quadrature components within the slit $\theta_O[k]$, $\zeta_O[k]$, $k = 0, ...19$, is formed by means of the **Polar-coordinates()** procedure. For the given realization of $\theta_O[k]$, $\zeta_O[k]$ the random distribution of $\theta_C[m]$, $\zeta_C[m]$ is then calculated as a cosine expansion linked to (7.74) in the form

$$\theta_C[15 + m] = \theta_C[15 + m] + \theta_O[k] \cdot \cos(\beta(m - 15)k) \quad ,$$

$$\zeta_C[15 + m] = \zeta_C[15 + m] + \zeta_O[k] \cdot \cos(\beta(m - 15)k) \quad ,$$

where m varies between -15 and 14, so that the value $m = 0$ can be associated with the central point of observation of the correlated field. Thereby, with the new integer variables $m = 15 - i$, and $n = 15 + i$, one particular contribution to the correlated field, variables $CF[i]$, is formed via summation as

$$CF[i] = CF[i] + \theta_C[m] \cdot \theta_C[n] + \zeta_C[m] \cdot \zeta_C[n] \quad ,$$

where i runs from 0 to 7.

On running the main loop, which contains $Max = 50000$ cycles, variables $CF[i]$ receive values, simulating the correlation function from a narrow slit. Here, it is the integer value $2i$ that models the distance Δx between two points within the correlated field, which are symmetrically with respect to the beam axis. After completion of the main loop, all the variable $CF[i]$ are normalized to the value of $CF[0]$ in order to form a normalized correlation function, which is assumed to be $\gamma_{12}(0) = \sin(\pi a \Delta x / z_0)/(\pi a \Delta x / z_0)$ (see eq. 7.26). Data associated with the obtained normalized correlation function are shown in Fig.7.29. Assuming that these data should be fitted by the function $\sin(\beta k_{\max}(2i))/(\beta k_{\max}(2i))$, modeling the function in (7.26), let us estimate the number $2i$ corresponding to the zero value of $\sin(\beta k_{\max}(2i))/(\beta k_{\max}(2i))$. For the given parameter $\beta = \pi/140$ and for $k_{\max} = 19$, the desired value of $2i$ is estimated to be between 7 and 8, which is in good agreement with the data presented in Fig.7.29.

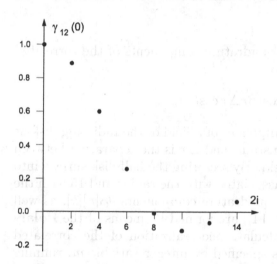

Figure 7.29 The function obtained when simulating the correlation field function of a slit. The integer numbers on the abscissa correspond to the argument Δx of the spatial correlation function $\gamma_{12}(0)$.

6.2 A disk

Procedure **CFDisk()** simulates the correlated field and correlation function of a circular source (Appendix 7.B). This procedure is similar to that in the previous case, with the exception of a simulation of the two-dimensional integration over the disk surface. In order to simplify the calculations, only one half of the surface is taken into account, due to the axial symmetry of the source (Fig.7.30,a).

We represent such a half disk by elements, each with an area of $r \Delta r \Delta \phi$ according to the polar coordinates r, ϕ. It follows from (7.30) that one

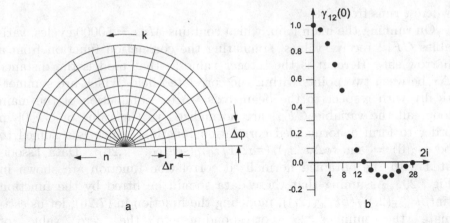

Figure 7.30. The sectoring of one half disk surface in terms of polar coordinates (a). The function obtained with simulating the spatial correlation function from a disk (b). Integer numbers on the abscissa ar assumed as being similar to the variable Δx of function $\gamma_{12}(0)$.

particular contribution to the quadrature components of the correlated field has to take the form

$$\theta_O \cdot r \cdot \cos(\beta r \Delta x \cos(\phi)) \quad ,$$

where θ_O is the quadrature component of a field of the radiating disk at the coordinates r, ϕ; β is a constant, and Δx is the separation between points within the correlated field. By sectoring the half disk surface into 10×15 elements, where 10 is associated with the radius and 15 with the polar angle, 150 values of the quadrature components $\theta_O[n][k]$, as well as $\zeta_O[n][k]$ ($n = 0,..9$; $k = 0,...14$), are formed by means of the **Polar-coordinates** procedure to calculate one realization of the correlated field θ_C, ζ_C. Assuming Δx, as specified by integer variable m, running from -15 to 14, the two-dimensional cosine expansion will have the from

$$\theta_C[15 + m] = \theta_C[15 + m] + \theta_O[n][k] \cdot \cos(\pi \beta mn \cos(\pi k/15)) \quad ,$$

$$\zeta_C[15 + m] = \zeta_C[15 + m] + \zeta_O[n][k] \cdot \cos(\pi \beta mn \cos(\pi k/15)) \quad ,$$

where the summations are performed by two variables n and k. As before, new integer variables $m = 15 - i$ and $n = 15 + i$ provide one particular contribution to the correlated field, variables $CF[i]$, which is formed via summation as

$$CF[i] = CF[i] + \theta_C[m] \cdot \theta_C[n] + \zeta_C[m] \cdot \zeta_C[n] \quad ,$$

where i runs from 0 to 14. After completion of the main loop of the procedure, $Max = 50000$, and after normalization to the value of $CF[0]$, 15 magnitudes of $CF[i]$ have to describe the profile of the function $2J_1(x)/x$, as in (7.32). These data are shown in Fig.7.30,b. Because the integer variable i specifies one half of the disc when calculating $CF[i]$ (the distance between two points within the correlated field), the diameter of the coherence area $(2i)_{ch}$ may be estimated by the first zero of the distribution of $CF[i]$ as

$$pi \cdot \beta \cdot (2i)_{ch} \cdot n_{max} \approx 1.22\pi \ .$$

Substitution of $n_{max} = 9$ and $\beta = 0.01$ gives 13.6 for the diameter of the coherence, which is in a good agreement with the observed data, where the first zero it is found at about 14.

6.3   A Doppler spectral line profile

Procedure **CFDoppler**(), Appendix 7.C, performs computations of the temporal correlation function in the case of a DOPPLER shaped spectral line. The general items of the procedure are similar to that of procedure **CFSlit**(), except new variables $sp[k]$, which model an amplitude spectrum related to the DOPPLER contour. When computing the quadrature components of the correlated fields, each cosine term includes the appropriate harmonics $sp[k]$, where the integer variable k corresponds to a certain frequency within the DOPPLER contour. Function $sp[k]$ is shown in Fig.7.31,a. Data obtained by the procedure are shown in Fig.7.31,b. The data display good approximation to a GAUSSian curve.

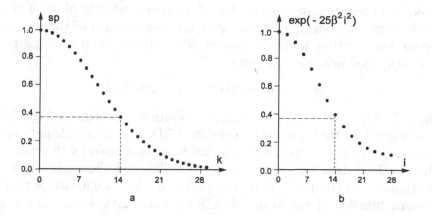

Figure 7.31. One half of the normalized amplitude distribution assigned to a DOPPLER contour (a). Simulated function $\exp(-\gamma_D^2\tau^2/4)$ (b).

In order to estimate the width of the contour in terms of the initial constants of the procedure, we should take into account that the variables $sp[k]$ describe the amplitude distribution over the spectral contour. Therefore, the spectral distribution of the average intensity should have a width which is twice as much as the width of $sp[k]$. Because $sp[k] = \exp(-0.005(k)^2)$, the width of the intensity distribution must be specified by the exponential factor in the form $\exp(-0.01(k)^2)$. This implies that the dimensionless constant associated with γ_D is equal to 10. The obtained data which simulates the function $\exp(-\gamma_D^2\tau^2/4)$, should obey the law: $\exp(-25 \cdot \tau^2)$, with a dimensionless variable τ. This variable has the discrete from: $\tau = \alpha \cdot i$, where the integer i is varying from 0 to 28 in the case under consideration, and α is a constant. On the other hand, in the phase $\beta \cdot (m - 20) \cdot k$ of the cosine expansion the integer k is associated with the circular frequency and $\beta \cdot (m - 20)$ with the variable τ. Hence, the required constant α is simply equal to β. Thus, the law which describes the obtained data has the form $\exp(-25 \cdot \beta^2 \cdot i^2)$. Let us estimate the value of i, for which $\exp(-25 \cdot \beta^2 \cdot i^2)$ is equal to $e^{-1} \approx 0.368$ (dashed lines in both graphs in Fig.7.31). It can be seen that $\exp(-25 \cdot \beta^2 \cdot i^2)$ has the value e^{-1} around $i = 14$. In turn, substitution of $\beta = \pi/210$ in the argument of the exponential function, together with the requirement $25 \cdot \beta^2 \cdot i^2 = 1$, gives i to be between 13 and 14, which is in good agreement with the estimation above.

6.4   A double-Doppler contour

We now compute the temporal correlation function for a double spectral line, consisting of two lines of identical contours. Due to the spectral symmetry the initial frequency of the spectral distribution has to be equal to zero; a cosine expansion of the amplitude distribution of the quadrature components will then give the correct form of the temporal correlation function under discussion. The amplitude spectral distribution $sp[k]$ used here has the form

$$sp[k] = \exp(-0.025 * (k - 20)^2)   ,$$

where $k = 0, \ldots 39$. The spectral distribution is shown in Fig.7.32,a. Procedure **CFDbDoppler**(), Appendix 7.D, performs calculations of data modeling the temporal correlation function caused by this contour. The obtained data are presented in Fig.7.32,b and show relaxation oscillations. According to (7.59) the period of the oscillations is that of a cosine function of the form $\cos(\Delta\Omega_0\tau)$, where $\Delta\Omega_0$ is one half of the separation of the centers of the spectral contours.

It follows from every cosine term $\cos(\beta \cdot m \cdot k)$ that τ should be associated with the variable $\beta \cdot k = 0.025 \cdot k$. $\Delta\Omega_0$, counted in terms of

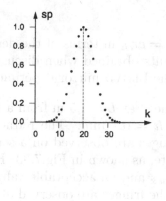
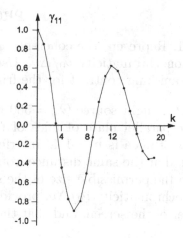

Figure 7.32. One half of the normalized amplitude distribution for a double-
DOPPLER coherence of that contour (a). Calculated temporal correlation function
(b).

integer numbers now, is equal to 20. It is seen from Fig.7.32 that the
period of oscillations is about $k = 12$. For the given values $\tau = 0.025 \cdot 12$
and $\Delta\Omega_0 = 20$ the product $\Delta\Omega_0\tau = 6$ is close to 2π, as the full period
of the cosine function should be.

SUMMARY

Correlation functions of the electromagnetic field are directly connected
with the visibility function of interference fringes. The basis for the calculation of the correlation functions are the properties of the radiating
surface of a light source. Most important properties are the statistical
independence of elementary sources on this surface and the assumption
that all elementary sources emit radiation with the same spectral composition. These properties allow that temporal and spatial correlation
functions can be calculated independently of each other.

The concept of a spatial mode allows a qualitative analysis of most
typical cases of interference. The total number of quanta within one
spatial mode of thermal radiation is always substantially less than unity,
which results in interference of single quanta. The treatment of thermal
radiation in terms of GAUSSian noise allows the computer simulation of
typical cases of field correlation.

PROBLEMS

7.1. Represent the coherence length $l_{ch} = c\tau_{ch}$ in terms of the degree of monochromaticity $\nu_0/\Delta\nu$. Use the results obtained when discussing the vanishing contrast for the fringes in the LLOYD mirror experiment.

7.2. A light source ($\lambda_0 = 550$ nm) is placed at $L = 1.5$ m from a thin plane parallel plate of mica of thickness $h = 0.1$ mm. The refractive index of mica is $n = 1.4$. Interference fringes are observed on a screen located at the same distance from the plate, as shown in Fig.7.33. Estimate the permissible size of the source ξ_{max} and an acceptable value of monochromaticity $\nu_0/\Delta\nu$, provided that the fringes are observed at the center of the screen, and that the angle of incidence is $\theta = 60°$.

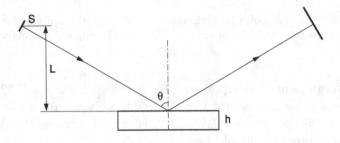

Figure 7.33.

7.3. The working plates of a FABRY-PEROT interferometer are separated by $h = 4$ mm. The reflectivity of the plates is $\mathcal{R} = 0.95$. Derive an expression for the temporal correlation function of light transmitted by the interferometer. Estimate the coherence length of the transmitted light.

7.4. Two plane parallel glass plates make a small dihedral angle $\alpha = 10^{-2}$ rad (Fig.7.34). A source of linear dimension $D = 2$ mm at a distance $L = 1$ m from the plates radiates quasi-monochromatic light ($\bar{\lambda} = 560$ nm) incident on the plates at an angle $\theta = 60°$. The surface of the upper plate is imaged by an objective inclined at the same angle θ to a screen. Estimate the number of fringes observed on the screen. Make an estimation for the effective range $\Delta\lambda$ of an optical filter required for such an observation.

7.5. It was found by measurements with the MICHELSON stellar interferometer that the angular diameter of Beteigeuze was $\theta = 0.047$ arc seconds. Calculate the value of the interferometer's baseline b needed to

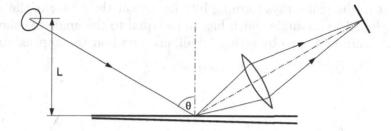

Figure 7.34.

achieve such a value with a wavelength $\lambda_0 = 550$ nm. Estimate the total number of quanta and the average light flux which form the interference fringes, provided that the total area of the diaphragms S_1, S_2 in front of the telescope objective have about $\Sigma = 125$ cm^2. The temperature of star's photosphere is approximately $T = 5000$ K. The radiation incident on the interferometer is made quasi-monochromatic by an optical filter, which provides a degree of monochromaticity of about 10^3.

SOLUTIONS

7.1. The disappearance of the fringes in Fig.7.37 is caused by the wide spectral distribution of the interfering light. This spectral width can be represented by the magnitude $\Delta\lambda$ or by the appropriate magnitude $\Delta\nu$ ($\Delta\nu = -\Delta\lambda\nu^2/c$). The visibility will approach the first zero value when $m_{max} = \lambda/\Delta\lambda$, where m_{max} is the maximal order of interference. Since the fringe of the zeroth order is located at the beginning of the interference pattern, m_{max} is also the total number of observed fringes before the first zero of the visibility. According to (7.71) $l_{ch} \approx \lambda_0 m_{max}$. On the other hand, $m_{max} = \lambda_0/\Delta\lambda = \nu_0/\Delta\nu$ has to be the degree of monochromaticity. Hence, the following relationships are found:

$$l_{ch} \approx \lambda_0 m_{max} = \lambda_0^2/\Delta\lambda = \lambda_0\nu_0/\Delta\nu = c/\Delta\nu \ .$$

7.2. Rays reflected from the upper surface of the plate can not provide an interference pattern on the screen, since these rays form a divergent pencil after reflection. The same statement is true for rays reflected from the lower surface of the plate. In order to produce an interference pattern located on the screen, one interfering ray must be reflected on the upper surface, and the second ray on the lower surface, as shown in Fig.7.35. Due to the symmetrical dispositions of source and screen with

respect to the plate, rays forming interference at the center of the screen
are included in an angle which has to be equal to the angle of coherence
Φ_{ch}. Assuming Φ_{ch} to be rather small, one can find the approximation

$$\Phi_{ch} = x_1 \cos\theta / L$$

for Φ_{ch}.

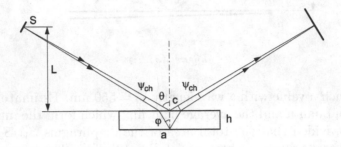

Figure 7.35.

In turn, $x_1/x = \sin(\pi/2 - \theta) = \cos\theta$, and $x_1 = x\cos\theta$. Since $x = h\tan\varphi$, where φ is the angle of refraction, for Φ_{ch} we get

$$\Phi_{ch} = \frac{h}{L}\tan\varphi\cos^2\theta \quad .$$

Because $n\sin\varphi = \sin\theta$, $\tan\varphi = \sin\theta/\sqrt{n^2 - \sin^2\theta}$. Now, for Φ_{ch}, we find

$$\Phi_{ch} = \frac{h}{L}\frac{\sin\theta\cos^2\theta}{\sqrt{n^2 - \sin^2\theta}} \quad .$$

Thus, the permissible dimension of the source can be estimated as

$$\xi_{max} \approx \frac{\lambda_0}{\Phi_{ch}} \leqslant \frac{\lambda_0 L\sqrt{n^2 - \sin^2\theta}}{h\sin\theta\cos^2\theta} \quad .$$

Substitution of all numerical values gives an estimation $\xi_{max} = 4.5$ cm.
In order to make an estimation for the acceptable value of monochromaticity $\nu_0/\Delta\nu$ we must find the path difference Δs between two interfering rays. Here we can use the expression found for the path difference
in the case of fringes of equal inclination, $\Delta s = 2nh\cos\varphi$. For the
given values h, n, and φ, the acceptable value of $\nu_0/\Delta\nu$ is given by the
condition $\Delta s = c\tau_{ch}$. According to the result of Problem 7.1 the coherence length can be represented in terms of monochromaticity $\nu_0/\Delta\nu$ and
$\lambda_0/\Delta\lambda$. Since $c\tau_{ch} = \lambda_0^2/\Delta\lambda$, the acceptable value of $\nu_0/\Delta\nu$ is given by

$$\frac{\nu_0}{\Delta\nu} = \frac{\lambda_0}{\Delta\lambda} = \frac{c\tau_{ch}}{\lambda_0} = \frac{2nh\cos\varphi}{\lambda_0} \quad .$$

Substitution of $\cos\varphi$ by $\left(\sqrt{n^2 - \sin^2\theta}\right)/n$ gives the expression

$$\frac{\nu_0}{\Delta\nu} = \frac{\lambda_0}{\Delta\lambda} \geq \frac{2h\sqrt{n^2 - \sin^2\theta}}{\lambda_0} \quad .$$

Substitution of numerical data gives us a needed degree of monochromaticity of $4 \cdot 10^2$.

7.3. According to the operating principle of the FABRY-PEROT interferometer its normalized spectral contour for transmitted light has to take the form of a LORENTZian curve (1.39):

$$\frac{I^{(t)}}{I_0} = \frac{1}{1 + (2\mathcal{F}\Delta\delta/\pi)^2} \quad ,$$

where $\Delta\delta$ is the phase deviation from a certain principal maximum, and $\mathcal{F} = \pi\sqrt{\mathcal{R}}/(1 - \mathcal{R})$ is the finesse. The phase deviation $\Delta\delta$ represented in terms of the angular frequency difference $\Omega = 2\pi(\nu_0 - \nu)$ is given as $\Delta\delta = 2h\Omega/c$. Thus, the LORENTZian contour becomes

$$\frac{1}{1 + (4h\Omega/(\varepsilon c))^2} \quad ,$$

where $\varepsilon = \pi/\mathcal{F} = (1 - \mathcal{R})/\sqrt{\mathcal{R}}$. The temporal correlation function associated with this contour has to take the normalized form

$$|\gamma_{11}(\tau)| = \exp(-\gamma\tau) \quad ,$$

where the constant is

$$\gamma = \frac{\varepsilon c}{4h} = \frac{c(1 - \mathcal{R})}{4h\sqrt{\mathcal{R}}} \quad .$$

We now estimate the coherence length to be $l_{ch} = 3c/\gamma$, which results in

$$l_{ch} = \frac{12h\sqrt{\mathcal{R}}}{(1 - \mathcal{R})} \quad ,$$

and, for the coherence time, we get

$$\tau_{ch} = \frac{12h\sqrt{\mathcal{R}}}{c(1 - \mathcal{R})} \quad .$$

Substitutions of numerical values gives $\tau_{ch} \approx 3.2 \cdot 10^{-9}$ s and $l_{ch} \approx 1$ m.

7.4. Fringes of equal thickness observed over the screen are caused by rays incident within the angle of coherence Φ_{ch}, as shown in Fig.7.36. We see that the region between the limits A and B, which confines the projections of the coherence area on the surface of the upper plate, should involve all fringes projected and observed over the screen. Therefore, the estimation of the total number of observable fringes has to take the form of the ratio $\overline{AB}/\delta x$, where δx is the fringe spacing on the surface of the upper plate. The path difference between two interfering rays in the case under consideration is estimated to be $2h\cos\theta$, provided that a thin air layer exists between the plates ($n = 1$). Hence, the requirement for maxima of the interference fringes becomes

$$2h\cos\theta = m\lambda_0 \quad .$$

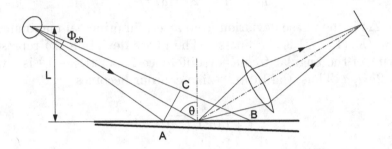

Figure 7.36.

Since the horizontal coordinate x orthogonal to the edge of the dihedral angle is connected to the approximate thickness of the air layer as $h = \alpha x$, the fringe spacing has to satisfy the relationship $2\delta x\cos\theta = \lambda_0/\alpha$, hence,

$$\delta x = \frac{\lambda_0}{2\alpha\cos\theta} \quad .$$

The linear dimension $\overline{AC}$ of the coherence area has to be estimated by $\overline{AC} = \Phi_{ch}L/\cos\theta$. For $\overline{AB}$, we then get

$$\overline{AB} = \overline{AC}/\cos\theta = \Phi_{ch}L/\cos^2\theta \approx \lambda_0 L/(D\cos^2\theta) \quad ,$$

if we assume $\Phi_{ch} \approx \lambda_0/D$. Taking into account the expression found for δx, the total amount of fringes between points A and B are found to be equal to

$$m_{\max} = \frac{\overline{AB}}{\delta x} \approx \frac{2L\alpha}{D\cos\theta} \quad .$$

Then the effective range $\Delta\lambda$ required to provide such observations is estimated according to λ_0/m_{max}:

$$\Delta\nu \leqslant \frac{\lambda_0}{m_{max}} = \frac{\lambda_0 D \cos\theta}{2L\alpha} .$$

Substitution of the numerical values gives $m_{max} \approx 20$ and $\Delta\lambda \leqslant 28$ nm.

7.5. According to (7.33) the separation d of the mobile mirrors which causes the first disappearance of the fringes in the case of a radiating disk, has to be equal to:

$$d = \frac{1.22\lambda_0}{\theta} ,$$

where $\theta = D/z_0$ is the angular diameter of the disk. Substitution of numerical values gives $d \approx 3$ m; hence, the radius of coherence is $r_{ch} = 1.5$ m. The total amount of quanta $\bar{n}$ within the coherence volume can be calculated assuming the star's photosphere emits like a black body. For the given temperature $T = 5000$ K and $\lambda_0 = 550$ nm one can find $\bar{n}$ to be

$$\bar{n} = \frac{1}{\exp(hc/(\lambda_0 k_B T) - 1)} \approx 6 \cdot 10^{-3} .$$

Since this number of quanta passes through the coherence area of radius $r_{ch} = 1.5$ m and, further, through the diaphragms of the area of $\Sigma = 125$ cm^2, the amount n of quanta, which form the interference fringes, it estimate to be $n = \bar{n}\Sigma/(\pi r_{ch}^2) \approx 10^{-5}$. For the given degree of monochromaticity provided by the filter, one can find the effective spectral width of the incident radiation from the relations $\Delta\lambda/\lambda_0 = \Delta\nu/\nu_0 = 10^{-3}$. Hence, $\Delta\nu = 10^{-3}\nu_0$, and the coherence time τ_{ch} is found to be $\tau_{ch} = 1/\Delta\nu = 10^3\lambda_0/c$. During such a period, n quanta form the observed interference fringes; therefore, the average light flux is estimated to be

$$\frac{n}{\tau_{ch}} = \frac{nc}{\lambda_0}10^{-3} \approx 5 \cdot 10^5 \text{ quanta/s} .$$

APPENDIX 7.A

```
CFSlit():
    x_in = 13724;  Max = 50000;  β = 3.1415926/140;
      for (M = 0; M < Max; M + +){
    for (m = 0; m < 30; m + +){
  θ_C[m] = 0.0; ζ_C[m] = 0.0;}
      for (k = 0; k < 20; k + +){
          Polar-coordinates();
          θ_O[m] = θ; ζ_O[m] = ζ;}
      for (m = −15; m < 15; m + +){
      for (k = 0; k < 20; k + +){
          θ_C[m + 15] = θ_C[m + 15] + θ_O * cos(β * (m − 15) * k);
          ζ_C[m + 15] = ζ_C[m + 15] + ζ_O * cos(β * (m − 15) * k);}}
      for (i = 0; i < 8; i + +){
      m = 15 − i, n = 15 + i
          CF[i] = CF[i] + θ_C[m] * θ_C[n] + ζ_C[m] * ζ_C[n];}
    }
    A = CF[0];
      for (n = 0; n < 8; n + +)CF[n] = CF[n]/A;
```

APPENDIX 7.B

```
CFDisk()
    Max = 50000;  pi = 3.1415926;  x_in = 1732;  β = 0.01;
      for {
      for (k = 0;  k < 15;  k + +){
      for (n = 0;  n < 10;  n + +){
      Polar-coordsinates();
  θ_O[n][k] = θ;  ζ_O[n][k] = ζ;}}
      for (m = −15;  m < 15;  m + +){
  θ_C[m + 15] = 0.0;  ζ_C[m + 15] = 0.0;
      for (k = 0;  k < 15;  k + +){
      z = pi * β * m * cos(k * pi/15.0);
      for (n = 0;  n < 10;  n + +){
          θ_C[m + 15] = θ_C[m + 15] + θ_O[n][k] * n * cos(z * n);
          ζ_C[m + 15] = ζ_C[m + 15] + ζ_O[n][k] * n * cos(z * n);}}}
      for (i = 0;  i < 15;  i + +){
      m = 15 − i; n = 15 + i;
          CF[i] = CF[i] + θ_C[m] * θ_C[n];
          CF[i] = CF[i] + ζ_C[m] * ζ_C[n];}
    }
    A = CF[0];
      for (n = 0; ,n < 15;  n + +)CF[n] = CF[n]/A;
```

APPENDIX 7.C

CFDoppler()
$Max = 50000;\ x_{in} = 173562;\ \beta = 3.1415926/(210.0);$
for $(k = 0;\ k < 40;\ k++)\quad sp[m] = \exp(-0.005 * (k^2));$
  for $(M = 0;\ M < Max;\ M++)\{$
  for $(k = 0;\ k < 30;\ k++)\{$
  $\theta_C[m] = 0.0; \zeta_C[m] = 0.0;$
  Polar-coordinates();
$\theta_O[k] = \theta;\ \zeta_O[k] = \zeta;\}$
  for $(m = -20;\ m < 20;\ m++)\{$
  for $(k = 0;\ k < 30;\ k++)\{$
  $\theta_C[m+20] = \theta_C[m+20] + \theta_O * sp[k] * \cos(\beta * (m-20) * k);$
  $\zeta_C[m+20] = \zeta_C[m+20] + \zeta_O[k] * sp[k] * \cos(\beta * (m-20) * k);\}\}$
  for $(i = 0;\ i < 20;\ i++)\{$
  $m = 20 - i;\ k = 20 + i;$
  $CF[i] = CF[i] + \theta_C[m] * \theta_C[k];$
  $CF[i] = CF[i] + \zeta_C[m] * \zeta_C[k];\}\}$
$\}$
$A = CF[0];$
  for $(n = 0;\ n < 15;\ n++)CF[n] = CF[n]/A;$

APPENDIX 7.D

CFDbDoppler():
$Max = 100000;\ x_{in} = 1723;\ \beta = 0.025;$
  for $(k = 0;\ k < 40;\ k++)\{$
  $z = (0.025 * (k - 20))^2;$
  $sp[k] = exp(-z);\}$
  for $(M = 0;\ M < Max;\ M++)\{$
  for $(k = 0;\ k < 40;\ k++)\{$
$\theta_C[m] = 0.0;$
$\zeta_C[m] = 0.0;$
  Polar-coordinates();
$\theta_O[k] = \theta\ \zeta_O[k] = \zeta;\}$
  for $(k = 0;\ k < 40;\ k++)\{$
  for $(m = 0;\ m < 40;\ m++)\{$
  $\theta_C[m] = \theta_C[m] + \theta_O * sp[k] * \cos(\beta * m * k);$
  $\zeta_C[m] = \zeta_C[m] + \zeta_O * sp[k] * \cos(\beta * m * k);\}\}$
  for $(k = 0;\ k < 20;\ k++)\{$
  for $(m = 0;\ m < 20;\ m++)\{$
$CF[k] = CF[k] + \theta_C[m] * \theta_C[m + k];$

$$CF[k] = CF[k] + \zeta_C[m] * \zeta_C[m+k]; \}\}$$
$$\}$$
$$A = CF[0];$$
for $(k = 0; \ k < 20; \ k++)CF[k] = CF[k]/A;$

Chapter 8

CORRELATION OF LIGHT INTENSITY

Early in 1950, serious attempts were made to observe the effects governed by the superposition of light beams emerging from independently radiating sources. Initially, experiments of this sort were intended to clarify the question, if two independent sources may effect a photodetector in a coherent manner. The first successful results were obtained by the American researchers A.FORRESTER, R.GUDMUNDSEN, and PH.JOHNSON in 1955.[7] In their experiments, the action of two σ-components of a ZEEMAN spectrum was investigated, which could be regarded as independent light sources since the spectral separation of the centers of the components was much larger than the width of each component, so they did not overlap. Light from both sources illuminated a photocell of a peculiar design. The photocurrent showed a weak peak at a frequency which was related to the frequency difference between the centers of the components.

Another experimental technique was suggested by R.HANBURY-BROWN and R.Q.TWISS in order to observe correlation effects between the photocurrents of two photodetectors illuminated by a partially coherent light beam which was believed to possess spatial coherence [8]. On one hand, the experiments were treated to be a definite proof for correlation in the photoemission under partially coherent light field conditions, and, on the other hand, as the existence of a correlation of intensities.

1. Correlation functions of intensity

Any phenomena of field correlation discussed throughout the previous chapter is accessible to measurements. We have seen that the coherence time and coherence area both can be estimated from the visibility function. Fluctuations of the instantaneous intensity can be expected to appear as a correlated noise within the coherence volume, or the spatial mode. Thus, it is anticipated that a measurement of these correlated fluctuations may be performed by means of two photo-detectors placed within the coherence volume of thermal radiation (Fig.8.1). For successful measurements, the relationship between the resolution time of the photo-detectors and the coherence time as well as the geometrical displacement of the photo-detectors have to be taken into account. The resolution time should be shorter than the coherence time.

From the physical point of view we define the problem of intensity correlation, in general, as: what results can be achieved with intensity measurements using two photodetectors at two arbitrary points within one coherent volume? (Fig. 8.1). In order to answer this question we have to solve two problems: We should first find the relationship between the mathematical form of the correlation function for the field and then the appropriate *correlation function for the intensity*, if such a relationship exists. The second problem is related to the properties of the photodetectors: They must have a suitable time resolution to allow the highest photon-counting rates possible.

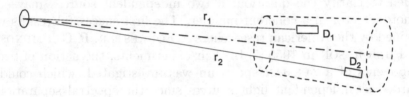

Figure 8.1. The instantaneous intensities measured by two photodetectors D_1 and D_2 may show a correlation effect of the photocurrents of the detectors, provided that the two detectors are located within a coherent volume (dashed lines).

Let us find a relationship between the correlation function of the intensity and the function $\Gamma_{12}(\tau)$ introduced in Chapter 7, assuming that the two concerned points are located within one coherence volume. We denote the complex amplitudes at the points of the first and second detectors by E_1 and E_2, respectively. In order to form a statistical average which is assigned to the correlation function of the intensity, we should consider the product $E_1 E_1^* E_2 E_2^*$. The term $E_1 E_1^*$ has to be associated with the instantaneous intensity I_1, and the term $E_2 E_2^*$ with I_2. Thus, $E_1 E_1^* E_2 E_2^*$ can be considered to describe the contribution to the correlation function under question in general. Due to the stochastic nature of the optical field radiated by any thermal source we can suppose that, in a long series of contributions, each having the form $E_1 E_1^* E_2 E_2^*$, contributions must be found which show no correlation. This fact, expressed in terms of statistical averaging, has to be written in the form $\langle E_1 E_1^* E_2 E_2^* \rangle \sim \langle I_1 \rangle \langle I_2 \rangle$. Besides these contributions, for the same reason, other sorts of products $E_1 E_2^* E_2 E_1^*$ must be found, where correlations between particular factors of the product $E_1 E_2^* E_2 E_1^*$ arise. These contributions has to provide an average

$$\langle E_1 E_1^* E_2 E_2^* \rangle = \langle E_1 E_2^* \rangle \langle E_2 E_1^* \rangle \sim \langle I_1 \rangle \langle I_2 \rangle \quad .$$

Hence, the complete form of the correlation function can be written as

$$\langle I_1 I_2 \rangle = \langle I_1 \rangle \langle I_2 \rangle + (\frac{c\varepsilon_0}{2})^2 \langle E_1 E_2^* \rangle \langle E_2 E_1^* \rangle \quad ,$$

and since $(c\varepsilon_0/2) \langle E_1 E_2^* \rangle = \Gamma_{12}(\tau)$ and $(c\varepsilon_0/2) \langle E_2 E_1^* \rangle = \Gamma_{12}^*(\tau)$, we finally find for the correlation function

$$\langle I_1 I_2 \rangle = \langle I_1 \rangle \langle I_2 \rangle + |\Gamma_{12}(\tau)|^2 \quad . \tag{8.1}$$

There are no other combinations of the quantities E_1, E_1^*, E_2, and E_2^* which could provide a real product of these four magnitudes. We emphasize that any correlation function of the field should be a function of the distance difference Δr under the uniformity condition of the field. This fact implies that $\langle I_1 \rangle = \langle I_2 \rangle$ is assumed. Hence, the function $\langle I_1 I_2 \rangle$ must be represented in terms of $\Gamma_{12}(\tau)$ with the requirement $\langle I_1 \rangle = \langle I_2 \rangle = \langle I \rangle$. It follows from (8.1) that the correlation function of the intensity represented in terms of the complex degree of coherence has to take the form

$$\langle I_1 I_2 \rangle = \langle I \rangle^2 \left(1 + \frac{|\Gamma_{12}(\tau)|^2}{\langle I \rangle^2} \right) = \langle I \rangle^2 \left(1 + |\gamma_{12}(\tau)|^2 \right) \quad . \tag{8.2}$$

It can be seen that any variations of $\langle I_1 I_2 \rangle$ must happen within one spatial mode, which means, within the limits of the appropriate coherence volume.

Let us introduce another correlation function, which should be a measure of the correlated fluctuations of the intensity. We denote $\Delta I_1 = I_1 - \langle I \rangle$ and $\Delta I_2 = I_2 - \langle I \rangle$. Then, the so-called *correlation function of the fluctuations* $\langle \Delta I_1 \Delta I_2 \rangle$ can be represented by $|\Gamma_{12}(\tau)|^2$ in the form

$$\langle \Delta I_1 \Delta I_2 \rangle = \langle I_1 I_2 - \langle I \rangle I_2 - \langle I \rangle I_1 + \langle I \rangle \langle I \rangle \rangle = \langle I_1 I_2 \rangle - \langle I \rangle^2 \quad ,$$

which, using 8.2, leads to

$$\langle \Delta I_1 \Delta I_2 \rangle = \langle I_1 I_2 \rangle - \langle I \rangle^2 = |\Gamma_{12}(\tau)|^2 = \langle I \rangle^2 |\gamma_{12}(\tau)|^2 \quad . \tag{8.3}$$

Using this expression we find that $\langle I_1 I_2 \rangle$ consists of two contributions: one is an invariable item, and the other is $\langle \Delta I_1 \Delta I_2 \rangle$:

$$\langle I_1 I_2 \rangle = \langle I \rangle^2 + \langle \Delta I_1 \Delta I_2 \rangle \quad , \tag{8.4}$$

and we see the fundamental role of the correlation of the fluctuations. Thus, we state that any variations of the correlation of the instantaneous intensity within a spatial mode must be caused by the function $\langle \Delta I_1 \Delta I_2 \rangle$. The spatial dependency of the functions $\langle \Delta I_1 \Delta I_2 \rangle$ and $\langle I_1 I_2 \rangle$ on variable Δr and on time interval τ has to be caused only by function $|\Gamma_{12}(\tau)|$, or by function $|\gamma_{12}(\tau)|$.

To sum up all features of the functions $\langle \Delta I_1 \Delta I_2 \rangle$ and $\langle I_1 I_2 \rangle$ found above, we can state for the radiation of all thermal sources, that the correlation of intensity fluctuations will cause observable correlation effects in photocurrents. Since this sort of radiation is subject to GAUSSIAN statistics, the fluctuations can achieve a value of average intensity $\langle I \rangle$ (see. 6.38):

$$\sigma_I = \left\langle (I - \langle I \rangle)^2 \right\rangle = \langle \Delta I^2 \rangle = \langle I \rangle^2 \quad . \tag{8.5}$$

It follows from (8.3) that, for a very small volume, where $\tau \to 0$ and $\Delta r \to 0$, the correlation function $\langle \Delta I_1 \Delta I_2 \rangle$ tends towards $\langle I \rangle^2$, since $|\gamma_{12}(\tau)|^2 \to 1$ and $\langle \Delta I_1 \Delta I_2 \rangle = \sigma_I$ for $\tau = 0$, $\Delta r = 0$, and $\langle I_1 I_2 \rangle = 2 \langle I \rangle^2$. In the other limiting case, where either $\tau > \tau_{ch}$, or $\Delta r > \Delta d_{ch}$, the fluctuations become uncorrelated, and $\langle \Delta I_1 \Delta I_2 \rangle \to 0$. If this is the case, the function $\langle I_1 I_2 \rangle$ takes the invariable value $\langle I \rangle^2$.

2.   Photocurrent statistics

2.1   The Mandel formula

In order to derive quantitative estimations for the effect of the counting time on the statistics of photocurrent pulses, we shall find a law which establishes probabilities for the occurrence of a required number of photocurrent pulses under quasi-monochromatic radiation. In the case of a monochromatic light wave, such probabilities are subject to POISSON statistics (Chapter 5). We note that for the given counting time T and the instantaneous intensity I, assumed here to be a constant value, the probability $P(N)$ for detecting N counts during interval T has the form of a POISSON distribution:

$$P(N) = \frac{\overline{N}^N}{N!} \exp(-\overline{N}) \quad ,$$

where $\overline{N} = \mu T$ is the power of the POISSON distribution. $\mu = q \Sigma I$, where q is the quantum efficiency, and Σ is the effective area of the photocathode.

Since we consider a quasi-monochromatic wave here, the instantaneous intensity should undergo random variations. Thus, the probability law mentioned above is believed to be true only for short time intervals ΔT, within which the instantaneous intensity remains invariable. Let the time resolution T_r of a photodetector be sufficiently short, so that $T_r < \Delta T$. Then, the average amount of photocurrent pulses found for one particular interval is $\overline{N} = q\Sigma I\Delta T$ as in the case of POISSON statistics, because I has a constant value for the interval ΔT. For the given value $\overline{N}$ a series of probabilities $P(N)$ can be calculated via the POISSON distribution. However, the average number $\overline{N}$, as well as the probabilities $P(N)$ found for the following interval ΔT may differ from those calculated before, because the instantaneous intensity can change. In turn, the amount of light energy W which passes through the area Σ during the interval ΔT will vary with I. Therefore, a particular probability $P(N)$ may now be regarded as a function of W. When changing W over a series of M intervals of duration ΔT, we may find the average probabilities $\overline{P(N)}$ in the form of an arithmetic average:

$$\overline{P(N)} = \frac{P(N, W_1) + ... + P(N, W_k) + ... + P(N, W_M)}{M} \quad , \qquad (8.6)$$

where the term $P(N, W_k)$ specifies one POISSON probability found at a certain value of light energy $W_k = \Sigma I_k \Delta T$. Let M be sufficiently large and let q_k be the total amount of terms in the sum $P(N, W_1) + P(N, W_2) + ... + P(N, W_k) + ... + P(N, W_M)$, for which W is within the same limits $W_k \leq W < W_k + \Delta W$. Hence, averaging in (8.6) can be performed with

$$\overline{P(N)} = \sum_{k=0}^{n} P(N, W_k) \frac{q_k}{M} \quad , \qquad (8.7)$$

where k ranges from 0 to some integer number n which specifies the maximum value of light energy. With M tending to infinity all the relative frequencies q_k/M will tend to the probabilities $P(N, W)$. Further, let us introduce the probability density function $f(W)$. Then the discrete averaging in (8.7) will take an integral form

$$\langle P(N) \rangle = \int_0^\infty P(N, W) f(W) dW \quad , \qquad (8.8)$$

where the upper limit is assumed to be infinity. Formula (8.8) is called the MANDEL *formula*, or the *photodetection equation*. Here, a particular average probability $\langle P(N) \rangle$ depends only on the distribution $f(W)$. This is also true in the general case, where the light energy has to be calculated over the counting time by means of integration:

$$W = \Sigma \int_0^T I dt \quad .$$

Due to the stochastic nature of thermal radiation the integrand function $I(t)$ can not be found in an explicit form. Nevertheless, explicit forms of the distribution $f(W)$ may be established, at least for limiting cases.

For example, as one limiting case we can assume that the random variation of the integral energy will follow the fluctuations of the instantaneous intensity. This gives us the so-called *inertialess photodetector*. In such a case, the coherence time, seen as the time interval for valuable variations of the intensity, is assumed to be much larger than the counting time. Therefore, the integral in (8.9) can be replaced by the simple dependency

$$W = \Sigma I T \quad . \tag{8.9}$$

This means that random variations of the energy W must occur, as well as variations of the intensity I. For this reason the distribution $f(W)$ can be replaced by the distribution of the instantaneous intensity, which has been found in Chapter 6. If this is the case, the average distribution in (8.8) can be calculated in its explicit form.

In another limiting case, so-called *inertial detection*, the counting time is much longer than the coherence time, which provides full smoothing of any fluctuations of the instantaneous intensity. Hence, W should have a constant value during the counting interval. If this is the case, each item of the summation in (8.7) will have a certain invariable magnitude, and we find: $\overline{P(N)} = P(N, W)$. In other words, for a given light energy W the quantity $\overline{P(N)}$ takes the form of a POISSON probability, as in the case of a monochromatic wave.

2.2   A computer model of inertialess detection

Let us discuss a model simulating the operating principle of an inertialess photodetector, where the counting time T is regarded to be much smaller than the coherence time. Algorithm **InertialessDetector()** presented in Appendix 8.A performs calculations of ten average probabilities stored in the variables $AP[N]$, with $N = 0, ...9$, which are related to the appearance of the required number N of photocurrent pulses during the counting interval. The first codes of the procedure are similar to that of procedure **CFDopler()**, discussed in Chapter 7. For the given amplitude spectral distribution in the form

$$sp[m] = \exp(-0.5(m/20)^2)$$

Table 8.1. Probabilities $\langle P(N) \rangle$ and $P(N)$ for $Q = 0.001$ and $Q = 0.002$.

N	$\overline{N} = 0.50, Q = 0.001$		$\overline{N} = 0.97, Q = 0.002$	
	$\langle P(N) \rangle$	$P(N)$	$\langle P(N) \rangle$	$P(N)$
0	0.6835	0.6054	0.5315	0.3776
1	0.2031	0.3038	0.2291	0.3677
2	0.0699	0.0762	0.1106	0.1790
3	0.0262	0.0127	0.0573	0.0581
4	0.0103	0.0016	0.0310	0.0141
5	0.0041	0.0001	0.0173	0.0027
6	0.0017	—	0.0098	0.0004
7	0.0007	—	0.0056	—
8	0.0003	—	0.0033	—
9	0.0001	—	0.0019	—

and the constant $\beta = 0.0125$ used for the cosine expansion, the dimensionless coherence time counted by integer numbers is found to be about 20. For one random realization of the field quantities, $\theta_C[k]$ and $\zeta_C[k]$, calculated by means of cosine expansions, are used to form 10 values of the instantaneous intensity as $\theta_C^2[k] + \zeta_C^2[k]$, where k runs from 0 to 9. Thus, each random value of the intensity is assumed to be calculated during some short counting interval. This interval represented by an integer is simply equal to 1, and, therefore, the requirement of inertialess detection is true. Now, assuming $W = I$, values of W must satisfy the relation $W[k] = \theta_C^2[k] + \zeta_C^2[k]$. Since each value $W[k]$ relates to an average number $\overline{N}$ of an appropriate POISSON distribution, 10 such distributions are calculated for each value $\overline{N} = Q \cdot W[k]$, where Q is a proportionality factor which simulates the quantum efficiency of the photo-cathode. Thus, one random realization of the correlated field gives rise to 10 different POISSON distributions. Further, only 10 probabilities $P(N)$, where N runs from 0 to 9, are taken into account for each distribution. The array $AP[N]$ is used to accumulate 10 averages, each related to a certain value of N.

When processing the main loop ($Max = 20000$), the values of $AP[N]$ tend to average magnitudes, which are then used for calculations of average probabilities $\langle P[N] \rangle$ by means of the simple relation $\langle P[N] \rangle = AP[N]/(10 \cdot Max)$. Here, factor 10 takes into account the fact that each $AP[N]$ was calculated 10 times for one random realization of the correlated field. It is clear that the summation of the partial contributions to the variables $AP[N]$ is similar to (8.6).

In order to compare the obtained magnitudes $\langle P[N] \rangle$ with a set of the appropriate POISSON probabilities the average amount of photo-counting pulses $\overline{N}$ is calculated using $\langle P[N] \rangle$:

$$\overline{N} = \sum_{1}^{9} N \cdot \langle P(N) \rangle \quad .$$

For the given number $\overline{N}$, 10 POISSON probabilities $P(N)$ are calculated as well. The calculations were performed for $Q = 0.001$ and $Q = 0.002$. The probabilities $\langle P(N) \rangle$ and $P(N)$ for $N = 0, \ldots 9$ are shown in Table 8.1, where "—" indicates that the appropriate value of $P(N)$ is smaller than 10^{-4}.

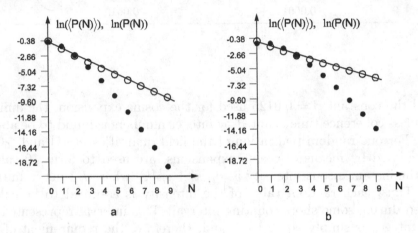

Figure 8.2. Logarithms of probabilities $\langle P(N) \rangle$ and $P(N)$ at Q=0.001 (a), and Q=0.002 (b) in the model of the inertialess detection. Hollow circles represent values of $\ln(\langle P(N) \rangle)$ and full circles of $\ln(P(N))$.

It is seen from Table 8.1 that with increasing N the probabilities $\langle P(N) \rangle$ decrease much more slowly than the appropriate values of POISSON probabilities. Emphasizing this peculiarity of $\langle P(N) \rangle$ and its different behavior from $P(N)$, Fig.8.2 shows the functions $\ln(\langle P(N) \rangle)$ and $\ln(P(N))$ obtained for $Q = 0.001$ (Fig.8.2,a) and $Q = 0.002$ (Fig.8.2,b). We suppose that in both cases the function $\ln(\langle P(N) \rangle)$ may be represented by a straight line, the slope of which depends on $\overline{N}$. Thus, the probabilities $\langle P(N) \rangle$ should satisfy an exponential law in the form

$$\langle P(N) \rangle = \frac{1}{Z} \exp(-\beta N) \quad , \tag{8.10}$$

Table 8.2. Probabilities $\langle P(N) \rangle$ and $P(N)$ for $Q = 0.001$ and $Q = 0.002$.

N	$\overline{N} = 0.52$, $Q = 10^{-4}$		$\overline{N} = 2.59$, $Q = 5 \cdot 10^{-4}$	
	$\langle P(N) \rangle$	$P(N)$	$\langle P(N) \rangle$	$P(N)$
0	0.6014	0.5925	0.1004	0.0750
1	0.2937	0.3101	0.2043	0.1942
2	0.0817	0.0811	0.2299	0.2516
3	0.0165	0.0141	0.1894	0.2172
4	0.0027	0.0018	0.1277	0.1406
5	0.0004	0.0001	0.0747	0.0728
6	0.0001	—	0.0394	0.0314
7	—	—	0.0191	0.0116
8	—	—	0.0087	0.0037
9	—	—	0.0037	0.0010

where β determines the slope of the function $\ln(\langle P(N) \rangle)$ and Z is a constant magnitude. For a given value of β and Z, the normalization condition has to be fulfilled:

$$\sum_{N=0}^{N=9} \langle P(N) \rangle = \frac{1}{Z} \sum_{N=0}^{N=9} \exp(-\beta N) = 1 \quad . \tag{8.11}$$

2.3 A computer model of inertial detection

It is easy to modify the code of the previous algorithm in order to simulate the operation of an inertial photodetector. In procedure **InertialDetector**(), see Appendix 8.B, the only variable W is used to add up values of the instantaneous intensity over the whole range of correlated field duration ($\tau_{ch} < T$). To intensify the expected effect, a range of 20 units was chosen, which is equal to the full range of variations of the correlated field. We also call attention to the code forming the probabilities $\langle P(N) \rangle$: the normalization of variables $AP[N]$ is calculated for integer numbers Max, because the light energy W is formed once for each realization of the correlated field. As before, Max is the total amount of such realizations. Probabilities $\langle P(N) \rangle$ and $P(N)$ obtained at two values of the quantity $\overline{N}$ and two values of factor Q are presented in Table 8.2.

The comparison of the data shows that the probabilities $\langle P(N) \rangle$ and $P(N)$ associated with the same value of $\overline{N}$ are close to each other.

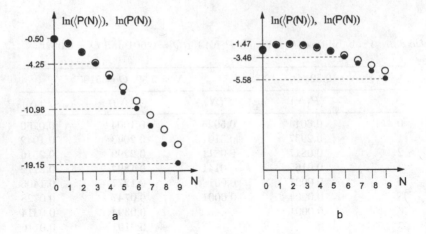

Figure 8.3. Logarithms of the magnitudes $\langle P(N) \rangle$ and $P(N)$ obtained in the model of the inertial detection for $Q = 10^{-4}$ (a), and $Q = 5 \cdot 0^{-4}$ (b). Bright circles denote values of $\ln(\langle P(N) \rangle)$ and dark circles of $\ln(P(N))$.

Figs. 8.3.a and 8,3,b, where the functions $\ln(\langle P(N) \rangle)$ (bright circles) and $\ln(P(N))$ (by dark circles) are shown, confirm the similarity.

Data found with these models show that fluctuations of the instantaneous intensity play a fundamental role in the formation of correlated intensities. When performing light detection by an inertialess photodetector, the statistics of the photocurrent pulses differ radically from POISSON's statistics of shot-noise. The reason is that the fluctuations can not be smoothed during the counting time interval T of a fast photodetector.

Supposing that any correlation of fluctuations has to exist within the coherence time τ_{ch}, the ratio τ_{ch}/T will be a qualitative measure of the speed of the photodetector. For an inertial photodetector the ratio τ_{ch}/T is about 1 or smaller. If this is the case, fluctuations will be smoothed by the photo-detector. Then, for any realization of a correlated field, such a photodetector will detect some average value of intensity, which must be a nearly constant quantity. Therefore, the statistics of the photocurrent pulses should have the form of a POISSON distribution, as in a case of a monochromatic wave.

We call attention to the fact that the new statistics of photocurrent pulses mentioned above is caused by the GAUSSian distribution of the quadrature components of the light field and by the exponential distribution of the instantaneous intensity (see (6.34) and (6.37)). This can be regarded to be a verification of all computer procedures simulating correlated fields used above. It has been proven that the noise of a GAUSSian distribution will also possess GAUSSian statistics after a linear filtering procedure. Moreover, the noise leaving such a linear filter must obtain correlation properties due to the finite band width of the filter, provided that the incident noise is regarded as being so-called "white noise", which means, that it is an absolutely uncorrelated noise. Thus, in all the computer models considered above, initially uncorrelated quadrature components of the GAUSSIAN noise take the form of correlated GAUSSIAN noise due to the linear filtering procedures used. Relationships (7.74) used in our procedures are typical examples of such a linear filtration.

3. The statistics of thermal radiation

3.1 An explicit form of the probabilities $\langle P(N) \rangle$

In one limiting case, in the case of inertialess detection, the distribution of the light energy $f(W)$ must be equal to the distribution of the instantaneous intensity (6.37), neglecting constant factors like the area of the photocathode Σ and the counting time T. In that case any variations of the instantaneous intensity are assumed to be much slower than the time-response of the photodetector. For this reason the light energy W has to be proportional to the instantaneous intensity in the form $W = T\Sigma I$, hence $\overline{N} \sim W$. A partial probability for the quantity W, found to be between the limits W and $W + dW$ is given by $f(W)dW$. This probability can be represented with the help of the distribution $f(I)$ from (6.37):

$$f(W) = \frac{1}{\langle W \rangle} \exp\left(-\frac{W}{\langle W \rangle}\right) .$$

Substitution for $f(W)$ in the MANDEL formula by the right-hand side of the expression above gives us

$$\langle P(N) \rangle = \frac{1}{\langle W \rangle} \int_0^\infty \frac{\overline{N}^N}{N!} \exp(-\overline{N}) \exp(-\frac{W}{\langle W \rangle})dW .$$

Since $\overline{N} \sim W$, it follows that $\langle N \rangle \sim \langle W \rangle$, hence, $W/\langle W \rangle = \overline{N}/\langle N \rangle$, and we can simplify the integrand:

$$\langle P(N) \rangle = \int_0^\infty \frac{\overline{N}^N}{N!} \exp\left\{-\left(\overline{N} + \frac{\overline{N}}{\langle N \rangle}\right)\right\} d\left(\frac{\overline{N}}{\langle N \rangle}\right) .$$

Now, by introducing a new variable $x = \overline{N} + \overline{N}/\langle N \rangle$, we get

$$\langle P(N) \rangle = \frac{\langle N \rangle^N}{(1 + \langle N \rangle)^{N+1}} \int_0^\infty \frac{x^N}{N!} \exp(-x)dx . \tag{8.12}$$

The first partial integration gives

$$\frac{1}{N!} \int_0^\infty x^N \exp(-x)dx = \frac{1}{(N-1)!} \int_0^\infty x^{N-1} \exp(-x)dx \quad .$$

Continuing the partial integration we will find a series of integrals as follows:

$$\frac{1}{(N-1)!} \int_0^\infty x^{N-1} e^{-x}\, dx = \frac{1}{(N-2)!} \int_0^\infty x^{N-2} e^{-x}\, dx = \dots$$

$$\dots = \frac{1}{0!} \int_0^\infty x^0 e^{-x}\, dx = 1\,.$$

Therefore, for the probability $\langle P(N) \rangle$, we finally find the expression

$$\langle P(N) \rangle = \frac{\langle N \rangle^N}{(1 + \langle N \rangle)^{N+1}} \quad . \tag{8.13}$$

The logarithm of $\langle P(N) \rangle$ takes the form of a linear function of argument N :

$$\ln\left(\langle P(N) \rangle\right) = -\ln\left(1 + \langle N \rangle\right) + N \cdot \left[\ln\left(\langle N \rangle\right) - \ln\left(1 + \langle N \rangle\right)\right] \quad , \tag{8.14}$$

which confirms our results found in the example of the computing simulation of inertialess detection.

Now we represent the probability $\langle P(N) \rangle$ in (8.13) by two coefficients:

$$\langle P(N) \rangle = \frac{1}{1 + \langle N \rangle} \left(\frac{\langle N \rangle}{1 + \langle N \rangle}\right)^N .$$

$\langle N \rangle / (1 + \langle N \rangle)$ is always less than unity, therefore $\langle P(N) \rangle$ can be represented in another way, for example by an exponential function $\exp(-\beta N)$:

$$\langle P(N) \rangle = \frac{1}{1 + \langle N \rangle} \exp(-\beta N) \quad , \tag{8.15}$$

where the constant factor β must satisfy the relation

$$\exp(-\beta) = \frac{\langle N \rangle}{1 + \langle N \rangle} \quad . \tag{8.16}$$

Thus, the statistical average of the amount N of photocurrent pulses obtained during the counting time T has to be represented by β as

$$\langle N \rangle = \frac{1}{\exp(\beta) - 1} \quad . \tag{8.17}$$

For the given average $\langle N \rangle$ and the explicit form of the probability $\langle P(N) \rangle$ important magnitudes can be calculated like the dispersion of N. By definition we find

$$\sigma_N = \langle (N - \langle N \rangle)^2 \rangle = \langle N^2 \rangle - \langle N \rangle^2 =$$

$$= \sum_0^\infty N^2 \langle P(N) \rangle - \langle N \rangle^2 = \frac{1}{1 + \langle N \rangle} \sum_0^\infty N^2 \exp(-\beta N) - \langle N \rangle^2 \quad . \quad (8.18)$$

Let us take into account that a geometric series of the factor $\exp(-\beta)$ gives the following sum:

$$\sum_0^\infty \exp(-\beta N) = \frac{\exp(\beta)}{\exp(\beta) + 1} \quad .$$

Differentiation twice with respect to β gives us

$$\sum_0^\infty N^2 \exp(-\beta N) = \frac{\exp(\beta)\,(1 + \exp(\beta))}{(\exp(\beta) - 1)^3} \quad .$$

The substitution of $1/(\exp(\beta) - 1)$ by $\langle N \rangle$ from (8.16) and of $\exp(\beta)$ by $(1 + \langle N \rangle)/\langle N \rangle$ according to (8.17) on the right-hand side of the last expression gives then

$$\sum_0^\infty N^2 \exp(-\beta N) = (1 + \langle N \rangle) \left(\langle N \rangle + 2 \langle N \rangle^2 \right) \quad ,$$

It then follows from (8.18) that the desired dispersion is given as

$$\sigma_N = \langle (N - \langle N \rangle)^2 \rangle = \langle N \rangle + \langle N \rangle^2 \quad . \quad (8.19)$$

Formula (8.19) establishes a fundamental property of inertialess detection: The dispersion of the amount of photocurrent pulses is a sum of two items, one is equal to $\langle N \rangle$ and one equal to the square of $\langle N \rangle$. We note that the analogous quantity found under the conditions of detection of a monochromatic radiation, contains only one term $\langle N \rangle$, as in the case of shot noise. Hence, in inertialess detection of quasi-monochromatic radiation the dispersion obtains an extra contribution which is caused by fluctuations of the instantaneous intensity. This is the "wave" noise, in contrast to the shot noise. The "wave" noise disappears by smoothing fluctuations by an inertial photodetector, which results in photocurrent pulses which show the shot noise of POISSON statistics.

3.2   Statistics of quanta within one spatial mode

Let a quasi-monochromatic wave propagate through a volume ΔV smaller than the coherence volume. At every time moment the field inside this volume is formed by monochromatic waves, each having a frequency distribution within a very narrow range around its carrier frequency ν_0. For every monochromatic component, the probability that a required number of photons is localized within the volume, is given by POISSON statistics as

$$P(n; \overline{n}) = \frac{\overline{n}^n}{n!} \exp(-\overline{n}) \quad , \tag{8.20}$$

where $\overline{n}$ is the density of photons around the point of observation.

Under these assumptions the fluctuations of the intensity of all monochromatic components within ΔV and within the frequency range of the quasi-monochromatic wave occur coherently. For these reaons, the quantity $\overline{n}$ of any monochromatic component has to be directly proportional to the instantaneous intensity:

$$\overline{n} \sim \frac{I}{ch\nu_0} \quad . \tag{8.21}$$

Therefore, the relationships

$$\langle n \rangle \sim \langle I \rangle \quad , \quad \text{and} \quad \frac{\overline{n}}{\langle n \rangle} = \frac{I}{\langle I \rangle}$$

are valid. We should remember that the volume of a spatial mode is the coherence volume. Thus, we are treating the smaller volume ΔV since we are interested in the photon statistics within the volume of one spatial mode.

We introduce the probabilities $\langle P(n) \rangle$ in a similar way as the probabilities $\langle P(N) \rangle$ for the photocurrent pulses before. These probabilities $\langle P(n) \rangle$ are given by the statistical averages

$$\langle P(n) \rangle = \int_0^\infty P(n) f(I) dI = \frac{\langle n \rangle^n}{(1 + \langle n \rangle)^{n+1}} \int_0^\infty \frac{x^n}{n!} \exp(-x) dx \quad ,$$

and this equation is similar to (8.12). Because the integral on the right-hand side has to be equal to 1, we find for $\langle P(n) \rangle$

$$\langle P(n) \rangle = \frac{\langle n \rangle^n}{(1 + \langle n \rangle)^{n+1}} \quad . \tag{8.22}$$

This result is quite analogous to (8.13), differing only in the notation n instead of N. The dispersion of the amount of quanta has to satisfy the formula (see (8.19))

$$\left\langle (n - \langle n \rangle)^2 \right\rangle = \langle n \rangle + \langle n \rangle^2 \quad . \tag{8.23}$$

Now an exponential representation of probabilities $\langle P(n) \rangle$ can be performed (see (8.15)):

$$\langle P(n) \rangle = \frac{1}{\langle n \rangle + 1} \exp(-\beta n) \quad , \tag{8.24}$$

where $\langle n \rangle$ must obey the relation

$$\langle n \rangle = \frac{1}{\exp(\beta) - 1}$$

as follows from (8.17).

With the assumption that such a quasi-monochromatic wave could appear due to the emission of a black body of suitable temperature T, we believe that the factor β must have the form $\beta = h\nu_0/(k_B T)$, where k_B is the BOLTZMANN constant. For this reason the probabilities $\langle P(n) \rangle$ and, respectively, the quantity $\langle n \rangle$ can be represented as

$$\langle P(n) \rangle = \frac{1}{1 + \langle n \rangle} \exp(-h\nu_0 n/(k_B T)) \quad \text{and} \tag{8.25}$$

$$\langle n \rangle = \frac{1}{\exp(h\nu_0 n/(k_B T)) - 1} \quad . \tag{8.26}$$

The quantity $\langle n \rangle$ obeys the law found for black body radiation, despite of the fact that such a quasi-monochromatic field can also be emitted by a non-equilibrium light source. For this reason T in (8.25) and (8.26) may be treated as the temperature of an equivalent black body light source. Nevertheless, formula (8.26), called the BOSE-EINSTEIN *distribution*, has an imperative sense for any optical radiation. The results concerning the probabilities $\langle P(n) \rangle$ in the forms (8.24), (8.25), the average amount of quanta per one spatial mode in the form of (8.26), and the dispersion in the form of (8.23) are often generally called BOSE-EINSTEIN *statistics* of radiation.

To emphasize the role of fluctuations in the formation of the dispersion (8.23) we compare two sorts of particles: molecules of an ideal gas and photons. In the model of an ideal gas, molecules collide, causing a change in the velocities and kinetic energies of the molecules. There are no other mechanisms for interaction between the molecules. Thus, the molecules are thought to be moving independently between two collisions. For this reason the average amount of molecules within a certain volume should be proportional to the size of the volume and to the density of the molecules:

$$\overline{N}_{mol} = \overline{n}V \quad .$$

Since the penetration of the volume by every molecule is assumed to be independent of other molecules, the probability that N molecules occur within the volume should obey the POISSON distribution. Therefore the dispersion of such classical particles takes the form

$$\langle (N - \overline{N}_{mol})^2 \rangle = \overline{N}_{mol} \quad ,$$

inherent to POISSON statistics.

For a given density of photons, the average amount of quanta within the volume takes a similar form:

$$\langle N \rangle_{phot} = \langle n \rangle V \quad .$$

In contrast to classical particles the penetration of the volume by photons should obey laws which describe probability waves. Due to the statistical independence of the elementary sources, superpositions should exist, in which the total energy is found to be a sum of the partial energies of the waves, in other words, where no interference effects can be observed. For all events of this sort, the dispersion could obey the usual classical statistics, and take the form of $\langle N \rangle_{phot}$. Nevertheless, due to the same primary reason, monochromatic waves should form superpositions, in which the total light energy within the limits of the volume is found to be very weak, or other superpositions, which provide a very high total energy. Superpositions of the last sort give rise to an extra increase of the dispersion by the term $\langle N \rangle_{phot}^2$.

We should also call attention to the fact that in optics, as soon as we discuss photons within one spatial mode, the magnitude $\langle n \rangle$ is equal to the degeneracy parameter δ. We have seen that under such conditions $\langle n \rangle$ has to be much less than unity. Therefore, the factor $\langle n \rangle$ affects the dispersion quite strongly as $\langle n \rangle^2$. In turn, it is the factor $\langle n \rangle^2$ which provides any correlation effects in experiments on correlation of the intensity. A relatively small value of $\langle n \rangle$ is the dominating reason that we are able to observe such correlation effects.

4. Optical beats experiment

We have mentioned above that in 1955 FORRESTER, GUDMUNDSEN, and JOHNSON first realized an experiment for observation of the superposition of optical fields caused by two independent light sources.

A simplified scheme of the experiment is shown in Fig.8.4. Two ZEEMAN σ-components of a green line of mercury ($\lambda_0 = 546$ nm) were used as two independent light sources. The frequency separation $\Delta\nu$ of the centers of the σ-components was controlled by the magnetic field strength B according to the formula for the ZEEMAN effect (see (1.49)):

$$\Delta\nu = \frac{eB}{4\pi m} .$$

The π component, emitted with the original frequency of the light source, was suppressed by a linear polarizer. In order to provide the independency of the two light sources from each other, the value of the magnetic field strength was high enough to generate non-overlapping contours, obeying the condition

$$\Delta\nu \gg \gamma_D , \tag{8.27}$$

where γ_D is the DOPPLER width of each ZEEMAN component. The basic idea of the experiment was to search for a beat effect, which would arise at frequency $\Delta\nu$. For the green mercury line used under these experimental conditions the width of both σ ZEEMAN components is about $\gamma_D = 10^9$ Hz , whereas the frequency separation, the so-called *beat frequency*, was varied around 10^{10} Hz, fulfilling requirement (8.27).

Light emitted by the source passed through an optical filter and a polarizer, selecting the desired radiation of both σ-components. Then the light beam was focused on the photocathode of a photodetector. A special design of the photodetector, which was called a photomixer, allowed a highly resonant amplification of the photoelectron beam by

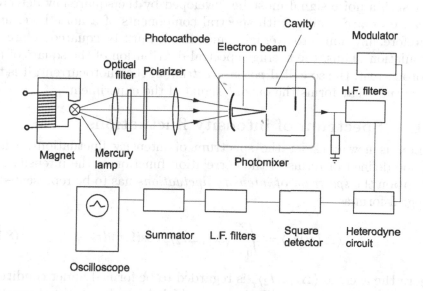

Figure 8.4. Scheme of the experiment of FORRESTER, GUDMUNDSEN, and JONSON.

means of a cavity. Due to the spherical shape of the photocathode the photoelectrons converged on the entrance slit of the cavity, whose resonant frequency Δf_R was chosen to be around the desired beat frequency: $\Delta f_R \approx \Delta \nu$. The output current of the cavity was led to a heterodyne circuit via a high-frequency radio filter in order to get a lower frequency signal. Such a signal was further processed by a square-law detector. Its output signal, after low-frequency filtration, was accumulated by a summator.

In order to analyze the operating principle of the experiment we should call attention to three important points of the transformations of the photocurrent into the accumulated data. At first we assume that photoelectrons arise due to photoemission during a very short time interval, thus the photocurrent i_{ph} is assumed to be directly proportional to the value of the instantaneous intensity I:

$$i_{ph} = QI \quad , \tag{8.28}$$

where the factor Q contains the quantum efficiency as well as the effective photosensitive area of the photocathode. Because of the chaotic processes of photoemission by thermal radiation the photocurrent i_{ph} must be regarded as noise. Secondly, we should take into account the finite spectral band of the cavity's resonant contour $R(\Omega)$, which provides a selective amplification of the photocurrent only within a certain narrow spectral band. The final important point of our consideration is that such a noise signal must be developed by the square-law detector, because we are dealing with spectral components of a noise here, and, therefore, any non-zero averaging magnitude must be squared before accumulation. Thus, the average spectral distribution of the square of the photocurrent, the so-called *power spectrum* of the photocurrent, it is the magnitude that forms the output signal of the experiment.

4.1   Spectrum of intensity fluctuations

Let us now consider the spectrum of intensity fluctuations, which can be defined in terms of the correlation function of the intensity. By definition the *spectrum of intensity fluctuations* has to be represented in integral form as

$$\langle \Delta I^2(\Omega) \rangle = \int_0^\infty \langle \Delta I_1 \Delta I_2 \rangle_\tau \cos(\Omega\tau)d\tau \quad , \tag{8.29}$$

where the average $\langle \Delta I_1 \Delta I_2 \rangle_\tau$ is regarded to be formed under conditions of temporal correlation, which means $\langle \Delta I_1 \Delta I_2 \rangle_\tau$ has to be a function of the parameter τ. The fact that any restriction of the spectral range associated with a distribution of intensity, as well as with intensity fluctuations, must cause a correlation effect, is similar to problems of spectral confinement in the case of an optical field. Actually, another integral form of the relationship between $\langle \Delta I_1 \Delta I_2 \rangle_\tau$ and $\langle \Delta I^2(\Omega) \rangle$ is

$$\langle \Delta I_1 \Delta I_2 \rangle_\tau = \int_0^\infty \langle \Delta I^2(\Omega) \rangle \cos(\Omega\tau) d\Omega$$

which confirms our considerations above. For this reason we introduce the spectrum of intensity fluctuations without a detailed discussion. In the case under consideration, the function $\langle \Delta I_1 \Delta I_2 \rangle_\tau$ takes the form of the square of the temporal correlation function $|\gamma_{11}(\tau)|$, which describes the correlated field from two nonoverlapping lines of GAUSSIAN shape (see (7.59)). Thus, for the spectrum of intensity fluctuations, we find

$$\langle \Delta I^2(\Omega) \rangle = \int_0^\infty \left[\exp(-\gamma_D^2 \tau^2/4) \cos(\Delta\Omega\tau/2) \right]^2 \cos(\Omega\tau) d\tau =$$

$$= \int_0^\infty \exp(-\gamma_D^2 \tau^2/2) \cos^2(\Delta\Omega\tau/2) \cos(\Omega\tau) d\tau \quad , \qquad (8.30)$$

where $\Delta\Omega$ is the separation of the centers of the lines. Since $\cos^2(\Delta\Omega\tau/2) = 0.5(1 + \cos(\Delta\Omega\tau))$ the integral in (8.30) can be written as a sum. One integrand varies with $\cos(\Omega\tau)$:

$$\frac{1}{2} \int_0^\infty \exp(-\gamma_D^2 \tau^2/2) \cos(\Omega\tau) d\tau \quad ,$$

the other is additionally modulated with high frequency by $\cos(\Delta\Omega\tau)$:

$$\frac{1}{2} \int_0^\infty \exp(-\gamma_D^2 \tau^2/2) \cos(\Delta\Omega\tau) \cos(\Omega\tau) d\tau \quad .$$

According to (7.55) the first integral has to be proportional to the function $0.5 \exp(-2\gamma_D^2 \Omega^2)$, and the second to the function $0.5 \exp(-2\gamma_D^2 (\Omega - \Delta\Omega)^2)$. Hence, the spectrum of the intensity fluctuations consists of two non-overlapping parts. One is presented by a DOPPLER contour of width $\sqrt{2}\gamma_D$, which peaks at $\Omega = 0$. The second part has a contour of the same shape and of the same width as the first one, but has a peak at the frequency $\Omega = \Delta\Omega$. The spectrum of intensity fluctuations is thus found in the form (Fig.8.5,b)

$$\langle \Delta I^2(\Omega) \rangle = 0.5 \exp(-2\gamma_D^2 \Omega^2) + 0.5 \exp(-2\gamma_D^2 (\Omega - \Delta\Omega)^2) \quad . \qquad (8.31)$$

When in this experiment the cavity is set to resonance with the beat frequency, the spectral composition of the output current of the cavity

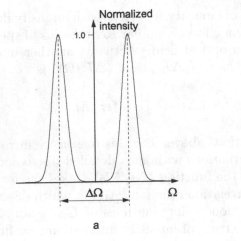
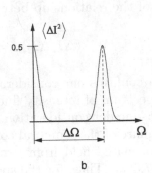

Figure 8.5. Normalized contour of a double GAUSSian line (a) and its spectrum of intensity fluctuations (b).

has to be associated with the spectral composition of the instantaneous intensity within a certain spectral range, and the output current is proportional to $I(\Omega - \Delta\Omega)$. Because of the action of the high-frequency filtering and the heterodyning, the signal becomes a lower frequency noise process, but it keeps the dependency of $I(\Omega)$ which is peculiar for the high-frequency part of the spectrum of intensity fluctuations. After passing the square-law detector the signal takes the form of a partial contribution to the accumulated average in the form $I^2(\Omega)$. When processing the average, a constant value assumed to be proportional to $\langle I \rangle^2$ can be filtered; therefore, the variable part of the average will be proportional to $\langle \Delta I^2(\Omega) \rangle$, the spectral distribution of intensity fluctuations around the beat frequency $\Delta\Omega$.

There were two ways for setting the beat frequency equal to the resonance frequency of the cavity. For a fixed cavity resonance contour $R(\Omega)$, the resonance could be achieved by a small change of the magnetic field B, which causes an appropriate change of the spectral separation $\Delta\Omega$. The other way was performed by tuning the peak frequency of the cavity to adjust it to a fixed value of the spectral separation $\Delta\Omega$. Progressing to resonance enhanced the output signal, whereas by moving away from resonance the output signal decreased. Additionally, a low-frequency scanning was performed under the resonance conditions. This provided the accumulation of spectral components within the narrow spectral band $\delta\Omega$, which was scanned from one side of the spectral contour to the other. With the requirement $\delta\Omega \ll \gamma_D$ the contour of the curve of $\exp(-2\gamma_D^2\Omega^2)$ could be determined with high spectral resolution.

Thus, the proper fast photodetection, performed with the help of the photo-mixer, allowed the photocurrent fluctuations to follow the fluctuations of the instantaneous intensity. Hence, the power spectrum of the photocurrent represented the spectrum of the intensity fluctuations. If no correlation would exist, the power spectrum would show a uniform distribution over the cavity resonance frequencies. If a correlation exists, a peak is obvserved when the resonance frequency matches the frequency splitting between both ZEEMAN components, and this is indeed the case. A beating of both light waves occours, and we can call the splitting frequency the *beat frequency.*

In spite of the fact that the accumulation of the signal was performed by scanning within a narrow band, which provided a sufficient gain in the signal-to-noise ratio, the "wave" noise was relatively small in the experiment under discussion due to the small value of the degeneracy parameter of the radiation under investigation. Nevertheless, this optical beat experiment showed that quasi-monochromatic waves can superimpose and perform correlated fluctuations of the instantaneous intensity, even if these waves were emitted by two independent sources. We call attention to the fact that any interference experiments with two sources, like the observation of interference fringes from a double star, are performed under the conditions of two independent sources. Since all correlation functions of the instantaneous intensity of thermal sources are represented in terms of the function $|\Gamma_{12}(\tau)|$ we can make explicit considerations of any effects of correlation of intensity. Thus, our discussion of the optical beat experiment should emphasize the important role of intensity fluctuations in the investigations of the correlation of intensity.

4.2 A computer model of the optical beat experiment

In the procedure **OpticalBeats**() presented in Appendix.8.C the simulation of correlated fields from a double line utilizes the ideas developed with the model of two spectral lines with DOPPLER contours (see 7.9.4). There, the amplitude spectrum $sp[k]$ assigned to the double line is calculated as ranging over 60 harmonics and has a peak at $k = 20$; thus it is associated with the peak frequency 40 of the spectrum of intensity fluctuations. Further, for 60 time points, each specified by the integer m, the values $\theta_C^2[m] + \zeta_C^2[m]$ of the instantaneous intensity are calculated. The magnitude of the light energy for each smallest interval takes the simple form $Q(\theta_C^2[m] + \zeta_C^2[m])$. For this reason the average amount $\overline{N}$ of photoelectrons is calculated by means of $\overline{N} = Q(\theta_C^2[m] + \zeta_C^2[m])$, where $Q = 0.05$. For the given value Q we assume that the maximal amount of photoelectrons, which may appear within any moment, will not exceed 19. Thus, for each $\overline{N}$ 20 POISSON probabilities $P[N]$ are calculated, where N runs from 0 to 19. To provide one value of N, which varies randomly between 0 and 19 and obeys POISSON statistics, a series of new variables is introduced, subjected to the rule $P[N] = P[N] + P[N - 1]$, where N varies from 0 to 19. Then, with $x = \textbf{Rnd}()$ and $0 \leq x < 1$, if $x \leq P[0]$ the amount of photoelectrons N is found to be 0, if $x > P[0]$, the next inequality $x \leq P[1]$ should be checked, and so on, until a certain inequality happens to be true and the appropriate number N is obtained. Repetition of such calculations 60 times allows the formation of 60 values of photoelectrons arising from the photocathode, which are stored in the array $Cathode[m]$. The power spectrum of the photocurrent $PwSp[i]$ is calculated as the spectral composition over the values $Cathode[k]$ by means of harmonics in the form $\cos(\beta ik)$. An appropriate code has the form

$$c = c + Cathode[k] \cos(\beta ik) \quad ;$$

$$PwSp[i] = c^2 \quad .$$

60 values of the function $PwSp[i]$ are shown in Fig.8.6. It can be seen that for the given parameter $Q = 0.05$ the power spectrum around the

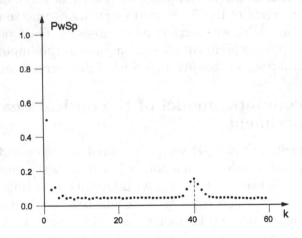

Figure 8.6. 60 values of the array $PwSp[i]$ obtained without actions of the cavity. This spectrum should be associated with the power spectrum of the photocurrent of the photocathode. A small peak around 40 specifies the presence of a double structure of light radiation.

desired beat frequency 40 looks like a small peak. We note that the parameter *ModeKey* must take 1 to perform the calculations above.

In the opposite case, *ModeKey* = 0, the procedure involves a code which takes into account the selective action of the cavity, applying a so-called pulse-response characteristics of the cavity to the current *Cathode*[k]. For a given resonance curve of the cavity, a LORENTZian contour in our example, the pulse-response characteristic is represented by the function $\exp(-0.01k)\cos((40+\delta)k\beta)$, where k specifies a certain moment. Here, the variable δ is regarded as a small frequency shift of the resonant frequency of the cavity. Thus, the output current of the cavity till moment i, stored in the array *Cavity*[i], is representing all preceding moments as follows:

$$Cavity[i] = Cavity[i] + Cathode[i - k]\exp(-0.01k)\cos((40 + \delta)k\beta) \quad .$$

We should emphasize that the summation above is similar to cutting down the spectral distribution of the input current by the spectral contour of the cavity. With a narrower spectral distribution, the decay of the pulse-response characteristic in time will be slower, which is determined in our code by the constant 0.01.

Further, one partial contribution to the spectrum of the intensity fluctuations for 20 harmonics, from 30 to 50, around a possible value of resonance frequency, is calculated. Here, the calculations of *PwSp*[i] are similar to those considered above, except that contributions to the power spectrum are performed by values of the array *Cavity*[k], which

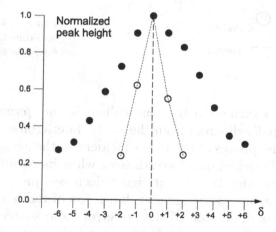

Figure 8.7. A set of values (solid circles), each indicating the peak spectral harmonics of the resonant curve of the cavity, found at 13 magnitudes of the detuning δ. One spectral contour of the cavity is shown by open circles for $\delta = 0$.

is the output current of the cavity instead of the photocurrent of the photocathode. Since the array $Cavity[k]$ is calculated at a certain magnitude of the frequency detuning δ, a set of distributions of the power spectrum may be found by varying δ. If only the peak harmonics of one particular distribution are taken into account, then a set of such peak values should show the influence of the detuning on the power spectrum. Figure 8.7 shows 13 peak values for a detuning δ running from -6 to +6. The height is normalized to the peak value at $\delta = 0$, mentioned above (solid circles). A number of harmonics of the resonant curve of the cavity associated with $\delta = 0$ are shown by open circles.

5.   Hanbury-Brown and Twiss experiments

Early in 1950 a new type of radio telescope was suggested by R.HAN-BURY-BROWN. In this radio telescope the signals from two aerials A_1 and A_2 were detected by square-law detectors independently (Fig.8.8). Then two low-frequency currents, each being proportional to an appropriate value of the instantaneous intensity, were multiplied together, thus recording the correlation function of the intensities.

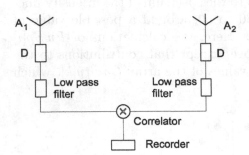

Figure 8.8 The stellar interferometer of intensities for radio waves.

It is clear that such an instrument, called the *interferometer of intensities*, is principally different from the stellar interferometer of MICHELSON, because the phases of the waves incident to the aerials are lost due to the quadratic detection. Nevertheless, when increasing the baseline of the interferometer, the output signal decreases progressively, which allows measurements of the angular dimension of radio sources. In this way, the angular dimension of the powerful radio source Cassiopeia-A was found to be about 2 minutes of arc and the dimension of another source, Cygnus-A, about 34 seconds of arc. For investigations of Cygnus-A an interferometer with a baseline of 3.9 km was used. The frequency of the radio waves was 125 MHz [8].

5.1   A laboratory experiment in the optical frequency range

In order to test this new idea in the optical frequency range, a laboratory experiment on intensity correlation was performed by HANBURY-BROWN and TWISS in 1956 [9]. The basic idea of the initial experiment was to search for the correlation of the photocurrents of two separated photocathodes when the light beams incident upon the two photodetectors are partially coherent, or, to confirm the idea that the correlation would be fully preserved in the process of photoelectric emission. For these reasons a laboratory experiment was carried out, which had some similarities with a MICHELSON interferometer. The original beam from a small aperture was split into two beams, as shown in Fig.8.9. A small rectangular aperture illuminated by a high–pressure mercury arc lamp acted as light source. The 435.8 nm mercury line was isolated by a system of filters, and the quasi-monochromatic beam was split by a semi-transparent mirror to illuminate the cathodes of two photomultipliers.

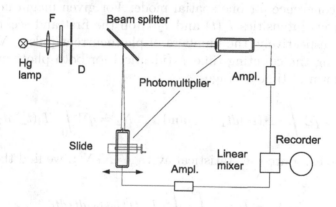

Figure 8.9. The apparatus for observation the intensity correlation effect with a mercury line of $\lambda_0 = 435.8$ nm.

In order to monitor the degree of coherence of the two light beams, the position of one photomultiplier could be varied in the horizontal plane in the direction transversal to the incident light beam. Two cathode apertures (as seen from the source through the beam splitter) were thus able to be superimposed or separated at any distance up to about three times their width. The fluctuations of the output currents from the photomultipliers were amplified within a bandpass of $3 - 27$ MHz and multiplied in a linear mixer, or a correlator. The average value of the product was recorded depending on the lateral movement of the mobile photomultiplier and gave the value of the correlation of intensity fluctuations.

When the photomultiplier was shifted across the light beam, the measurements of the intensities of the quasi-monochromatic field showed a high signal level on the smaller background of the shot noise. In such way a relation between the partial coherence in two beams and the correlation of the probabilities for emission of photoelectrons by two separated photodetectors had been obtained. The progress of these correlation measurements was predetermined by the circumstances that the counting time of the photodetectors, acting as the resolution time, was close to the coherence time (in early stages of experimental investigations of this effect the counting time was even longer than the coherence time).

5.2   Correlation function of photocurrents

Now we derive a relationship between the output signal in the HANBURY-BROWN– TWISS experiment and the degree of coherence $\gamma_{12}(\tau)$, assuming that the function $\gamma_{12}(\tau)$ has the form $\gamma_{12}(\tau) = \gamma_{12}(0)\gamma_{11}(\tau)$ as the degree of coherence for one spatial mode. For given magnitudes of the instantaneous intensities $I_1(t)$ and $I_2(t)$ on the first and second photocathodes, respectively, the numbers of photocurrent pulses N_1 and N_2 found during the counting time T (identical for both photomultipliers) have the form of time-averages:

$$N_1 = q\Sigma \int_0^T I_1(t_1)dt_1 \quad , \quad \text{and} \quad N_2 = q\Sigma \int_0^T I_2(t_2)dt_2 \quad .$$

In order to compute the statistical average $\langle N_1 N_2 \rangle$, we find the product $N_1 N_2$ as

$$N_1 N_2 = (q\Sigma)^2 \int_0^T \int_0^T I_1(t_1)I_2(t_2)dt_1dt_2 \quad .$$

Thus, the statistical average related to the output signal of the experiment is given as

$$\langle N_1 N_2 \rangle = (q\Sigma)^2 \int_0^T \int_0^T \langle I_1(t_1)I_2(t_2) \rangle dt_1 dt_2 \quad .$$

Since the average $\langle I_1(t_1)I_2(t_2) \rangle$ has to depend only on the difference $t_1 - t_2$, it should be represented by $|\gamma_{12}(\tau)|^2$ from (8.2). Hence, for $\langle N_1 N_2 \rangle$ we find

$$\langle N_1 N_2 \rangle = (q\Sigma \langle I \rangle)^2 \int_0^T \int_0^T \left[1 + |\gamma_{12}(t_1 - t_2)|^2 \right] dt_1 dt_2 =$$

$$= (q\Sigma \langle I \rangle)^2 \int_0^T \int_0^T dt_1 dt_2 + (q\Sigma \langle I \rangle)^2 \int_0^T \int_0^T |\gamma_{12}(t_1 - t_2)|^2 dt_1 dt_2 \quad.$$

$$(8.32)$$

The first integration gives the invariable magnitude

$$(q\Sigma \langle I \rangle)^2 \int_0^T \int_0^T dt_1 dt_2 = \langle N \rangle^2 \quad,$$

where $\langle N \rangle$ is the average amount of photocurrent pulses, which is assumed to be the same for both photocathodes. Because $|\gamma_{12}(t_1 - t_2)|^2$ is an even function, the integration of the second integral in (8.32) can also be performed over the non-negative range of the variable $t_1 - t_2$ ($t_1 - t_2 \geqslant 0$). The new non-negative variable $\tau = t_1 - t_2$ provides a new form of the integrals in (8.32):

$$\langle N_1 N_2 \rangle = \langle N \rangle^2 + 2(q\Sigma \langle I \rangle)^2 \int_0^T \int_0^T |\gamma_{12}(\tau)|^2 d\tau d(t_1 - \tau) \quad,$$

where the factor 2 takes into account the integration over one half of the total range of variables due to the even properties of function $|\gamma_{12}(\tau)|^2$. For a fixed τ the integration over variable t_1 gives a line integral over the variable τ:

$$\langle N_1 N_2 \rangle = \langle N \rangle^2 + 2(q\Sigma \langle I \rangle)^2 \int_0^T (T - \tau)|\gamma_{12}(\tau)|^2 d\tau \quad.$$

Since $|\gamma_{12}(\tau)|^2 = |\gamma_{12}(0)|^2 |\gamma_{11}(\tau)|^2$ is true, and since $(q\Sigma \langle I \rangle)^2 = \langle N \rangle^2 / T^2$, for $\langle N_1 N_2 \rangle$ we find the expression

$$\langle N_1 N_2 \rangle = \langle N \rangle^2 \left[1 + |\gamma_{12}(0)|^2 \frac{2}{T^2} \int_0^T (T - \tau)|\gamma_{11}(\tau)|^2 d\tau \right] \quad. \quad (8.33)$$

Formula (8.33) describes the output signal in the HANBURY-BROWN–TWISS experiment in general, provided that the effective apertures of the photocathodes are much smaller than the coherence area of the light beams, and that polarized light is detected. In any case, the requirement of spatial coherence must be fulfilled. On the other hand, the photodetectors should provide the detection of the intensities over a frequency band as wide as possible.

Let us assume that the coherence time is shorter than the counting time T. The function $|\gamma_{11}(\tau)|^2$ will then have non-zero values within the interval T, and the variable τ can be assumed to be shorter than T in the term $(T - \tau)$. Thus, the integral in (8.33) allows the following approximation:

$$\int_0^T (T - \tau)|\gamma_{11}(\tau)|^2 d\tau \approx T \int_0^T |\gamma_{11}(\tau)|^2 d\tau = T\tilde{\tau}_{ch}/2 \quad, \quad (8.34)$$

where $\widetilde{\tau}_{ch}$ is defined by the integral

$$\widetilde{\tau}_{ch} = \int_{-T}^{T} |\gamma_{11}(\tau)|^2 d\tau \qquad (8.35)$$

when the inequality $\widetilde{\tau}_{ch} < T$ is true. Substituting of the integral in (8.34) by $T\widetilde{\tau}_{ch}/2$ gives

$$\langle N_1 N_2 \rangle = \langle N \rangle^2 \left[1 + |\gamma_{12}(0)|^2 \frac{\widetilde{\tau}_{ch}}{T} \right] =$$

$$= \langle N \rangle^2 + \langle N \rangle^2 |\gamma_{12}(0)|^2 \frac{\widetilde{\tau}_{ch}}{T} \qquad . \qquad (8.36)$$

In turn, by using the effective spectral width of a spectral line in the form $\Delta\Omega = 1/\widetilde{\tau}_{ch}$ and the effective band width of the detecting apparatus in the form $\Delta\Omega_{\text{det}} = 1/T$, we find another simple approximation for the output signal:

$$\langle N_1 N_2 \rangle = \langle N \rangle^2 + \langle N \rangle^2 |\gamma_{12}(0)|^2 \frac{\Delta\Omega_{\text{det}}}{\Delta\Omega} \qquad . \qquad (8.37)$$

The term $\langle N \rangle^2 |\gamma_{12}(0)|^2 \widetilde{\tau}_{ch}/T$ in (8.36) and the term $\langle N \rangle^2 |\gamma_{12}(0)|^2 \Delta\Omega_{\text{det}}/\Delta\Omega$ in (8.37) are equivalent to each other. Both terms approximate how much the useful signal exceeds the background of shot noise. The latter is given by $\langle N \rangle^2$. This is obvious, since the term $\langle N \rangle^2$ in (8.36) and (8.37) is neither dependent on the spatial nor on the temporal degree of coherence, which means, that it relates to the shot noise.

Besides the magnitude $\langle N_1 N_2 \rangle$, the correlation function of the fluctuations of the photocurrent pulses $\langle \Delta N_1 \Delta N_2 \rangle$, is of interest (see (8.3)), which is represented by

$$\langle \Delta N_1 \Delta N_2 \rangle = \langle N \rangle^2 |\gamma_{12}(0)|^2 \frac{\widetilde{\tau}_{ch}}{T} \qquad . \qquad (8.38)$$

For a given value of average intensity $\langle I \rangle$, the average amount of photocurrent pulses per one spatial mode, which must be associated with the degeneracy parameter δ, is directly proportional to the magnitude $\langle I \rangle \widetilde{\tau}_{ch}$. In turn, $\langle I \rangle T$ has to be proportional to $\langle N \rangle$. Thus, the ratio $\widetilde{\tau}_{ch}/T$ in (8.38) can be replaced by $\delta/\langle N \rangle$. Finally, we obtain

$$\langle \Delta N_1 \Delta N_2 \rangle = \langle N \rangle \delta |\gamma_{12}(0)|^2 \qquad . \qquad (8.39)$$

5.3   The signal-to-noise ratio of the intensity interferometer

Let us discuss the signal-to-noise ratio of an intensity interferometer, where the formation of the output signal is performed by means of a correlator, which includes a coincidence circuit C as shown in Fig.8.10. Photocurrent pulses received from two photomultipliers P_1 and P_2 arrive at counters T. Each counter records an amount of incident pulses during the counting period T. Let N_1 be the amount of pulses detected by the first counter and N_2 by the second one. The shorter the period T is, the greater the fluctuations of both amounts above will be. To take into account the fluctuations, and further to form an output magnitude, the counted pulses are accumulated by two integrators, producing the so-called evaluative means $\overline{N}_1$ and $\overline{N}_2$. Subtraction of the actual numbers of pulses from the evaluative mean gives two actual fluctuations: $\Delta N_1 = \overline{N}_1 - N_1$ and $\Delta N_2 = \overline{N}_2 - N_2$.

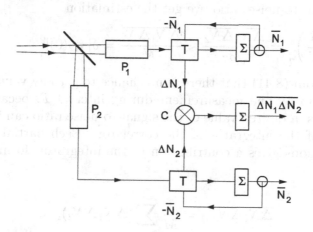

Figure 8.10. An intensity interferometer designed with a coincidence circuit for photocurrent pulses.

The coincidence circuit receives the two fluctuations and forms the product $\Delta N_1 \Delta N_2$, which corresponds to one measurement during the counting interval T. Such a value comes to the integrator of the correlator, where the evaluative mean $\overline{\Delta N_1 \Delta N_2}$ is collected. By definition, the signal-to-noise ratio of the magnitude $\overline{\Delta N_1 \Delta N_2}$ is represented in terms of the mean square deviation $\sqrt{\sigma_T}$ as

$$\left(\frac{S}{N}\right)_T = \frac{\overline{\Delta N_1 \Delta N_2}}{\sqrt{\sigma_T}} = \frac{\overline{\Delta N_1 \Delta N_2}}{\sqrt{\overline{(\Delta N_1 \Delta N_2)^2} - \left(\overline{\Delta N_1 \Delta N_2}\right)^2}} \quad , \quad (8.40)$$

where the subscript T emphasizes the fact that σ_T as well as the signal-to-noise ratio relates to one interval T.

To simplify (8.40) we assume that ΔN_1 and ΔN_2 are independent, at least to a first approximation, since fluctuations of the instantaneous intensity are primarily caused by shot-noise rather than by wave-noise, due the very small value of the degeneracy parameter δ. This assumption allows the simplification

$$\sigma_T = \overline{(\Delta N_1 \Delta N_2)^2} - \left(\overline{\Delta N_1 \Delta N_2}\right)^2 = \overline{\Delta N_1^2} \cdot \overline{\Delta N_2^2} - \left(\overline{\Delta N_1}\right)^2 \left(\overline{\Delta N_2}\right)^2 \quad ,$$

and, since $\overline{\Delta N_1} = 0$ and $\overline{\Delta N_2} = 0$, the right-hand side of the last expression becomes equal to $\overline{\Delta N_1^2} \cdot \overline{\Delta N_2^2}$. Because both magnitudes $\overline{\Delta N_1^2}$ and $\overline{\Delta N_2^2}$ are formed by shot-noise, we can use $\overline{\Delta N_1^2} \approx \overline{N_1}$ and $\overline{\Delta N_2^2} \approx \overline{N_2}$ for the dispersions of the shot-noise. Finally, for $\overline{N_1} \approx \overline{N_2} = \overline{N}$, the estimation of $\sqrt{\sigma_T}$ takes the simple form $\sqrt{\sigma_T} \approx \overline{N}$. Thus, using (8.39) for the signal-to-noise ratio, we get the estimation

$$\left(\frac{S}{N}\right)_T = \frac{\overline{\Delta N_1 \Delta N_2}}{\sqrt{\sigma_T}} \approx \frac{\overline{\Delta N_1 \Delta N_2}}{\overline{N}} = \delta |\gamma_{12}(0)|^2 \quad . \tag{8.41}$$

It follows from (8.41) that there is no chance to get any valuable signal-to-noise ratio by one measurement during interval T, because $\delta \ll 1$. Nevertheless, a desired value of the signal-to-noise ratio can be achieved by means of the integrator of the correlator. Each partial product of the fluctuations gives a contribution to the integrator, forming a time-average

$$\overline{\Delta N_1 \Delta N_2}_M = \frac{1}{M} \sum_{i=1}^{i=M} (\Delta N_1 \Delta N_2)_i \quad ,$$

where M is the total amount of the partial evaluative averages found during the time t of observation $t = MT$. Now, for the signal-to-noise ratio for the time $t = MT$ we obtain

$$\left(\frac{S}{N}\right)_t = \frac{\overline{\Delta N_1 \Delta N_2}_M}{\sqrt{\sigma_t}} \approx \frac{\sqrt{\sigma_T}}{\sqrt{\sigma_t}} \left(\frac{S}{N}\right)_T \approx \frac{\sqrt{\sigma_T}}{\sqrt{\sigma_t}} \delta |\gamma_{12}(0)|^2 \quad ,$$

where $\sqrt{\sigma_t}$ is the root-mean-square of the magnitude $\overline{\Delta N_1 \Delta N_2}_M$. Assuming all evaluative magnitudes $(\Delta N_1 \Delta N_2)_i$ as independent from each other, the value of $\sqrt{\sigma_t}$ will be smaller than $\sqrt{\sigma_T}$ by a factor $\sqrt{M}$. Hence, the obtained signal-to-noise ratio after a time of observation $t = MT$ is

$$\left(\frac{S}{N}\right)_t = \sqrt{M} \left(\frac{S}{N}\right)_T = \sqrt{\frac{t}{T}} \delta |\gamma_{12}(0)|^2 \quad . \tag{8.42}$$

This formula shows that for the given counting interval T the time t needed to form a required value of the signal-to-noise ratio is dependent on the degeneracy parameter δ and on the value of the spatial degree of coherence.

Let the required value of the signal-to-noise ratio be equal to 10, and let us estimate the effective band width of the electrical circuits of the intensity interferometer to be about 25 MHz. Then, for the superimposition of the photocathodes, where $|\gamma_{12}(0)|^2 \approx 1$, and for $\delta \sim 10^{-4}$, the time t needed to achieve the required value of the S/N ratio is estimated to be

$$t \geq 10T\delta^{-2} = \frac{10^9}{(2\pi)25 \cdot 10^6} \approx 7 \text{ s} \quad .$$

However, to achieve the same value of S/N ration, for example, at $|\gamma_{12}(0)|^2 = 0.3$, the required time should be enlarged by a factor of about 10, because $t \sim T|\gamma_{12}(0)|^{-4}$. Then t is found to be about 70 s. It follows from (8.37) and (8.42) that, with all other factors the same, a wider bandwidth of the circuit for detecting the photocurrent pulses permits a suitable signal-to-noise ratio for a shorter time of observation t.

5.4   The stellar interferometer of intensities

After successful evidence of intensity correlation in the laboratory experiment, the idea of such measurements was utilized in a new type of optical stellar interferometer by HANBURY-BROWN and TWISS in 1956. A simple scheme of the *stellar interferometer of intensities* is shown in Fig.8.11. Two mirrors separated by a desired baseline focus the incident light on two photomultipliers. The photocurrent pulses generated by the photomultipliers pass through two radio circuits, each containing a delay unit and an amplifier. The band widths of the amplifiers are about $5 - 45$ MHz. The delay units permit equalization of a delay caused by the difference in the arrival time of the light from a star at the two mirrors. The time delay required to compensate such a time difference is estimated to be about $\Delta t \approx 1/\Delta f_{circ}$, where Δf_{circ} is the effective band width of the circuits processing the photocurrents.

We would like to point out that, in principle, this arrangement is a stellar interferometer of MICHELSON's type working in the time domain. The time difference between the two interfering rays must be compensated more accurately than the coherence time, that is, up to $\Delta t \ll 1/\Delta\nu$, where $\Delta\nu$ is the effective width of the light radiation. It follows from estimations of these time delays that the interferometer of intensities allows a much larger baseline than the MICHELSON stellar interferometer, because the accuracy requirements for the adjustment

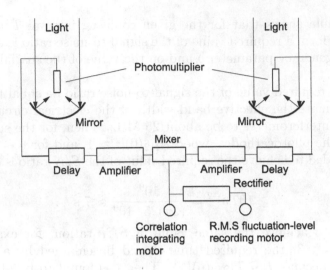

Figure 8.11.   The stellar interferometer of intensities.

are around $\Delta\nu/\Delta f_{circ}$. Moreover, the photocurrents are proportional to the instantaneous intensities. Therefore, any wave phases are absent in these photocurrents, thus, atmospheric disturbances along the light propagation in the photocurrents does not perturb the intensity signals.

The outputs from both amplifiers are multiplied together in a linear mixer and the accumulated average value of the product is recorded on the revolution counter of an integrating motor (its rotational speed depends on the actual value of input current). The reading of this counter gives a direct measure of the correlation between the intensity fluctuations in the light incident on the two mirrors. The input of the correlation motor was fed via a rectifier to a second integrating motor in order to get a value directly proportional to the root-mean-square of the fluctuations of this input. This is done to eliminate an uncertainty in the gain by expressing all results as the ratio of the integrated correlation to the root-mean-square fluctuations. Data obtained by HANBURY-BROWN and TWISS when measuring the correlation from Sirius (α Canis Majoris A) are presented in Table 8.3.[10].

It is seen from Table 8.3 that the diameter of the coherence area from Sirius could be estimated to be about 9 meters, which allowed an estimation of the angular diameter of Sirius to be about 0.0068 arc seconds. We call attention to the duration of the observation, which were about 5 – 6 hours to permit an acceptable signal-to-noise ratio.

Table 8.3. Data obtained with measurements of correlation on Cirius.

Base in m	2.5	5.54	7.27	9.2
Observing time (min)	345	285	280	170
S/N	8.5	3.59	2.65	0.83
$\gamma_{12}^2(0)$	0.84 ± 0.07	0.64 ± 0.12	0.52 ± 0.13	0.19 ± 0.15

Table 8.4. Data obtained with the computer model of the HANBURY-BROWN – TWISS experiment.

Base	Mcurr1	Mcurr2	OutS1	OutNorm	DF	SNM
0	1.744	1.741	3.075	1.01	8.765	1.037
2	1.583	1.822	2.343	0.81	9.147	0.774
4	1.485	1.762	1.063	0.41	8.571	0.363
6	1.500	1.569	0.200	0.08	8.592	0.068

5.5 A computer model of the Hanbury-Brown - Twiss experiment

Algorithm **IntensityInterferometer**(), see Appendix 8.D, simulates the operating principle of the apparatus in the HANBURY-BROWN – TWISS experiment in its principal points. At the beginning of the algorithm two series of variables $\theta_C[m]$ and $\zeta_C[m]$ are calculated to deliver simulated quadrature components of the correlated field, as in the case of a narrow slit (see 7.9.1), where $\beta = \pi/140$. Further, a set of variables $W[0]$, $W[1]$, $W[2]$, $W[3]$ associated with the light energy are calculated. These variables describe the photocurrents at four spatial points of observation. Variables $W[0]$ and $W[1]$ are associated with two points separated by the actual baseline of the interferometer, whereas variables $W[2]$, $W[3]$ belong to two points at the periphery of the coherent area. The fluctuations of light intensity can be assumed as having an appreciable correlation at points specified by $W[0]$ and $W[1]$, and correlation of nearly zero at the peripheral points.

To convert the light energies into photocurrents, a code similar to that used in the model of the optical beat experiment is utilized, where the maximal amount of photoelectrons is restricted to 27. Thus, four random variables: $curr[0]$, $curr[1]$, $curr[2]$, $curr[3]$ represent the actual amount of photoelectrons at the four points mentioned above. Then, four mean values of photocurrent pulses are all formed by means of a recurrent code:

$$Mcurr = (Mcurr * (M-1))/M + curr/M \quad ,$$

where M is the actual value of the number of outer loops completed from the beginning, and $curr$ is the actual value of a current. A set of sums are then calculated in order to represent the results in terms of a signal-to-noise ratio similar to that discussed above. One sum stored in the variable $OutS1$ is formed by the terms

$$OutS1 = OutS1 + (curr[0] - Mcurr1) * (curr[1] - Mcurr2) \quad ,$$

representing fluctuations of two photocurrent pulses at points where a correlation is assumed to exist. The other sum, variable $OutS2$, contains terms

$$OutS2 = OutS2 + (curr[2] - Mcurr3) * (curr[3] - Mcurr4) \quad ,$$

and represents the contribution provided by two peripheral points. Along with these sums, the summation of terms in the form of the square of one particular term of the previous sum is performed and stored in variable DF:

$$DF = DF + ((curr[2] - Mcurr3) * (curr[3] - Mcurr4))^2 \quad .$$

After completion of the inner loop a set of means normalized to the value of Max are calculated: $OutS1/M$, $OutS2/M$, DF/M. Thus, the final value of calculations takes the form of a signal-to-noise ratio:

$$SNM = \frac{OutS1/M}{\sqrt{DF/M - (OutS2/M)^2}} \quad .$$

For the given values of $Mcurr1$, $Mcurr2$, and $OutS1$ the output signal is calculated in the normalized form

$$OutNorm = \frac{OutS1}{Mcurr1 * Mcurr2}$$

Data found for four values of the baseline of the interferometer, variable $Base = 2*base$, are presented in Table.8.4. Normalized values of the output signal are shown in Fig.8.12, where the variable $Base$, the distance between two photodetectors, is represented by the integer numbers.

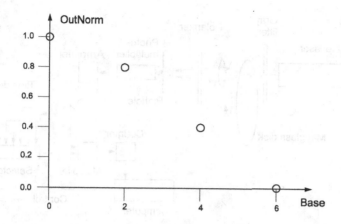

Figure 8.12. Normalized values of output signal *OutNorm*, depending on *Base*, in the model of the HANBURY-BROWN – TWISS experiment.

6. Correlation of pseudo-Gaussian light

Already in an early state of the description of light by statistical methods, properties of laser light were investigated. It was found that laser radiation associated with one single mode of the laser resonator shows statistical properties similar to that of a monochromatic wave. Such laser radiation, detected by a photodetector, should cause a photocurrent noise with POISSON statistics. In particular, this fact implies that no correlation of the laser intensity can be found. Thus, POISSON statistics can be applied to laser radiation in the single mode regime as the simplest model which is sometimes called the model of an "ideal" laser. The demonstrational experiment considered below was first realized by L.MANDEL in 1963 [11].

Let us consider a setup which is able to demonstrate the statistical properties of such an "ideal" laser. All principle items of this setup are shown in Fig.8.13. A He-Ne gas laser radiates under conditions of the transversal TEM_{00} mode at $\lambda = 632.8$ nm. A short optical resonator of the laser, having a length of only 15 cm, assures that only one longitudinal mode is located under the DOPPLER profile of the laser medium, thus the laser emits single mode radiation (containing only one sharp frequency).

The laser beam is weakened by an optical gray filter and then passes through a pinhole onto the photocathode of a photomultiplier, operating under photon counting conditions. In such a way, the weak light flux is represented by a sequence of photocurrent pulses which arrive at the input of a time-delay line in random order. The time-delay line

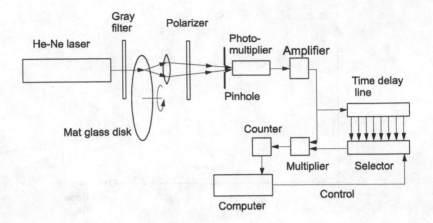

Figure 8.13. Initiation of pseudo-GAUSSian light of with long-time correlation.

contains eight outputs. The first output reproduces each incoming photocurrent pulse, while the remaining seven outputs form pulses retarded
with respect to the incoming pulses. The retardation is increased from
one output to the next by the same interval τ. After amplification, each
original photocurrent pulse is led to the first input of a coincidence circuit. The second input of this circuit is switched by a commutator to
one input of the time-delay line in series. The resolution time of this
circuit was set to the duration of the photocurrent pulses, about 30 ns.

The pulses from the coincidence circuit are counted during each period ΔT by means of a counter. The period $\Delta T = 1$ s was chosen in
order to eliminate any effect of instabilities of the laser intensity. Using
the output data of the coincidence circuit for ΔT, seven magnitudes,
each corresponding to a certain magnitude of the normalized temporal
correlation function of the photocurrent pulses, are formed as follows:

$$\frac{\overline{I(t)I(t+\tau)}_{\Delta T}}{\overline{I(t)}^2_{\Delta T}} \quad , \quad \frac{\overline{I(t)I(t+2\tau)}_{\Delta T}}{\overline{I(t)}^2_{\Delta T}} \quad , \quad .... \quad , \quad \frac{\overline{I(t)I(t+7\tau)}_{\Delta T}}{\overline{I(t)}^2_{\Delta T}} \quad .$$

We assume that such normalized magnitudes of the correlation function
are not subjected by slow variations of the intensity of the incident laser
beam. These magnitudes are further collected as arithmetic means for
each of the seven retardation periods $n\tau$ by a computer during the interval of observation, $t = 10$ min. The accumulated arithmetic means
represent 7 points of the correlation function under consideration.

The fact that "ideal" laser light shows no correlation of its intensity is
illustrated in Fig.8.14,a, where the magnitudes of the correlation function may be well approximated by a horizontal line drawn at 1. This

means that the correlation function is nearly independent of the time delay $n\tau$ ($\tau = 8.5$ μs), which confirms that the POISSON statistics is valid for such radiation.

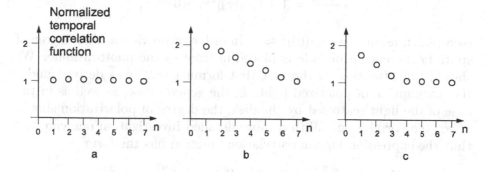

Figure 8.14. The data found after accumulation of pulses from the scheme of coincidance during period $t = 10$ min; (a) – without the mat glass disk; (b) at linear velocity $v = 50$ cm/s; (c) at linear velocity $v = 180$ cm/s.

The situation will be essentially changed when introducing a mat glass disk, which is rotating with uniform angular velocity, into the laser beam. Now the correlation function depends on the time delay $n\tau$, and on the angular velocity of the disk. The correlation functions found for two different velocities of the disk are shown in Fig.8.13,b,c. These functions show a maximum value of about 2 at $\tau = 1$ and their value is gradually reduced with time-delay $n\tau$ down to 1. A dependence of this sort is inherent to GAUSSian light. The generation of the *pseudo*-GAUSSian *light* from the laser beam is caused by scattering of the beam by the micro-structure of the rotating mat glass disk. Let us consider an area of the disk illuminated by the laser beam at some instant t. We assume a huge number of scattering centers of the mat glass to be within this area, each being specified by a certain optical path length for the given wavelength λ. The mat glass disk has a chaotically varying thickness within this area, which causes wavelets emitted by the scattering centers to have chaotic phases. A superposition of the wavelets at a distant point therefore results in a random field, which might show GAUSSian nature. The chaotic distribution of the scattering centers stays unchanged until this area is replaced by another one due to rotation of the disk. For a given effective width d of the laser beam and for the given linear velocity v of the area mentioned above, a period d/v specifies, roughly, intervals during which one chaotic superposition can be regarded as static. This implies that the period d/v estimates the coherence time of such chaotic radiation. It follows from Fig.8.13,b,c that, the greater the linear velocity

is, the shorter the coherence time will become. That is also confirmed
by using a normalized correlation function of intensity

$$\frac{\langle I_1 I_2 \rangle}{\langle I \rangle^2} = 1 + |\gamma_{11}(\tau)|^2 |\gamma_{12}(0)|^2 \quad ,$$

(see (8.2)), assuming $|\gamma_{12}(0)|^2 \approx 1$. In order to provide a high degree of
spatial coherence a pinhole is placed in front of the photocathode. We
should call attention to the fact that formula (8.2) was derived under
the assumption of polarized light. In the general case, as well as in the
case of the light scattered by the disk, the degree of polarization should
affect the correlation. With a more detailed investigation one can find
that the expression for the correlation function has the form

$$\langle I_1 I_2 \rangle = \langle I \rangle^2 \left[1 + 0.5(1 + P^2)|\gamma_{11}(\tau)|^2 |\gamma_{12}(0)|^2 \right] \quad ,$$

where P is the degree of polarization. To provide totally polarized light
in the scheme under consideration a polarizer is used.

Let us discuss the parameters of the experiment. It is assumed that
the GAUSSian light scattered by the disk is reliably formed if the amount
of the scattering centers distributed over the laser spot is rather large.
For example, this amount will be about $10^5 - 10^4$ for a diameter of the
laser beam of $d = 1$ mm, when the average size of a scattering center is
$1 - 10$ μm. The coherence time can be estimated by

$$\tau_{ch} = \frac{d}{v} \quad , \tag{8.43}$$

where v is the linear velocity of the scattering centers within the laser
spot. Corresponding to two velocities $v_1 = 50$ cm/s and $v_2 = 180$
cm/s, the coherence time can be estimated to be $\tau_1 \approx 0.1/50 = 200$
ms and $\tau_2 \approx 0.1/180 = 56$ ms, respectively. Values of the coherence
time found from the experimental data are $\tau_1 = 60$ ms and $\tau_2 = 17$ ms,
respectively, that agrees with the estimated value expect from a factor
3.

SUMMARY

Light quanta, or photons, obey the BOSE-EINSTEIN statistics. In a particular case of thermal radiation the probability for the localization of a required number of quanta within a spatial mode takes the form of Eq.(8.24). This probability differs essentially from a POISSON probability. In the case of thermal radiation the BOSE-EINSTEIN statistics is responsible for the wave noise of photocurrent pulses. Because of wave noise there exists the possibility of detection of the correlation of intensity fluctuations. For this detection is needed to provide an acceptable resolution time of the photodetector: the resolution time should be less than the coherence time of the optical radiation under investigation. This condition allows to measure the spatial correlation of instantaneous intensities within one spatial mode.

PROBLEMS

8.1. Radiation from the sun, processed by a filter with $\overline{\lambda} = 500$ nm and a band width of 0.5 nm $(\Delta\lambda/\overline{\lambda} = 10^{-3})$, is detected by a photodetector with a sensitive area $\Sigma = 1$ mm^2 during recording intervals $T = 10^{-3}$ s. The sun is observed under an angular diameter $\theta_s = 9.2$ mrad. Its radiation is regarded to be black body radiation at $T = 5300$ K. Find the relative root-mean-square error of the detected signal.

SOLUTIONS

8.1 Using the BOSE-EINSTEIN distribution in the form of Eq. (8.28) one can find that the mean amount of quanta per one spatial mode $\langle n \rangle$ is much smaller than 1, using the parameters of the current problem. It can be estimated to be $\langle n \rangle \approx \exp(-h\nu/(k_B T))$. This implies that the detecting noise can be regarded as only being shot noise when neglecting the "wave" noise of such radiation. Thus, the root-mean-square error of the detected signal should be equal to the value calculated for the amount of quanta within the detecting volume ΣTc, where c is the light velocity. Since $\langle n \rangle$ is the amount of quanta within one coherence volume, the mean amount of detected quanta within the volume ΣTc is equal to $\overline{N} = \langle n \rangle \Sigma Tc/V_{ch}$, where $V_{ch} = S_{ch}l_{ch}$. In turn, $l_{ch} = \overline{\lambda}^2/\Delta\lambda$ and $S_{ch} \approx \pi(1.22 \cdot \overline{\lambda}/\theta_s)^2/4 \approx (\overline{\lambda}/\alpha_s)^2$, hence $V_{ch} \approx (\overline{\lambda}/\theta_s)^2\overline{\lambda}(\overline{\lambda}/\Delta\lambda)$. Finally, one gets for $\overline{N}$ the relation

$$\overline{N} = \frac{\langle n \rangle \Sigma Tc}{V_{ch}} = \exp(-h\nu/(k_B T))\frac{\Sigma Tc}{(\overline{\lambda}/\theta_s)^2\overline{\lambda}(\overline{\lambda}/\Delta\lambda)} \ .$$

Because for shot noise $\sqrt{\overline{\Delta N^2}}/\overline{N} = 1/\sqrt{\overline{N}}$, we get for the desired magnitude of the detected signal

$$\frac{1}{\sqrt{\overline{N}}} = \exp(h\nu/(2k_BT))\frac{\overline{\lambda}}{\theta_s}\sqrt{\frac{\overline{\lambda}}{\Delta\lambda}\frac{\overline{\lambda}}{STc}} .$$

Substitution of the numerical values gives $1/\sqrt{\overline{N}}$ to be about $7 \cdot 10^{-5}$. We call attention to the fact that $1/\sqrt{\overline{N}}$ decreases with increasing time T as $\sqrt{T}$.

APPENDIX 8.A

InertialessDetector()
$\beta = 0.0125$, $Q = 0.001$, $Max = 20000$, $x_{in} = 1732$;
  for $(m = 0;\ m < 20;\ m + +)\{z[m] = \exp(-0.5(m/20)^2)\}$
  for $(M = 0;\ M < Max;\ M + +)\{$
  for $(m = 0;\ m < 20;\ m + +)\{$
$\theta_C[m] = 0.0$;
$\zeta_C[m] = 0.0;\}$
  for $(m = 0;\ m < 20;\ m + +)\{$
  Polar-coordinates();
  for $(k = 0;\ k < 20;\ k + +)\{$
$\theta_C[m] = \theta_C[m] + \theta * z[k] * \cos(\beta mk)$;
$\zeta_C[m] = \zeta_C[m] + \zeta * z[k] * \cos(\beta mk);\}\}$
  for $(m = 0;\ m < 10;\ m + +)\{$
$W[m] = \theta_C[m] * \theta_C[m] + \zeta_C[m] * \zeta_C[m]$;
$\overline{N} = Q * W[m]$;
$P[0] = \exp(-\overline{N})$;
$AP[0] = AP[0] + P[0]$;
  for $(N = 1;\ N < 10;\ N + +)\{$
$P[N] = (P[N-1]\overline{N})/N$;
$AP[N] = AP[N] + P[N];\}\}$
$\}$
  for $(N = 1;\ N < 10;\ N + +)\{$
$\langle P[N] \rangle = AP[N]/(10 * Max);\}$

APPENDIX 8.B

InertialDetector():
$\beta = 0.0125$, $Q = 0.0001$, $Max = 80000$; $x_{in} = 1735$;
  for $(m = 0;\ m < 20;\ m + +)\{$
$z[m] = \exp(-0.5(m/20)^2)\}$
  for $(M = 0;\ M < Max;\ M + +)\{$
  for $(m = 0;\ m < 20;\ m + +)\{$
$\theta_C[m] = 0.0$;
$\zeta_C[m] = 0.0;\}$
  for $(m = 0;\ m < 20;\ m + +)\{$
  Polar-coordinates();
  for $(k = 0;\ k < 20;\ k + +)\{$
$\theta_C[m] = \theta_C[m] + \theta * z[k] * \cos(\beta mk)$;
$\zeta_C[m] = \zeta_C[m] + \zeta * z[k] * \cos(\beta mk);\}\}$
$W = 0$;
  for $(m = 0;\ m < 20;\ m + +)\{$
$W = W + \theta_C[m] * \theta_C[m] + \zeta_C[m] * \zeta_C[m];\}$
$\overline{N} = Q * W$;

$P[0] = \exp(-\overline{N})$;
$AP[0] = AP[0] + P[0]$;
  for $(N = 1; N < 10; N++)\{$
$P[N] = (P[N-1]\overline{N})/N$;
$AP[N] = AP[N] + P[N];\}$
$\}$
  for $(N = 1; \ N < 10; \ N++)\{$
$\langle P[N] \rangle = AP[N]/Max;\}$

APPENDIX 8.C

OpticalBeats():
  $x_{in} = 1732; \ Max = 100000; \ \beta = 0.025; \ Q = 0.05; \ ModeKey = 1; \ \delta = 0;$
  for $(i = 0; \ i < 60; \ i++)\{$
$sp[i] = exp(-0.25*(i-20)^2)$;
$PwSp[i] = 0.0;\}$
  for $(M = 0; \ M < Max; \ M++)\{$
  for $(i = 0; \ i < 60; \ i++)\{$
$Cathode[i] = 0.0; Cavity[i] = 0.0; \theta_S[i] = 0.0; \zeta_S[i] = 0.0;\}$
  for $(i = 0; \ i < 60; \ i++)\{$
  Polar-coordinates() ; $\theta_O[i] = \theta; \zeta_O[i] = \zeta;\}$
  for $(k = 0; \ k < 60; \ k++)\{$
  for $(i = 0; \ i < 60; \ i++)\{$
$\theta_S[k] = \theta_S[k] + sp[i]*\theta_O[i]*cos(i*k*\beta)$;
$\zeta_S[k] = \zeta_S[k] + sp[i]*\zeta_O[i]*cos(i*k*\beta);\}\}$
  for $(m = 0; \ m < 60; \ m++)\{$
$\overline{N} = Q*(\theta_S^2[m] + \zeta_S^2[m]); P[0] = exp(\overline{N})$;
  for $(k = 1; \ k < 20; \ k++)P[k] = P[k-1]/k$;
  for $(k = 1; \ k < 20; \ k++)P[k] = P[k] + P[k-1]/k$;
$x = \mathbf{Rnd()}$;
  for $(k = 0; \ k < 20; \ k++)\{$
  if $(x <= P[k])\{Cathode[m] = k; k = 20;\}\}\}$
  if $(ModeKey == 1)\{$
  for $(i = 0; \ i < 60; \ i++)\{$
$c = 0.0$;
  for $(k = 0; \ k < 60; \ k++)\{$
$c = c + Cathode[k]*cos(i*k*\beta);\{$
$PwSp[i] = PwSp[i] + c;\}\}$
  else $\{$
  for $(i = 59; \ i >= 0; \ i--)\{$
  for $(k = 0; \ k < 60; \ k++)\{$
$Cavity[i] = Cavity[i] + Cathode[i-k]*exp(-k*0.01)*cos(40+\delta)*\beta*k)$;
$\}\}$
  for $(i = 30; \ i < 50; \ i++)\{$
$c = 0.0$;
  for $(k = 0; k < 60; k++)\{$

$c = c + Cavity[k] * \cos(i * k * \beta);$}
$PwSp[i] = PwSp[i] + c;$}}
 if $(ModeKey == 1)${
 for $(i = 0;\ i < 60;\ i + +)${
$PwSp[i] = PwSp[i]/Max;$}}
 else {
 for $(i = 30;\ i < 50;\ i + +)${
$PwSp[i] = PwSp[i]/Max;$}}

APPENDIX 8.D

IntensityInterferometer()
 $Max = 50000;\ x_{in} = 17325;\ \beta = \pi/140;$
 for $(M = 0;\ M < Max;\ M + +)${
 for $(m = 0;\ m < 4;\ m + +)W[m] = 0.0;$
 for $(m = 0;\ m < 20;\ m + +)\{\theta_C[m] = 0.0; \zeta_C[m] = 0.0;$ }
 for $(k = 0;\ k < 20;\ k + +)$ {
 Gauss $(1.0);\ \theta_O[k] = \theta; \zeta_O[k] = \zeta;$ }
 for $(m = -10;\ m < 10;\ m + +)${
 for $(k = 0;\ k < 20;\ k + +)${
$\theta_C[m + 10] = \theta_C[m + 10] + \theta_O[k] * \cos(\beta * (m - 10) * k);$
$\zeta_C[m + 10] = \zeta_C[m + 10] + \zeta_O[k] * \cos(\beta * (m - 10) * k);$}}
$W[0] = \theta_C[n - base] * \theta_C[n - base] + \zeta_C[n - base] * \zeta_C[n - base];$
$W[1] = \theta_C[n + base] * \theta_C[n + base] + \zeta_C[n + base] * \zeta_C[n + base];$
$W[2] = \theta_C[n - 4] * \theta_C[n - 4] + \zeta_C[n - 4] * \zeta_C[n - 4];$
$W[3] = \theta_C[n + 4] * \theta_C[n + 4] + \zeta_C[n + 4] * \zeta_C[n + 4];$
 for $(n = 0;\ n < 4;\ n + +)${
$AN = Q * W[n]; P[0] = \exp(-AN);$
 for $(N = 1;\ N < 27;\ N + +)P[N] = (P[N - 1] * AN)/N;$
 for $(N = 1;\ N < 27;\ N + +)P[N] = P[N] + P[N - 1];$
 for $(N = 0;\ N < 27;\ N + +)${
$x =$ **Rnd()**;
 if $((x <= P[N]))\{curr[n] = N;$ **break;**}}}
$Mcurr1 = (Mcurr1 * (M - 1))/M + (curr[0])/M;$
$Mcurr2 = (Mcurr2 * (M - 1))/M + (curr[1])/M;$
$Mcurr3 = (Mcurr3 * (M - 1))/M + (curr[2])/M;$
$Mcurr4 = (Mcurr4 * (M - 1))/M + (curr[3])/M;$
$OutS1 = OutS1 + (curr[0] - Mcurr1) * (curr[1] - Mcurr2);$
$OutS2 = OutS2 + (curr[2] - Mcurr3) * (curr[3] - Mcurr4);$
$DF = DF + ((curr[2] - Mcurr3) * (curr[3] - Mcurr4))^2$
}
$OutS1 = OutS1/M; OutS2 = OutS2/M;$
$DF = DF/M; DF = DF - OutS2 * OutS2;$
$SNM = OutS1/\sqrt{DF};$
$OutNorm = OutS1/(Mcurr1 * Mcurr2);$

References

[1] Joseph W. Goodman. Introduction to Fourier Optics. McGraw-Hill Inc. 1968

[2] L. Boltzmann. The analytical prove of the second law of thermodynamics by means of theorems of "alive force". Wien. Ber. 63, 712, 1871

[3] L.Boltzmann. Further investigation of thermal equilibrium between molecules of a gas. Wien. Ber. 66, 275, 1872

[4] A. Einstein. Strahlungs-Emission und -Absorption nach der Quantentheorie. Verhandl. Dtsch. Phys. Ges. 18, 318, 1916

[5] A. Einstein. Über einen die Erzeugung und Verwandlung des Lichtes betreffenden heuristischen Gesichtspunkt. Ann. d. Phys. 17, 132, 1905 A. Einstein. Zur Theorie der Lichterzeugung und Lichtabsorption. Ann. d. Phys. 20, 199, 1906

[6] M.Planck. Über eine Verbesserung der Wien'schen Spektralgleichung. Verh. d. Dtsch. Phys. Ges. 2, 202, 1900 M.Planck. Zur Theorie des Gesetzes der Energieverteilung im Normalspektrum. Verh. d. Dtsch. Phys. Ges. 2, 237, 1900 M.Planck. On the Law of Distribution of Energy in the Normal Spectrum. Ann. d. Phys. 4, 553, 1901

[7] A.T.Forrester, R.A.Gudmundsen, and P.O.Johnson. Photoelectric Mixing of Incoherent Light. Phys. Rev. 99, 1691, 1955

[8] R.Hanbury-Brown, R.C.Jennison, and M.K. Das Gupta. Apparent Angular Sizes of Discrete Radio Sources: Observations at Jodrell Bank, Manchester. Nature 170, 1061, 1952

[9] R.Hanbury-Brown, and R.Q.Twiss. Correlation between photons in two coherent beams of light. Nature, No 4497, 27, 1956

[10] R.Hanbury Brown, R.Q.Twiss. A test of a new type of stellar interferometer on Sirius. Nature, 178, 1046, 1956

[11] L.Mandel. Letter to the Editor. J.Opt. Soc.Amer. 52, 1407, 1962

Index